Economics and Ageing

José Luis Iparraguirre

Economics and Ageing

Volume IV: Political Economy

José Luis Iparraguirre
Age UK
London, UK

ISBN 978-3-030-29012-2   ISBN 978-3-030-29013-9 (eBook)
https://doi.org/10.1007/978-3-030-29013-9

© The Editor(s) (if applicable) and The Author(s) 2019
This work is subject to copyright. All rights are solely and exclusively licensed by the Publisher, whether the whole or part of the material is concerned, specifically the rights of translation, reprinting, reuse of illustrations, recitation, broadcasting, reproduction on microfilms or in any other physical way, and transmission or information storage and retrieval, electronic adaptation, computer software, or by similar or dissimilar methodology now known or hereafter developed.
The use of general descriptive names, registered names, trademarks, service marks, etc. in this publication does not imply, even in the absence of a specific statement, that such names are exempt from the relevant protective laws and regulations and therefore free for general use.
The publisher, the authors, and the editors are safe to assume that the advice and information in this book are believed to be true and accurate at the date of publication. Neither the publisher nor the authors or the editors give a warranty, expressed or implied, with respect to the material contained herein or for any errors or omissions that may have been made. The publisher remains neutral with regard to jurisdictional claims in published maps and institutional affiliations.

Cover illustration: © Darren Nakata / EyeEm

This Palgrave Macmillan imprint is published by the registered company Springer Nature Switzerland AG.
The registered company address is: Gewerbestrasse 11, 6330 Cham, Switzerland.

Contents

Part I	**Economics of Happiness and Quality of Life**	1
1	**Conceptualisations and Measurement**	3
1.1	Utility	6
1.2	Welfare	9
1.3	Subjective and Objective Well-Being	10
	1.3.1 Subjective Well-Being	10
	1.3.2 Objective Well-Being	14
1.4	Psychological Well-Being	15
1.5	Optimal Well-Being	16
1.6	Happiness	17
1.7	Human Development	18
1.8	Quality of Life	19
	1.8.1 CASP-19	23
	1.8.2 Long-Term Care	23
	1.8.3 Health Care	26
	1.8.4 Community Quality of Life	31
References		32
2	**Theories and Empirical Findings**	45
2.1	Comparison Theories	48
2.2	Easterlin Paradox	66
2.3	Set Point, Adaptation Level, and the Hedonic Treadmill	67

2.4		Personality Traits and Genetics	74
2.5		Needs-Based Theories	76
2.6		Inequality and Happiness	79
2.7		Happiness Along the Life Cycle	79
References			85

3 Happiness and Policy 101

3.1		Gross National Happiness	106
3.2		National Accounts of Well-Being	107
	3.2.1	Time-Based National Well-Being Accounts	108
3.3		Happy Life Expectancy or Happy Life Years	109
	3.3.1	Inequality of Happiness	111
3.4		Closing Thought	111
References			116

Part II Inequality and Poverty 119

4 Inequality 121

4.1		Introduction	121
4.2		Distribution and Moral Theory	123
	4.2.1	Value Claims	124
	4.2.2	Moral Desert	126
	4.2.3	Utilitarianism	128
	4.2.4	Libertarianism	128
	4.2.5	Contractualism	128
	4.2.6	Capabilities	129
	4.2.7	Consequentialist and Deontological Approaches	129
4.3		Measurement of Distribution and Inequality	129
	4.3.1	Inequality of Whom?	129
	4.3.2	Inequality of What?	133
	4.3.3	Measures of Inequality	135
4.4		Population Ageing and Distributional Issues	157
4.5		Intergenerational Transmission of Inequality	161
References			165

5	**Poverty, Deprivation, and Social Class**		175
	5.1	An Embarrassment of Definitional Riches?	177
		5.1.1 Equivalisation	179
		5.1.2 Absolute Poverty	182
		5.1.3 Relative Poverty	187
		5.1.4 Multidimensional Approaches	192
		5.1.5 Mortality-Adjusted Poverty Rates	202
		5.1.6 Subjective Poverty	204
		5.1.7 Financial Distress	205
		5.1.8 Financial Security	207
		5.1.9 Chronic or Persistent Poverty	207
	5.2	Theories of Poverty	214
		5.2.1 Individualist Approaches	218
		5.2.2 Structuralist Approaches	222
		5.2.3 Intergenerational Income Elasticity	223
		5.2.4 Equal Burden-Sharing	227
		5.2.5 The Great Gatsby Curve	229
		5.2.6 Anti-poverty Role of Pension Income in Low-Income Developing Countries	230
	5.3	Social Class and Later Life	231
	References		240
6	**Some Questions of Intergenerational Economics**		255
	6.1	Intergenerational Transfers	259
	6.2	Intergenerational Mobility	261
		6.2.1 Multigenerational Mobility	267
	6.3	Justice Between Generations	272
		6.3.1 Prudential Lifespan	273
		6.3.2 Fair Innings	275
	6.4	Equity, Solidarity, Conflict, and Ambivalence	275
		6.4.1 Generational Equity	276
		6.4.2 Generational Interdependence and Solidarity	279
		6.4.3 Generational Contract	283
		6.4.4 Intergenerational Ambivalence	284
	6.5	Indices of Intergenerational Fairness	286
		6.5.1 The Intergenerational Fairness Index	286
		6.5.2 Intergenerational Justice Index	288
	References		291

7 Ageing, House Prices, and Economic Crises — 301
- 7.1 Introduction — 301
- 7.2 Residential Mobility in Later Life — 303
  - 7.2.1 Ageing, Moving, and House Prices — 303
  - 7.2.2 Housing-Related Financial Products — 307
- 7.3 Housing and Poverty in Later Life — 314
  - 7.3.1 Housing and Risk in Later Life — 316
- 7.4 Housing and Pensions — 318
  - 7.4.1 Housing Income and Public Spending on Older People — 318
  - 7.4.2 Housing and the Retirement Decision — 320
- References — 324

Part III Behavioural Economics and Ageing — 335

8 Behavioural Economics and Individual Ageing — 337
- 8.1 Prospect Theory — 340
- 8.2 Framing Effects — 346
- 8.3 Anchoring Effect — 349
- 8.4 Priming — 350
- 8.5 Sunk Cost Effect — 350
- 8.6 Mental Accounting — 352
- 8.7 Myopia — 354
- 8.8 Lack of Willpower — 354
- 8.9 Complexity — 356
- 8.10 Same Findings, Other Approaches — 358
- References — 360

9 Behavioural Economics and Policy — 369
- 9.1 Libertarian Paternalism — 370
  - 9.1.1 Nudge — 371
- 9.2 Constitutionally Constrained Paternalism — 373
- 9.3 Autonomy-Enhancing Paternalism — 373
- 9.4 Asymmetric Paternalism — 374
- 9.5 The Save More Tomorrow™ Programme — 375
- References — 382

Part IV Political Economy — 387

10 Economics and the Political Economy of Ageing — 389
- 10.1 Introduction — 389
- 10.2 Political Economy of Ageing: The Orthodox Economics View — 391
  - 10.2.1 Population Ageing and the Median Voter Model — 393
  - 10.2.2 The Age of Policy Makers — 406
  - 10.2.3 Elderly Power and Fiscal Leakage — 408
  - 10.2.4 Interest Group Models — 418
- References — 421

11 Gerontological Views — 429
- 11.1 Political Gerontology — 429
  - 11.1.1 Political Participation — 430
  - 11.1.2 Age-Related Franchise Limits — 432
  - 11.1.3 Realignment, Cognitive Mobilisation, and Regret — 434
- 11.2 Social Gerontology and the Political Economy of Ageing — 435
- References — 439

Part V The Silver Economy — 443

12 The Silver Economy — 445
- 12.1 Introduction — 445
- 12.2 The 'Ageing' Consumer — 447
  - 12.2.1 Demand-Driven Market Segmentations — 448
  - 12.2.2 Other Market Segmentations — 452
- 12.3 The Retirement-Consumption Puzzle — 453
- 12.4 Ageing and the Consumer Society — 456
  - 12.4.1 Successful Ageing and the Consumer Society — 457
  - 12.4.2 Affluenza — 460
- References — 461

Part VI Postscript	471
References	472
Glossary: Volume IV	473
Index	477

List of Figures

Fig. 2.1	Mediator and moderator variables	60
Fig. 3.1	Feeling of happiness among population aged 50 or over. (**a**) 2005–2009. (**b**) 2010–2014. *Source: World Values Survey*	103
Fig. 3.2	Life expectancy and happy life years, 2005–2014. *Source: VeenhovenHLYdata*	110
Fig. 4.1	Lorenz curve. *Source: Figure is illustrative, prepared with mock data*	145
Fig. 4.2	Gini, Paglin-Gini, and Age-Gini coefficients of Family Income Distributions United States of America, 1947–1972. *Source: Paglin (1975, Table 3)*	148
Fig. 5.1	Income distribution (after housing costs) and poverty lines of the whole UK population and pensioner households. *Source: Estimation by the author from the Family Resources Survey, United Kingdom (2015/2016)*	191
Fig. 5.2	Subjective poverty line. *Source: Figure is illustrative, prepared with mock data*	204
Fig. 8.1	Prospect theory value function. *Source: Figure is illustrative, prepared with mock data*	343
Fig. 8.2	Prospect theory value function with different loss aversion and risk aversion parameters	344
Fig. 8.3	Prospect theory value function for older people (Watanabe and Shibutani 2010, fig. 3)	344
Fig. 8.4	Sunk cost effect	345
Fig. 10.1	Single-peaked preference of voter A. *Source: Figure is illustrative, prepared with mock data*	396
Fig. 10.2	Single-peaked preferences of two voters. *Source: Figure is illustrative, prepared with mock data*	397

Fig. 10.3 Single-peaked preferences of five voters. *Source: Figure is illustrative, prepared with mock data* 398

Fig. 10.4 Single-peaked preferences of five voters. *Source: Figure is illustrative, prepared with mock data* 399

Fig. 10.5 Median before and after the introduction of a salient dimension. *Source: Figure is illustrative, prepared with mock data* 411

List of Tables

Table 5.1	Different poverty trajectories	208
Table 6.1	Intergenerational mobility matrices	263
Table 6.2	Male occupational intergenerational mobility matrix, Norway 1960–1980	264
Table 6.3	Father-son intergenerational transition probability matrix, Norway 1960–1980	265
Table 6.4	Wealth stock of two generations over time	274
Table 6.5	Conceptual dimensions of intergenerational solidarity	280
Table 7.1	Sources of housing income	308
Table 10.1	Probability of paradox of voting	407

Part I

Economics of Happiness and Quality of Life

1

Conceptualisations and Measurement

> **Overview**
>
> This chapter presents the rationale for the studies of happiness and related concepts in economics and discusses definitions and alternative conceptualisations. Topics covered include utility, welfare and well-being, happiness, human development, and quality of life (the latter particularly in connection with health and long-term care).

It's official: this part has a 'feel good' factor to it. Research shows that paying attention to happiness either by responding to questions about happiness or by self-monitoring happiness levels as part of a study *increases* happiness (honestly; see Bakker et al. 2015; Ludwigs et al. 2017).[1] You may object that answering a few questions or filling a form about how happy you are is not the same as reading about the academic output on the topic. Good point. And I must admit that I do not know of any research that has looked into the relative happiness levels of the students and scholars in the field: do they know something no one else does? Anyway, your focus will be placed on happiness over the two chapters, let's hope that the effect of thinking about happiness is not a question of the means to corral the attention to the topic, but of the actual thinking, irrespective of whether it is by a questionnaire, an interview, or, as in this case, two chapters in a book.

OK, I cheated: this part is not just about happiness (feeling less good already?). The Article 25 of the Universal Declaration of Human Rights, proclaimed and adopted by the United Nations General Assembly in 1948, states: 'Everyone has the right to a standard of living adequate for the health and well-being of himself and of his family...' (Nations 1948). In addition, the signatories of the Declaration of Independence of the United States of America held as one of the 'self-evident truths' that the pursuit of *happiness* is one of the unalienable rights all individuals are endowed with. A right to well-being,[2] a right to pursue happiness...Are these one and the same thing? And what about utility, life satisfaction, welfare, wellness, and quality of life? These are all familiar, everyday terms and are often used interchangeably. They do not need any explanation, do they? In fact, they do. Usually taken as synonymous with no further ado, these are still fairly ill-defined concepts, despite having been subject to thorough analysis and extensive discussion for...centuries.[3] Sometimes economists equate all the various terms seemingly to 'move on'. For example, the opening line in Easterlin (2005b, p. 29) reads:

> I take the terms "well-being", "utility", "happiness", 'life satisfaction", and "welfare" to be interchangeable and measured by the answer to a question such as that asked in the United States General Social Survey (GSS): "Taken all together, how would you say things are these days – would you say that you are very happy, pretty happy, or not too happy?"

Feldman (2010, p. 9) remarked that no philosopher would treat evaluative terms such as well-being like descriptive terms such as utility—a distinction that was lost among almost all economists working in this field![4]

Seeking clarification in the scope and aims of academic journals specialised in these topics renders more heat than light. The *International Journal of Well-Being* defines its scope as 'the general subject or topic of what makes a life go well for someone', so well-being would be what makes life go well. In turn, the *Journal of Happiness Studies* is introduced as 'devoted to scientific understanding of subjective well-being', from which it transpires that 'subjective well-being' is another way of saying 'happiness'. Finally, the *Applied Research in Quality of Life* journal opted for an enumeration, long but still not exhaustive, of connected concepts 'directly related to quality of life and social indicators [which] include happiness, subjective wellbeing, life satisfaction, the good life, the good society, economic wellbeing, family wellbeing, quality of work life, community quality of life, spiritual wellbeing, leisure wellbeing, social wellbeing, emotional wellbeing, psychological wellbeing, and quality of home life, among others'. Among others! Oh, yes, morale, pleasure, instant utilities, psychological wealth, human development, and even 'ophelimity'[5] and 'wantability',[6] for example...

There is more than conceptual quibbling behind this terminological cacophony according to Frawley (2015): it serves a semiotic and ideological purpose as it gives happiness 'a perhaps unique ability to simultaneously act as, at first glance, a free-floating signifier, meaning many things to many people, while at the same time accomplishing its normative alignment' [p. 114].

Alas, this is not the place to settle the definitional conundrums, and—as Machlup (1943, p. 26) noted:

> To complain about the continuous changes of concepts and terms, about the conversion of ideas and names, is to misunderstand the nature of growth of a body of knowledge – at least in the social sciences…When the family of related concepts – related by meaning or by name or both – is large, an occasional examination of the family record and a probing of the relationships is helpful.

This chapter presents such an examination and probing. Notwithstanding, Feldman (2010, p. 6) remarked that some of the contrasting results in happiness studies arise from conceptual misunderstandings:

> One of the greatest sources of misunderstanding in the happiness literature is the fact that two seriously different sorts of enquiry are confused under the title "the attempt to determine what happiness is". On the one hand, there is a question about the most promising causes of happiness…[On the other,] views about what is likely to cause happiness, or to increase happiness; or views about what happiness is "based upon" or how we can achieve happiness.

Research sometimes investigates distinct concepts, although terming the variables of interest the same, so that two papers may have, say, 'happiness' in their title but one is about life satisfaction and the other one tackles subjective well-being. And it is true that with blurry definitions often come fuzzy measurements, results, and interventions. As Poon and Cohen-Mansfield (2011, p. 4) wrote: 'There are countless theories and debates on the definition, meaning, measurement, antecedent, consequences, and impact of well-being'. (And not just of well-being, but, you know, happiness, life satisfaction, etc.) In some studies, despite the fact that the same theoretical framework and definitions are used, the variables are measured using different instruments (e.g. asking disparate questions).

Another potential methodological pitfall is that the instruments used in this area of specialisation are more liable to manipulation and influence by context than those in other areas of social research where the measurement and data collection is done by comparatively more reliable methods (Deaton and Stone 2016), which requires a careful consideration of design, implementation, validation, reliability, and analysis of results. As Blanchflower and Oswald (2011b, p. 7) warned:

> It is likely that debates about the right interpretation of subjective measures will continue throughout the 21st century and beyond

Therefore, this chapter also swiftly reviews the most common quantitative instruments used to measure happiness and related concepts, as a familiarity with how these concepts are operationalised helps understand and assess the findings in the literature.

One last point before we embark on this journey: Gilhooly et al. (2005) commented that in gerontology there has been a shift away from the use of 'happiness' and 'well-being' in favour of 'quality of life'. Little (1950) commented that classical economists were concerned about 'wealth', then an interest in 'happiness' and 'satisfaction' ensued, and it was only with the publication of Pigou (1932) that the term 'welfare' was popularised. In turn, we are witnessing a recent shift within the economics profession back towards 'happiness': the 'economics of happiness' is one of the hottest fields of specialisation in terms of publications, for instance. As we will see, other economists work on 'quality of life' and still others prefer to use somewhat related terms of equally widespread currency such as welfare, well-being, and human development. Frawley (2015, p. 4) proposed that among all the different signifiers (quality of life, subjective well-being, etc.), it was 'happiness' which became 'the main "sign vehicle" through which claims about the existence of a new social problem initially made their way onto the public stage', …in a context of declining beliefs about the perfectibility of society and where the unchallenged ascendency of the market has left the sphere of the individual and interpersonal among the few sites open to change [p. 17].

Let's see, now, what these concepts mean.[7]

1.1 Utility

Utility has two main meanings in economics. In one interpretation, it corresponds with the satisfaction that an economic agent derives from the goods and services she consumes (what is also known as the 'value-in-use' or 'use value' of a commodity). In this view, it is 'the capacity of a good to satisfy desire'

(Viner 1925, p. 369) or want. The alternative interpretation in economics adopts a behavioural view of utility, according to which it corresponds to the desire or preference for a choice rather than the actual satisfaction derived from it.

Both conceptualisations, nevertheless, assume that utility derives from decisions. In other words, according to mainstream economics, decision utility is the only type of utility, the only game in town. In contrast, behavioural science has distinguished between decision, predicted, and experienced utility (which is divided into instant and remembered utility) (Kahneman et al. 1997). Predicted utility refers to the expected pleasure (or pain) derived from future goods, services, activities, and situations. Experienced utility is the hedonic and affective quality of the goods, services, activities, and situations (see also Berridge and ODoherty 2014). Each of these options have a concurrent, immediate level of hedonic and affective quality (instant utility) and a retrospective, time-prolonged quality (remembered utility) as when someone remembers fondly that particular afternoon.[8] Experienced utility comes closer than decision utility to the notion of satisfaction. Carter and McBride (2013) showed that the average experienced utility function is S-shaped: the relationship between net expected payment levels and average satisfaction is not linear but fits an S-shaped curve.

Economists also distinguish between cardinal and ordinal utility. Cardinality defines utility from a hedonistic point of view as a measurable psychic quantity: the levels of satisfaction that an economic agent derives from the goods and services she consumes. An understanding of utility as the capacity of satisfaction interprets it as a hedonistic experience. An understanding of utility as the desire for an action interprets it as ordinal. The hedonistic conceptualisation is also known as 'experience utility', whilst the alternative view is known as 'decision utility' (Dolan and Kahneman 2008; Kahneman et al. 1997). The cardinal, hedonistic, experience approach posits that utility can be measured, just like height, temperature or waist circumference. Instead, the ordinal, decision approach posits that utility is subject to ranking rather than measurement—see Strotz (1953) and Van Praag (1991). Ordinality assumes that comparisons between options, alternatives, choices and bundles of commodities can be made regarding the utility they generate, but measurements of the utility the goods generate are assumed to be either impossible, meaningless, or unnecessary: for ordinality, whether utility is also cardinal or not is irrelevant; what would be relevant is which option is better, not by how much (Samuelson 1938). The discussion around cardinal and ordinal utility has been active since the nineteenth century. Suffice to say here that it has ramifications in most branches of economics; for example, as Van Praag (1991, p. 71) pointed out:

> ...the whole literature on income inequality and poverty would be reduced to a sterile exercise if we do not accept the implicit cardinal utility measurement and interpersonal comparability on which these concepts are based.

It is worth noting that not only income can be treated as a cardinal or ordinal magnitude: so can status (Bilancini and Boncinelli 2008), affect (Pham et al. 2015), health (Wouters et al. 2015), and so on.

Two usual measures of utility and preferences are income and QALYs—see Chap. 5 in Volume II. Income (either at the individual or household level) is used to appraise the distributional impact of policy measures. Utility is equated by income (as a proxy), and individual lifetime utility is represented by the discounted sum of flow or instantaneous (i.e. one-period) utility over the life course of an economic agent. The assumption is that income (money in general) is subject to diminishing marginal utility: the utility of an additional amount of money is worth more the less money an individual or household has or earns. The assumption, then, is that, for example, the utility of a person earning 100 dollars would grow more after a 10 dollar increase in her income than that of someone earning 1000 dollars after an increase in her income by 10 dollars.

Behavioural economics has identified several decision mechanisms, heuristics, and that provide ample evidence that economic agents do not make rational decisions (see Part III in this volume). Consequently, even if they had enough income to choose the bundle of goods and services that would maximise their decision utility, they would not maximise their experienced utility (Kahneman et al. 2004b). Furthermore, a number of findings in the happiness literature suggest that income and happiness or satisfaction do not correlate very highly (see below).

From a different theoretical perspective, Saint-Paul (2011) went as far as to consider the equality between utility and happiness a 'fallacy'. In fact, this author made the distinction that utility is about what economic agents want whilst happiness is about how they feel:

> Feeling well is one of their goals, but they have other goals as well. Status, ambition, survival, wealth accumulation, building a family, testing one's physical and intellectual limits, and achieving fame are other goals that are equally respectable, and often more respectable, than just feeling well. Only the individual can make a statement about the relative importance of those goals. In identifying utility

with happiness, researchers are making their own paternalistic statement that individuals should care more about feeling well, relative to other goals, than they actually do.

(Saint-Paul 2011, p. 57)

1.2 Welfare

Closely related with the concept of utility is one of the two meanings of 'welfare' in economics. Welfare is used either to refer to the total amount of utility experienced from the consumption of goods or services—as in a branch of microeconomics known as 'welfare economics'—or to the public subsidies and transfers towards 'health, education, housing and employment as well as social security' (DSS—The Welfare Reform Green Paper Consultation Team 1998, p. 9), such as in the 'welfare state' or the 'welfare system'. This second meaning of the word 'welfare' is not connected with the meanings of happiness, quality of life, and the like. Regarding the first meaning, a common conceptualisation of welfare is to assume the existence of a mathematical function of individual income, either cross-sectionally or over time, either for a given society ('social welfare') or an individual (e.g. in the concept of welfare as 'lifetime utility'). These conceptualisations reflect a cardinal view of utility. Utilitarian, Rawlsian, and other social welfare functions (in the sense of total amount of experienced utility) entail the assumption that utility (and welfare) can be measured and that such measurement is meaningful.

An implication underlying this understanding of welfare is that, by assuming that each individual is rational and that a rational individual seeks to maximise her welfare and that she knows what to choose in order to do so, the utility function can be equated to preferences and the choices to individual welfare, so that if an economic agent prefers *a* to *b*, her welfare is higher with *a* than with *b*. This is the basic tenet of the 'revealed preferences' approach. One problem with revealed preferences is that choices may not always translate into higher welfare. Scitovsky (1976, p. 4) observed that this equating of choice and what is best for an agent 'seemed to rule out—as a logical impossibility—any conflict between what man chooses to get and what will best satisfy him.' An agent may act altruistically on behalf of others despite knowing that her actions will reduce her own utility. Moreover, some may take steps now to preclude getting what she knows she will prefer in the future despite not being good for her, that is to preclude any temporal inconsistency in her preferences (as when a friends asks you: 'do not give me that second drink when I ask for it' Schelling

1984, p. 1). However, others may not exhibit such a strong self-determination and self-control, of course. That is why some economists prefer to separate preferences from welfare on the basis that if an economic agent prefers (or chooses) a to b, it does not necessarily follow that $W(a) > W(b)$ (where $W(.)$ denotes welfare). In connection with this point, Ng (1979) distinguished between self-concerned, self-minded, self-attending, self-regarding, and self-centred agents. Self-concerned agents only concern for the welfare of others if their own welfare is affected. The welfare of self-minded individuals is not influenced by the welfare of other agents, but their preferences may be influenced by other agents' welfare. The welfare of self-attending agents is not affected by the activities of other agents. Self-minded and self-attending agents are self-regarding. Self-centred agents are, at the same time, self-minded and self-concerned.

Given the many mismatches between choice and preferences that vitiate the theoretical construct of revealed preferences, Easterlin (2005b) highlighted the superiority of direct, self-reported measurement instruments. In contrast, other authors defend the revealed preferences approach as it allows comparisons between the utility derived from choices that have been made and that from decisions not taken (Saint-Paul 2011).

1.3 Subjective and Objective Well-Being

Well-being may refer to a mental state or to living conditions. In the case of a mental state, it is usually termed 'subjective well-being'; when it refers to living conditions, sometimes it is known as objective well-being. Objective living conditions are tenuously associated with subjective well-being, which provides justification for the separation of both concepts (Walker and Lowenstein 2009).

1.3.1 Subjective Well-Being

Subjective well-being has been defined as "a broad category of phenomena that includes people's emotional responses, domain satisfactions, and global judgements of life satisfaction" (Diener et al. 1999, p. 277) or as 'Good mental states, including all of the various evaluations, positive and negative, that people make of their lives, and the affective reactions of people to their experiences' (*OECD Guidelines on Measuring Subjective Well-being* 2013, p. 10) (see also Shmotkin 2011). According to Diener et al., subjective well-being is less a concept than a whole research field in which affective reactions and cognitive evaluations of life satisfaction are studied in various domains (work, family, health, the self, etc.). Other authors maintain the distinction between affective and cognitive-judgmental components but as parts of subjective well-being inasmuch as a theoretical construct, not as a separate field of research. Under the banner of subjective well-being, two concepts are usually confounded: emotional well-being and life satisfaction. Emotional well-being (also known as experienced happiness and hedonic well-being) refers to the frequency and intensity of positive and negative emotions in everyday life (e.g. how often and how intensely an individual feels anger or joy). Life satisfaction (also known as life evaluation) refers to what an individual thinks about or how she assesses her life; hence, it is considered to be the cognitive component of subjective well-being. These two concepts suggest that the perception and evaluation of life situations are key determinants of subjective well-being (Rudinger and Thomae 1990).

Several instruments to measure subjective well-being have been developed based on self-reported indicators of psychological states, including[9]:

- The Positive and Negative Affect Schedule (PANAS), introduced by Watson et al. (1988), which focuses on affect and measures emotional well-being; the Satisfaction With Life Scale (SWLS) presented by Diener et al. (1999), which focuses on life satisfaction; the Self-Anchoring Striving scale or 'ladder of life' (Cantril 1966; Cantril and Roll 1971; Kilpatrick and Cantril 1960), which combines both components of subjective well-being; and the Multidimensional Personality Questionnaire ((MPQ), which although measures various aspects of personality, includes a well-being trait dimension (see Tellegen and Waller 2008[10]). PANAS consists of a list of twenty feelings and emotions—ten of positive affect and ten of negative affect. Positive affect feelings and emotions include interested, alert, excited, inspired, strong, determined, attentive, enthusiastic, active, and proud. Negative affect feelings and emotions include irritable, distressed, ashamed, upset, nervous, guilty, scared, hostile, jittery, and afraid. Respondents are invited to express along a scale from 1 (denoting very slightly or not at all) to 5 (extremely) the extent to which they feel right now or had felt over the previous week regarding each feeling and emotion in the list.
- the Personal Well-Being Index (PWI). The PWI, which was developed as one of the two elements of the Australian Unity Wellbeing Index, measures satisfaction levels in seven different personal life domains (Cummins et al. 2003): standard of living, health, achievements in life, personal relationships, feeling of safety, community connectedness, and security about the

future. This index was used by Rodriguez-Blazquez et al. (2011) and Rojo-Perez et al. (2014) among older people in Spain.
- The SWLS consists of five statements; respondents are invited to indicate their agreement along a scale ranging between 1 (denoting strong disagreement) and 7 (strong agreement). The statements are:
  - In most ways my life is close to my ideal
  - The conditions of my life are excellent
  - I am satisfied with my life
  - So far I have got the important things I want in life
  - If I could live my life over, I would change almost nothing
- The Self-Anchoring Striving scale invites respondents to describe what life would be like if they were to imagine their future in the best possible light. This is followed by an open-ended question about what it would be like under the worst possible light and it is accompanied by a drawing of a ladder with eleven rungs numbered from 0 to 10, which respondents are told represents all their hopes and aspirations (where 0 corresponds to the worst state of affairs and 10 the best). Respondents are then asked to indicate where they stand along the ladder in terms of their aspirations at the moment and where they think they will in five years' time. The 'self-anchoring' element in the title of the instrument refers to the feature that any given score is comparable among people of different backgrounds, chronological ages, and so on, even though their hopes and aspirations may (and certainly do) differ: their sense of accomplishment regarding hopes and aspirations would be the same if they placed themselves on the same rung. This instrument allows to elicit the substance of people's hopes and fears, which may be construed as the basis for a conceptualisation of quality of life with subjective weighting factors. Imagine, for example, that 95 per cent of respondents said that life would be best if they had excellent health; this would give support, then, to include indicators of health as part of an operational definition of quality of life, and would give health a greater weight than another domain which, say, 72 per cent of respondents agreed would be best they had or experience, and so on.
- The well-being trait in the MPQ scale includes twenty-three items. Individuals who exhibit higher scores on well-being tend to have a cheerful happy disposition; feel good about themselves; see a bright future ahead; be optimist; report they live interesting, exciting lives; and say they enjoy the things they are doing.
- the Scale of Positive and Negative Experience (SPANE). It is a measure of affect (Diener et al. 2010) that asks respondents to score from 1 (very rarely

or never) to 5 (very often or always) how much they experienced each of a number of feelings, including positive and negative, good and bad, happy and sad, and so on, during the previous four weeks. The PWB (also known as the Flourishing Scale—see also Huppert and So 2013) is formed of a list of statements about which respondents are asked to indicate their agreement along a 7-category scale from 1 (strongly disagree) to 7 (strongly agree). The statements are:

- I lead a purposeful and meaningful life
- My social relationships are supportive and rewarding
- I am engaged and interested in my daily activities
- I actively contribute to the happiness and well-being of others
- I am competent and capable in the activities that are important to me
- I am a good person and live a good life
- I am optimistic about my future
- People respect me

- the Gallup-Sharecare Well-Being Index (previously known as Gallup-Healthways Well-Being Index (GHWBI)) (Gallup 2018) is a measurement instrument that combines both elements of subjective well-being as it includes questions about emotional well-being and the Self-Anchoring Striving scale to measure life satisfaction. It is obtained by averaging indicators across five domains: purpose, supportive relationships, finances, community, and health.
- the Life Satisfaction Index (LSI-A) (Neugarten et al. 1961) consists of five components, each measured on a five-point scale, giving a rating between 5 (lowest) and 25 (highest). (There is a shorter version, the LSI-Z). The five components are:

  - Zest (as opposed to apathy): the extent to which an individual takes pleasure from daily activities
  - Resolution and fortitude: how much an individual considers her life as meaningful and accepts what how her life has been
  - Congruence between desired and achieved goals: the extent to which an individual feels she has achieved her goals in life
  - Positive self-concept: the extent to which an individual holds a positive image and conception of herself
  - Mood tone: how optimistic an individual is

  The LSI-A and LSI-Z scales have been used to assess life satisfaction among sub-groups of older adults, for example, people with different levels of frailty, housebound, recently bereaved, nursing home residents, and so on, and also its association with social support, functional ability, self-reported health status, retirement and work trajectory, ageism, and so on.

Longitudinal studies found that LSI in later life diminishes with increased financial stress (Chou and Chi 1999, 2002)—see Aquino et al. (1996) for a similar finding but using a different measure of life satisfaction—and with poverty and deprivation (Cheung and Chou 2017).

- the CASP-19 scale (see below)

In policy-focused and macro studies, subjective well-being is measured by asking about levels of life satisfaction and reporting percentages of respondents who say they have high levels, and so on. Several surveys and studies ask a single question about current overall life satisfaction (e.g. 'how satisfied are you with your life nowadays?'), which respondents are invited to score along a scale. Questions such as this elicit individuals' cognitive assessments of life in general with no specific life conditions or time frames being primed. Depending on the scope of the study and the framework guiding the research, questions about how happy, how worried, and how depressed the respondent feels at the moment or felt the previous day are also included. Sometimes, instead of, or in addition to, a single overall question, satisfaction with specific areas of life (e.g. personal relations, financial situation, area of residence) is also surveyed.

1.3.2 Objective Well-Being

As I already pointed out, when well-being refers to living conditions and life chances and events (Oishi 2010), it is sometimes termed 'objective well-being'. Objective well-being is either quantified by direct measures such as life expectancy at birth and gross national product per capita or by indirect measures such as implicit pricing.

The Organisation for Economic Development (OECD) developed a framework for measuring well-being, which combines indicators from two domains: 'economic well-being' or material living conditions (including income and wealth, availability and quality jobs, earnings, housing) and indicators of quality of life defined as 'the set of non-monetary attributes of individuals [that] shapes their opportunities and life chances, and has intrinsic value under different cultures and contexts' [p. 5], including: 'health status, work-life balance, education and skills, social connections, civic engagement and governance, environmental quality, personal security and subjective well-being' (OECD 2013a, p. 1). The OECD's framework is based on the capability approach developed by the Indian economist Amartya Sen (Sen 2001), in which subjective well-being is only one element of a wider conceptualisation of well-being. Well-being is a multidimensional concept: in a study of objective well-being in later life in the United Kingdom, Green et al. (2017) found forty significant indicators—the most important ones being participation in creative and cultural activities, physical activity, cognitive skills, mental well-being, and educational attainment.

The implicit price (also known as hedonic or shadow price) of an amenity or characteristic of a good or an area is equivalent to the price differential between two goods or areas only differing in that particular amenity or characteristic. If any two goods almost equal except for one aspect differ in price, that price difference is known as the implicit, hedonic, or shadow price. For example, the difference in price between fair trade and non-fair trade products or the property prices of exactly the same house in two neighbourhoods equivalent in terms of parks, crime rates, transport links, cultural activities, and so on, but differing in the quality of the primary schools. In all these cases, the price differential corresponds, implicitly, to the value placed by consumers on the distinguishing characteristic. Srinivasan and Stewart (2004) applied implicit pricing to estimate quality of life in fifty-five counties in England and Wales. They tested the relationship between a set of 'amenities' and 'inconveniences' and market prices that would reflect location decisions by firms and households, in order to estimate the implicit prices of those area characteristics. The findings suggest that differentials in house prices and hourly wages are associated mostly with differences in air pollution and population density.

1.4 Psychological Well-Being

Psychological well-being combines subjective well-being with positive psychological functioning. Drawing from different theoretical perspectives (lifespan development, self-actualisation, personal growth models, and positive mental health), Ryff (1989) developed the conceptualisation of psychological well-being (PWB) as an alternative to subjective well-being. PWB measures socio-psychological prosperity and consists of the following six domains, which would be 'the key goods in life central to positive human heath', according to Ryff and Singer (1998, p. 3):

- purpose in life, or meaning in life: having a sense of direction
- the quality of connections with others: trusting, close relationships

- self-acceptance or self-regard: to hold a positive image and attitude of oneself, and to accept good and bad aspects of one's personality
- environmental mastery: the capacity to mould the world around us creatively in a way that suits one's psychic condition
- autonomy: inner self-control and regulation
- personal growth: to aim and strive to develop and expand one's potential

The PWB scale is a metric developed by Ryff and Keyes (1995) and Ryff and Singer (1996) to assess these domains. Ryff (1989) compared young, middle-aged, and older adults along four of these domains: environmental mastery, personal growth, purpose in life, and autonomy. Compared to young adults, older people exhibit higher scores in environmental mastery and lower scores in the other three domains. However, against middle-aged adults, older people showed significant reduced ratings in purpose of life and personal growth only.

Demakakos et al. (2008) looked into well-being among people aged 50 or over in England between 2004–2005 and 2008–2009. Their definition of well-being comprised life satisfaction (measured with the SWLS), quality of life (CASP-19), negative affect (measured with a depression scale), and a loneliness scale. The authors did not attempt to combine these indicators into a single index; instead, they considered each one at a time. The findings showed that wealth was associated with the four components of well-being and that life satisfaction and quality of life deteriorated and loneliness increased during the period, independently from wealth. Another important result was that limitations with activities of daily living (see Volume II, Chap. 6) significantly impacted on all four measures.

1.5 Optimal Well-Being

Optimal well-being (OWB) is the combination of high subjective and psychological well-being. Keyes et al. (2002) showed empirical evidence that despite some overlapping, SWB and PWB are two distinct constructs. Consequently, a valid typology resulted from the cross-classification of both measures. These authors reported that the older the individual (in a US-based sample of adults aged 25–74 years), the higher the profiles of SWB (Cantril's ladder) and PWB (Ryff's et al. scale), hence the more likely she would exhibit optimal levels of well-being (the odds also increased with educational attainment). The probability of having high SWB and low PWB also increased with chronological age: later life would be associated with greater contentment and satisfaction with life but lower thriving than youth.

1.6 Happiness

Another concept related to utility, welfare, and well-being is that of happiness. Ng (1979, p. 2) stated that 'individual welfare may be taken as an individual's well being, or more explicitly, his happiness, taking happiness to subsume both sensual pleasure and pain and spiritual delights and sufferings'. One of the leading authors in the field, the Dutch sociologist Ruut Veenhoven, defined happiness as 'the degree to which an individual judges the overall quality of his life-as-a-whole favorably' (Veenhoven 1984a, p. 22), and went on to say that happiness is 'an essentially experiential phenomenon which cannot be identified with particular external conditions or with a way of life' [p. 38]. Nevertheless, years later, this same author went on to state that 'happiness can also be called "life-satisfaction"' (Veenhoven 1991, p. 2) and also that 'in the end, happiness is determined by need gratification' (Veenhoven 2011, p. 7). Combining these statements, happiness can be defined as 'a subjective evaluation of how satisfied a person is with her life in all its dimensions and facets as a result of the degree that her needs are met'. This approximation to a definition of happiness is not dissimilar to the understanding of utility by the founders of the Austrian School in the nineteenth century (Eugen Böhm von Bawerk and Carl Menger, in particular) in terms of 'a psychic reality, a feeling that was evident from introspection, independent of any external observation', as Schumpeter (1994, p. 1026) explained.

The branch in economics known as the 'economics of happiness' focuses on this conceptualisation of happiness akin to a subjective measure of life satisfaction. However, some psychologists distinguish between life satisfaction and happiness: the latter would be transitory and more changeable, dependable on external circumstances, whilst life satisfaction would be more stable and temporally reliable. Happiness would reflect the effect of positive and negative life experiences, even ongoing and transient ones, and consequently would capture affective aspects; life satisfaction would reflect a relationship between lifetime aspirations and realisations in general (George 1981; Larsen et al. 1985).

With roots in Greek classic philosophy, happiness and well-being are usually classified into hedonistic (or hedonic) and eudaimonic (or eudaemonist)[11] (Delle Fave et al. 2011). From Socrates onwards, a basic tenet of ancient Greek moral philosophy is that happiness is the final end of all human actions—what Vlastos (1991, p. 203) termed the 'Eudaemonist Axiom'—and that any disagreement among Greek classical philosophers on this topic hovered around the reasons for this to be so, that is, around the link between virtue and happiness (Vlastos 1991) (see also Shiner and Jost 2003). Philosophers of a hedonistic persuasion, following Epicurus, opined that virtue had only an instrumental value insofar as it would lead to pleasure and absence of pain. Other philosophers, following Aristotle, opined that happiness is not the only thing individuals should desire for its own sake—honour, pleasure, and other virtues would also have intrinsic value. Finally, the Cynics and the Stoics opined that virtue is equal to happiness, its only component. Drawing on this distinction, psychologists and economists have classified happiness into hedonic and eudaimonic. The concept of hedonic happiness is based on avoiding pain and maximising pleasure, on positive emotions and, therefore, it finds its philosophical grounding in utilitarianism. The concept of eudaimonic happiness is based on functionings, fulfilment, and interactions, and on meaning and purpose; it finds its philosophical grounding in virtue ethics. Several authors have attempted to integrate both approaches: for example, Seligman (2002) proposed three paths to happiness—pleasure, engagement, and meaning—and Keyes (2002) proposed the notion of flourishing and the mental health continuum, combining emotional, social, and psychological indicators. Seligman's approach is particularly relevant to studies of happiness in later life (see below) because it classifies positive emotions into those directed to the past, the present, and the future (Seligman 2002, Appendix). Positive emotions directed to the past include satisfaction, contentment, pride, and serenity. Positive emotions directed to the present comprise two categories: pleasures and gratifications. Positive emotions directed to the future are optimism, hope, confidence, trust, and faith.

1.7 Human Development

According to Alkire (2002, p. 182), human development comprises 'human flourishing in its fullest sense—in matters public and private, economic and social and political and spiritual' and consists of well-being attainment, agency—that is, 'what [individuals] are able to do about the causes they follow'—and 'nonindividualist aspects of social living that are of utmost importance'. In turn, the United Nations defined 'human development' as the 'process of enlarging people's choices' among which it mentioned 'to live a long and healthy life, to be educated and to have access to resources needed for a decent standard of living …, political freedom, guaranteed human rights and personal self-respect' (UNDP 1990, p. 1). Both definitions denote that the concept of human development is rooted in the capability approach (Deneulin and Shahani 2009; Land 2015).

Human development has been operationalised as the combination of life expectancy, education, and income. The most well-known composite measure is the Human Development Index (HDI), which the United Nations compiles annually (UNDP 2016). The HDI is calculated at a national level and is composed of three variables: life expectancy (as a measure of a 'long and healthy' life), years of schooling (as a proxy for knowledge), and gross national income per head (as a proxy for standard of living). The United Nations also compiles a life course and gender-gap version of the HDI. For older people, the indicators are life expectancy at age 50, and old-age pension recipients. The gendered version of the HDI for older people combines female life expectancy at age 50 and the percentage of women in receipt of old-age pensions as a proportion of all recipients.

A related concept is sustainable development: a process that 'requires total capital—that is, economic capital, human and social capital and environmental capital—to be non-decreasing' (DETR 1999, p. 12). Sustainable development expands the concept of sustainable growth in economics by adding the requirement that human, social, and environmental capital grow at a 'sustainable' rate as well as the economy.[12] Therefore, sustainable development can be also defined as 'time-consistent quality of life' (Iparraguirre 1992, p. 11).

1.8 Quality of Life

Quality of life in general, and in later life in particular, is possibly the least clearly defined and, at the same time, the most broadly encompassing concept related to happiness and well-being. To illustrate: the United Kingdom's Economic and Social Research Council (ESRC) allocated £3.5 million in 1998 (just over £5 million in 2017 prices) to a research programme of which one topic area that comprised five projects was devoted exclusively to the definition and measurement of quality of life in old age (Walker 2005; Walker and Hagan Hennessy 2004). The complexity and multidimensionality of the concept of quality of life is reflected in the following two well-known definitions:

- The World Health Organisation defined quality of life as (WHO 1994, p. 17):

  an individual's perception of his or her position in life in the context of the culture and value system where they live, and in relation to their goals, expectations, standards and concerns. It is a broad ranging concept, incorporating in a complex way a person's physical health, psychological state, level of independence, social relationships, personal beliefs and relationship to salient features in the environment.

- Lawton (1991, p. 6) offered this structural definition:

    Quality of life is the multidimensional evaluation, by both intrapersonal and social-normative criteria, of the person-environment system of an individual in time past, current and anticipated.

Quality of life 'is an irreducible network of interwoven parts, encompassing elders themselves (mind, body, and spirit), their animate and inanimate environment, their life experiences in space and time, and the functions or powers created by the interwoven parts'. This irreducibility has, almost by necessity, turned quality of life into a multidimensional, amorphous, complex, and holistic construct (Mollenkopf and Walker 2007), applicable to all facets and domains of life. For example, in a regression analysis of self-reported global quality of life among people aged 65 or over in Great Britain, Bowling et al. (2002) found that the variables with highest explanatory power were social comparisons and expectations, personality and psychological characteristics, health and functional status, and social capital.

In a survey of the literature, Brown et al. (2004) identified nine different models of quality of life:

- Objective indicators (based on measures of standard of living, health, housing, education, etc.)
- Subjective indicators (based on measures of life satisfaction and psychological well-being, happiness, morale, etc.)
- Satisfaction of human needs (based on a combination of objective measures of deprivation and self-actualisation and social belonging)
- Psychological models (based on measures of cognitive ability, personality traits, social competence, etc.)
- Health and functioning models (based on measures of health status, disability, depression, etc.)
- Social health models (based on measures of social networks and integration with local community)
- Social cohesion and social capital (mainly based on measures of community and area-level resources, and trust, criminality, etc.)
- Environmental models (based on place of residence and factors that influence participation and independence)

- Ideographic or individualised approaches (based on values, perceptions, satisfaction, etc.)

'Quality of life' can be seen, then, as an umbrella term denoting what such indicators show about a population or subpopulation or about a region or country. In academia, the concept of quality of life began in the 1970s, though in policy circles it appeared by the 1950s mostly in relation with environmental and urban living conditions (Katz and Gurland 1991). Attempts were made to produce a summary figure or index, while other initiatives opted for a list of components or scorecard, generally without considering their interrelationships. As with any other single figure formed by aggregating different elements, there is a gain in the simplicity behind aggregate indices. However, it is easy to forget the complex weave of indicators that are linked together, sometimes implicitly, by any given conceptual model: the simplicity of a single index is more apparent than real. The scorecard approach differs from aggregated indices in that it does not amalgamate a number of selected variables into a unique construct. Instead, it presents a list of indicators to scrutinise the level and evolution of quality of life in its various, predefined dimensions albeit with no attempt to weigh either the indicators within the dimensions or the dimensions themselves. Here the risk is, as Katz and Gurland (1991, p. 340) noted, that without taking into account the interconnectedness between the components of quality of life, 'understanding is incomplete and, too often, of limited use for life's important decisions'.

The domains and indicators can be elicited from opinions from experts, and reviews of and findings in the academic literature, or from the population under study (e.g. older people in a country). These definitions are known as 'expert' and 'lay' definitions, respectively. To measure quality of life across geographical areas, the estimation of the implicit prices of given characteristics within a region or country is also used, as was described in Sect. 1.3.2 above. Most of these measures bring together elements from different domains of life, including social, economic, institutional, and environmental, as well as subjective assessments of life satisfaction (which, according to Bond and Corner (2004), is the most important domain of quality of life). Depending on the focus of the study, personality traits are also included. Studies of quality of life in older people tend to focus on functioning or competence (physical, social, cognitive, emotional, and behavioural; sometimes, including sexual) and perceptions about health and satisfaction with life (Arnold 1991). In some instances, there is a bias or excessive emphasis towards the negative aspects of old age. For example, Netuveli and Blane (2008, p. 114) opined that there was value in studying the quality of life of older people given that

> ...the elderly are peculiarly vulnerable due to (1) declining physical and mental capabilities; (2) exit from labour market with greater dependence on pensions; (3) break down of extended families; and (4) isolation due to death of contemporaries, especially that of spouse or partner

The expert opinions approach is related to the 'social indicators' movement that started in sociology in the 1960s in reaction to the then prevalent view in policy circles that economic growth was always conducive to a better life. Eventually, some social gerontologists advanced the notion of the importance of lay definitions, on the premise that older people themselves would be best positioned to evaluate their own quality of life (Bowling and Gabriel 2007; Farquhar 1995). Epistemologically, researchers who support the use of lay definitions opine that survey-based studies fail to give the respondents truly a voice and treat older people merely 'as passive vessel of answers...repositories of facts and the related details of experience' (Holstein and Gubrium 1995, pp. 7–8). The lay definitions approach seeks to actively engage older people in the co-production of knowledge about, in this case, the quality of later life.

Two examples of this approach are Bowling (1995a) and Bowling et al. (2003), which have tried to elicit the domains of quality of life in later life, as well as their relative importance, from responses by older people themselves living in Great Britain. The lists resulting from both studies are not wholly comparable due to differences in the methods, but provide an overall indication of the constituents of a life of quality according to older people (in the mid-1990s and mid-2000s in Britain). The 1995 study identified own health, family relationship, health of a close person, standard of living, and social activities as the five most important areas of life. The 2003 study listed social relationships, social roles and activities, other activities done alone, health and psychological outlook and well-being (i.e. optimism/pessimism, enthusiasm/frustration, etc.) among the five most relevant realms. Researchers who replicated these studies over time reported slight changes in the rankings of items but fairly consistent compositional elements of quality of life according to older people. Bond and Corner (2004) concluded that these studies suggest that subjective aspects of life are more important than objective indicators. In the same vein, in a study of correlates with responses to the open-ended question 'What is quality of life for you?' asked to a sample of older people in Göteborg, Sweden, Wilhelmson et al. (2005) found that social relations was the most important domain followed by health, activities (travel, walks, 'a good book', etc.), and functional ability (albeit with marked differences in the rankings for men and women).

1.8.1 CASP-19

Based on the needs satisfaction theory, Hyde et al. (2003) proposed the CASP-19 as a measure of quality of life in old age. It encompasses nineteen items over the following four domains (defined by):

- control
- autonomy
- pleasure
- self-realisation

Control is 'the ability to actively intervene in one's environment' and autonomy is 'the right of an individual to be free from the unwanted interference of others' (McGee et al. 2011, p. 268). These two items would reflect 'prerequisites of free participation in society'. Pleasure is 'the sense of happiness or enjoyment derived from engaging with life', and self-realisation has been defined as 'the fulfilment of one's potential' (McGee et al. 2011, p. 268). These two domains would capture 'the more active and reflexive dimensions of being old in an increasingly complex society in which "post-materialist" values prevail' (Wiggins et al. 2004, p. 696).

Netuveli et al. (2006) applied the CASP-19 measure to data from people aged 50 or over in England in 2002 and reported that it was negatively associated with depression, poor perceived financial situation, mobility limitations, disability, and limiting long-standing illness. In turn, it was positively associated with trusting relationships and frequent contacts with family members and friends, positive perception of the living environment, and having two cars in the household. The authors concluded that quality of life in later life could be improved by tackling financial hardship, functional limitations, trusting personal relationships, and the quality of neighbourhoods.

1.8.2 Long-Term Care

Other meanings of 'quality of life', not related to happiness, life satisfaction, and the rest, are used in specific domains: the so-called focused definitions of quality of life. Two notable areas of extensive development and research on quality of life are long-term care and health. Some authors have recommended the use of measures of overall quality of life and domain-specific quality of life for economic evaluations on the grounds that they do not capture the same factors and are complementary to each other—see, for example, Davis et al. (2013) for such a proposal in the context of falls prevention for older people.

Volume II, Chaps. 5 and 6, introduced some instruments to measure quality of life in health and social care settings. For example, the CarerQol stated preferences method used to measure the monetary value of caregiving services (Brouwer et al. 2006; Hoefman et al. 2017) and the SF-36 and EQ-5D indices of health-related quality of life. Various tools have been designed to measure long-term or social care-related quality of life (SCRQoL). Most instruments focus on the quality of life of the recipient of the service, although some measure the quality of life of the caregivers who provide the services. Besides these objective indicators of quality of life in long-term care institutional settings, a number of studies have used direct methods to elicit the opinions of residents (and sometimes their relatives) about the elements, characteristics, activities, or functionings that influence the quality of their lives the most. Some domains consistently feature in these definitions of quality of life, such as the quality and ethos of the care services, activities, and therapies, aspects of personal identity and agency, and connectedness to family and the wider community (Cooney et al. 2009; Kwong et al. 2014; Murphy et al. 2007; Yeung and Rodgers 2017).

ASCOT

One SCRQoL is the Adult Social Care Outcomes Toolkit (ASCOT), which led to the NI 127, a composite indicator of quality of life among recipients of social care produced by England's National Health Service. According to the Department of Health, this 'gives an overarching view of the quality of life of users based on outcome domains of social care related quality of life' (DoH 2011, p. 8).[13] Iparraguirre and Ma (2015) used this measure of quality of life as a proxy for care outcome to benchmark quality of care service provision across local authorities in England. The NI 127 covers eight different domains:

- Control: how much control the individual has over her daily life
- Personal care: keeping clean and presentable in appearance
- Food and nutrition
- Accommodation: how clean and comfortable the respondent's home or care home is

- Feeling of safety
- Social participation: how much contact the individual has with people she likes
- Occupation (how the individual spends her time)
- Dignity: the way the individual is helped and treated

In addition, some studies have focused on the perceptions and opinions about the quality of life of the residents by their family members (Robichaud et al. 2006) and staff (Thapa and Rowland 1989), providing additional facets to the topic, but which have proved challenging so far to combine or incorporate into a cost-effectiveness framework. Tools that measure carers' quality of life contribute with additional intelligence on the costs of care provision to, especially, informal and unpaid carers. Therefore, their inclusion in cost-effectiveness studies has been recommended (Goodrich et al. 2012; Hoefman et al. 2013; Al-Janabi et al. 2011). One such valuation tool for assessing the quality of life of informal care providers is the CarerQol instrument (Brouwer et al. 2006; Hoefman et al. 2017).

CarerQol

CarerQol consists of two parts, the CarerQol-7D and the CarerQol-VAS. The CarerQol-7D is a list of seven statements about the burden (the instrument solely captures negative effects of caregiving) derived from providing informal care services: fulfilment from carrying out tasks, relational problems with care receiver, own mental health problems, juggling care responsibilities with other day-to-day activities, financial problems resulting from caregiving, support to carry out care tasks, and physical health problems. Each response has three categories: 'no', 'some', and 'a lot of'. The CarerQol-VAS is a one-item scale from 0 to 10 about how the carer feels at the moment. Brouwer et al. (2006) reported that both parts are closely and negatively associated: the average CarerQol-VAS scores goes down as the severity of problems increases.

The CarerQoL tool has been designed to complement economic evaluations of caregiving and its inclusion can change the main conclusions derived from QALY studies that do not incorporate the strain on informal carers of the services they provide (Goodrich et al. 2012; Krol et al. 2015).

1.8.3 Health Care

The World Health Organisation has defined quality of life as an individual's 'perception of his or her position in life in the context of the culture and value system where they live, and in relation to their goals, expectations, standards and concerns' (Organization 1997, p. 1). This broad ranging definition is—rather curiously, considering the organisation that produced it—not exclusively or primarily health related (although its operationalisation is).

However, there are many health-related definitions of overall quality of life and also specific measures of quality of life by disease (e.g. cancer, dementia, aphasia, depression, arthritis), condition (e.g. frailty, end-of-life), or treatment (e.g. renal replacement therapy, palliative care).

Generally speaking, quality of life in the health domain signifies health status plus 'widely valued aspects of life …that are not generally considered as "health", including income, freedom, and quality of the environment' (Guyatt et al. 1993, p. 622) (see also Volume II, Chap. 5). In certain academic and policy circles, this is the only meaning given to this concept. For example, the International Society for Quality of Life Research (ISOQOL) announces that its goal is to promote "the rigorous investigation of health-driven quality of life measurement from conceptualisation to application and practice".[14]

The main health-related measures of quality of life are:

ICECAP

ICECAP is a measure of quality of life inspired by the capability approach. It consists of two instruments, the ICECAP-O for older people and the ICECAP-A for the general adult population. The ICECAP-O comprises five attributes:

- Attachment (love and friendship)
- Security (thinking about the future without concern)
- Role (doing things that make you feel valued)
- Enjoyment (enjoyment and pleasure)
- Control (independence)

EQ-5D

The EQ-5D index is a five-item questionnaire that describes 243 unique health states according to mobility, self-care, usual activities, pain, anxiety, and depression (Dolan 1997; Szende et al. 2007).

SF-36

The Medical Outcomes Study 36-Item Short Form Health Survey (SF-36) (Ware et al. 1993) is a form with thirty-six questions that produces an eight-scale profile, with four factors corresponding to physical health and the other four factors to mental health. Physical health factors include physical functioning, social role functioning, bodily pain, and self-rated general health. The mental health factors are vitality, social functioning, emotional role functioning, and self-rated mental health.

This instrument and the WHOQOL described below have been used extensively in developing countries. See, for example, the application of the SF-36 survey by Wyss et al. (1999) to study quality of life among the adult population in urban areas in Tanzania or Lera et al. (2013), who used it to assess the quality of life of older people in Chile. Furthermore, Tourani et al. (2018) carried out a meta-analysis and systematic review of twenty-five studies of quality of life in later life using the SF-36 in Iran alone.

WHOQOL

The World Health Organisation's Quality of Life (WHOQOL) assessment tool is organised around six domains and several sub-domains or facets. The domains are:

- Physical
- Psychological
- Level of independence
- Social relationships
- Environment
- Spirituality/religion/personal beliefs

An add-on module was developed specifically for older people: the WHOQOL-OLD (Power et al. 2005). It comprises the following six facets:

- Sensory Abilities
- Autonomy
- Activities
- Social participation
- Death and dying
- Intimacy

As mentioned above, this instrument has also been widely applied to assess quality of life of older people in developing countries. For example, Campos et al. (2014) and Miranda et al. (2016) in Brazil; Nguyen et al. (2018) in Vietnam; Khan and Tahir (2014) in Malaysia; Kumar et al. (2014) in India; Oo et al. (2015) in Myanmar; and Dorji et al. (2017) in Bhutan.

MDS-HSI

The MDS Health-Status Index (MDS-HSI) score is an assessment instrument that consists of a core set of questions and additional modules according to different healthcare settings (post-acute, mental health, home care, etc.). The MDS-HSI comprises six attributes: sensation (vision, hearing, speech), mobility, emotion, cognition, self-care, and pain (Wodchis et al. 2003) (see also Lam and Wodchis 2010).

SEIQOLDW

The Schedule for the Evaluation of Individual Quality of Life (SEIQOLDW) (McGee et al. 1991; OBoyle 1994) is an instrument based on the notion that quality of life is ultimately an individual concept in nature and that any two people may have different views about which aspects of life are important (Browne et al. 1994). Therefore, SEIQOLDW favours the elicitation of individual judgements by patients and does not incorporate assessments by physicians or carers. It includes a general question of happiness and satisfaction in life and then asks for the five most important areas of life (or cues) at present, which the respondents are then invited to rate in importance. The five areas of life vary according to the population in which the instrument is administered. For example:

- Among healthy older people in the Republic of Ireland, health, living conditions, and physical, social, and emotional functioning were identified (OBoyle 1994).
- A study of hospitalised older people in Scotland reported the following most important cue levels: family, health, leisure activities, home, money, relationship with spouse, and friends (Mountain et al. 2004).
- Leisure activities, family, relationships, social life, independence and peace and contentment were the cues rated as most important by older people living in care homes in London in the United Kingdom (Hall et al. 2011).
- Octogenarians in Aberdeen, in the United Kingdom, reported health, family, relationships, finances, and social pastimes as the most important areas in life (Seymour et al. 2008).

OPQOL

The Older People's Quality of life Questionnaire (OPQOL) is based on qualitative research carried out in Britain (Bowling 1995a,b). This tool comprises the following main domains: psychological well-being and positive outlook, having health and functioning, social relationships, leisure activities, neighbourhood resources, adequate financial circumstances, and independence. It performed well in comparisons with the CASP-19 and the WHOQOL-OLD in three separate samples of older people (Bowling and Stenner 2011).

HUI

The Health Utilities Index (HUI) (Furlong et al. 2001; Horsman et al. 2003) is a patient-reported instrument that includes a utility score measurement tool (based on states preference scores) to classify and describe health status. It consists of two classification systems, the HUI Mark 2 (HUI 2) and the HUI Mark 3 (HUI3), which can describe over one million health states. HUI2 includes six attributes: sensation, mobility, emotion, cognition, self-care, and pain. HUI3 consists of eight attributes: vision, hearing, speech, ambulation, dexterity, emotion, cognition, and pain. Despite similarity in the names of the attributes, each classification system measures something different. For example, emotion in HUI2 captures distress and anxiety, whilst in HUI3 this attribute is concerned with the poles happiness-depression.

A joint study of older people in the United States of America and Canada reported a significant statistical association between HUI3 scores and household income, after controlling for a number of individual and household variables (Huguet et al. 2008). The authors surmised that access to healthcare and socio-economic inequalities could explain this relationship. Furthermore, McGrail et al. (2009) found that income inequality could explain *half* of HUI3 inequalities in each country. A similarly strong association between income and HUI3 scores was found in a study on French- and English-speaking adult (not only older) Canadians (Kopec et al. 2000).

Complementarity Between Health-Related Quality of Life Measurement Instruments

Several other measures of health-related quality of life have been proposed, including the Leiden-Padua (LEIPAD) questionnaire (De Leo et al. 1998), the Manchester Short Assessment (MANSA) (Priebe et al. 1999), the Herdecke Quality of life questionnaire (Ostermann et al. 2005), the Assessment of Quality of Life (AQoL) (Hawthorne et al. 1999), the Lancashire Quality of Life Profile (LQLP) (Oliver 1991), the Sickness Impact Profile (Bergner et al. 1981; Gilson et al. 1975), the Reintegration to Normal Living Index (Wood-Dauphinee and Williams 1987), and so on.

This proliferation of measurement instruments of health-related quality of life should not be seen as a waste of effort and talent by wheel-reinventing academics. Instead, comparative studies show that the most widely used tools are in fact complementary to one another and there is agreement in the benefits of using more than one measurement tool at the same time. To illustrate with a few examples:

- Davis et al. (2013) compared the EQ-5D and ICECAP-O instruments in the context of a fall prevention programme for older people in Canada. They found that both tools provided complementary information, therefore capturing disparate components of quality of life.
- In a systematic review of measurement instruments for economic evaluations of health and social care-related quality of life outcomes in older people, Bulamu et al. (2015) recommended the EQ-5D to obtain a measure of QALYs, in combination with the ASCOT or the ICECAP-O.
- Leeuwen et al. (2015) compared the EQ-5D, ICECAP-O, and ASCOT measures among frail older people participating in a geriatric care model in

the Netherlands. The authors also found complementarities among these tools: in particular, the ASCOT was more responsive to changes in quality of life broadly defined, the EQ-5D to changes in physical limitations, and the ICECAP-O to changes in mental health and disability.

1.8.4 Community Quality of Life

Sirgy and Cornwell (2001) and Sirgy et al. (2000) developed the notion and measurement tools of satisfaction with community-based services, which they proposed as a domain of quality of life (see also Tonon 2017 for a series of studies of community quality of life in Spain and Latin American countries). In this literature, a distinction is made between community and neighbourhood. Neighbourhood is the geographical area psychologically nearest to an individual beyond her home, whilst community is related to the geographical levels in which 'reasoned policy choice' is made—for example, a town or municipality. Satisfaction with the community comprises the evaluation of services in three domains: government, business, and non-profit services.

Green et al. (2017) included satisfaction with local leisure, medical and transport services, and with local shopping facilities in its index of well-being in later life for the United Kingdom. All these indicators were found to be significantly and positively associated with overall scores of well-being.

Forjaz et al. (2011) developed a community well-being index to measure the level of satisfaction with the local place of residence among older adults in Spain (see also Giraldez-Garcia et al. 2013). This index combines a self-reported overall satisfaction with the place of residence and satisfaction with eleven different items in the community: economic situation, environment, social conditions, distribution of wealth, health services, social services for older people, support to families, trust in people, leisure services, belonging, and security. Three sub-scales were identified: community services, community attachment, and physical and social environment.

Kim and Lee (2014) also presented a community well-being index for older people. Applied to data from Korea, this index combines individual and local-area data, although it is methodologically cumbersome: for example, indicators of satisfaction with family relationships, medical services, and the civil service are combined with objective health indicators and with intergenerational class mobility, or number of hours of TV viewing.

Notes

1. Even better news: Haucap and Heimeshoff (2014) reported that studying *economics* also boosts your happiness!
2. A rather finicky note on spelling. The editors of the *International Journal of Wellbeing* explain that 'wellbeing' was chosen in the title of the publication instead of 'well-being' as the latter expression would stand as the antonym of 'ill-being'. I am mindful of this potential connotation, but I have chosen the hyphenated version—except when quoting authors who used the alternative spelling—because it is more widely used throughout the English-speaking world.
3. For a good survey of the recent history of happiness studies and related terms in various social sciences (albeit not economics), see Angner (2011), and for a history of happiness in philosophy, see McMahon (2006).
4. See MacKerron (2012) for a concise introduction to economics of happiness.
5. Pareto (1896).
6. Fisher (1918).
7. Phillips (2006) presents a classification of measures similar to the one that follows (see also Gasper 2010).
8. ...'no doubt unforgettable and, yet, already forgotten', as Argentine writer Jorge Luis Borges put it in his poem 'Las cosas'.
9. See Andrews and Robinson (1991), Larsen et al. (1985), and McNeil et al. (1986), for comparisons between different measures.
10. See also https://www.upress.umn.edu/test-division/mpq.
11. From the Greek $\varepsilon \dot{v} \delta \alpha \iota \mu o v \acute{\iota} \alpha$, literally a good spirit, soul, or inner self (Vittersø 2016)—although there is disagreement among philosophers about what, for example, Aristotle exactly meant; see Shiner and Jost (2003).
12. This is closely aligned with the so-called Golden Rule either in its biblical form of doing onto others as you would have them do onto you (Matthew 7:12; Luke 6:31) or in the literature on optimal economic growth as a consumption-maximising steady state growth path.
13. See also Caiels et al. (2010) and Netten et al. (2009, 2010).
14. ISOQOL's Mission Statement, available at http://www.isoqol.org/.

References

Alkire, Sabina (2002). "Dimensions of human development". In: *World Development* 30.2, pages 181–205.

Andrews, Frank M. and John P Robinson (1991). "Measures of Subjective Well-being". In: *Measures of Personality and Social Psychological Attitudes. Volume 1 of Measures of social psychological attitudes* Edited by John P Robinson, Philip R. Shaver, and

Lawrence S. Wrightsman. San Diego, CA: United States of America: Academic Press, pages 61–114.

Angner, Erik (2011). "The evolution of eupathics: The historical roots of subjective measures of well-being". In: *International Journal of Wellbeing* 1.1, pages 4–41.

Aquino, Juan A et al. (1996). "Employment status, social support, and life satisfaction among the elderly". In: *Journal of Counseling Psychology* 43.4, pages 480–489.

Arnold, Sharon B (1991). "Measurement of Quality of Life in the Frail Elderly". In: *The concept and measurement of quality of life in the frail elderly* Edited by James E Birren et al. San Diego, CA: United States of America: Academic Press, pages 50–73.

Bakker, Arnold et al. (2015). *Happiness raised by raising awareness: effect of happiness using the happiness indicator* EHERO Working Paper 1. Rotterdam: The Netherlands: Erasmus Happiness Economics Research Organisation, Erasmus University Rotterdam.

Bergner, Marilyn et al. (1981). "The Sickness Impact Profile: development and final revision of a health status measure". In: *Medical care* 19.8, pages 787–805.

Berridge, Kent C and John P ODoherty (2014). "From experienced utility to decision utility". In: *Neuroeconomics. Decision Making and the Brain* Edited by Paul W Glimcher and Ernst Fehr. London: United Kingdom: Academic Press, pages 335–351.

Bilancini, Ennio and Leonardo Boncinelli (2008). "Ordinal vs cardinal status: two examples". In: *Economics Letters* 101.1, pages 17–19.

Blanchflower, David G and Andrew J Oswald (2011b). "International happiness: A new view on the measure of performance". In: *The Academy of Management Perspectives* 25.1, pages 6–22.

Bond, John and Lynne Corner (2004). *Quality of Life and Older People* Rethinking Ageing. Maidenhead: United Kingdom: Open University Press.

Bowling, Ann (1995a). "The most important things in life. Comparisons between older and younger population age groups by gender. Results from a national survey of the public's judgements". In: *International Journal of Health Sciences* 6, pages 169–176.

—— (1995b). "What things are important in people's lives? A survey of the public's judgements to inform scales of health related quality of life". In: *Social Science & Medicine* 41.10, pages 1447–1462.

Bowling, Ann and Zahava Gabriel (2007). "Lay theories of quality of life in older age". In: *Ageing & Society* 27.6, pages 827–848.

Bowling, Ann and Paul Stenner (2011). "Which measure of quality of life performs best in older age? A comparison of the OPQOL, CASP-19 and WHOQOL-OLD". In: *Journal of Epidemiology & Community Health* 65.3, pages 273–280.

Bowling, Ann et al. (2002). "A multidimensional model of the quality of life in older age". In: *Aging & Mental Health* 6.4, pages 355–371.

Bowling, Ann et al. (2003). "Let's ask them: a national survey of definitions of quality of life and its enhancement among people aged 65 and over". In: *The International Journal of Aging and Human Development* 56.4, pages 269–306.

Brouwer, Werner et al. (2006). "The CarerQol instrument: a new instrument to measure care-related quality of life of informal caregivers for use in economic evaluations". In: *Quality of Life Research* 15.6, pages 1005–1021.

Brown, Jackie, Ann Bowling, and Terry Flynn (2004). *Models of quality of life: A taxonomy and systematic review of the literature review* Review. Sheffield: United Kingdom.

Browne, JP et al. (1994). "Individual quality of life in the healthy elderly". In: *Quality of life Research* 3.4, pages 235–244.

Bulamu, Norma B, Billingsley Kaambwa, and Julie Ratcliffe (2015). "A systematic review of instruments for measuring outcomes in economic evaluation within aged care". In: *Health and quality of life outcomes* 13:179.1. https://doi.org/10.1186/s12955-015-0372-8.

Caiels, James et al. (2010). *Measuring the outcomes of low-level services: final report* PSSRU Discussion Paper 2699. Canterbury: United Kingdom: Personal Social Services Research Unit, University of Kent at Canterbury.

Campos, Ana Cristina Viana et al. (2014). "Aging, Gender and Quality of Life (AGEQOL) study: factors associated with good quality of life in older Brazilian community-dwelling adults". In: *Health and quality of life outcomes* 12.1. https://doi.org/10.1186/s12955-014-0166-4.

Cantril, Albert Hadley (1966). *The Pattern of Human Concerns* New Brunswick, NJ: United States of America: Rutgers University Press.

Cantril, Albert Hadley and Charles W Roll (1971). *Hopes and fears of the American people* New York, NY: United States of America: Universe Books.

Carter, Steven and Michael McBride (2013). "Experienced utility versus decision utility: Putting the 'S' in satisfaction". In: *The Journal of Socio-Economics* 42, pages 13–23.

Cheung, Kelvin Chi-Kin and Kee-Lee Chou (2017). "Poverty, deprivation and life satisfaction among Hong Kong older persons". In: *Ageing & Society* pages 1–19. https://doi.org/10.1017/S0144686X17001143.

Chou, K-L and Iris Chi (1999). "Determinants of life satisfaction in Hong Kong Chinese elderly: A longitudinal study". In: *Aging & Mental Health* 3.4, pages 328–335.

Chou, Kee-Lee and Iris Chi (2002). "Financial strain and life satisfaction in Hong Kong elderly Chinese: Moderating effect of life management strategies including selection, optimization, and compensation". In: *Aging & Mental Health* 6.2, pages 172–177.

Cooney Adeline, Kathy Murphy and Eamon OShea (2009). "Resident perspectives of the determinants of quality of life in residential care in Ireland". In: *Journal of advanced nursing* 65.5, pages 1029–1038.

Cummins, Robert A et al. (2003). "Developing a national index of subjective wellbeing: The Australian Unity Wellbeing Index". In: *Social indicators research* 64.2, pages 159–190.

Davis, Jennifer C et al. (2013). "A comparison of the ICECAP-O with EQ-5D in a falls prevention clinical setting: are they complements or substitutes?" In: *Quality of Life Research* 22.5, pages 969–977.

De Leo, Diego et al. (1998). "LEIPAD, an internationally applicable instrument to assess quality of life in the elderly". In: *Behavioral Medicine* 24.1, pages 17–27.

Deaton, Angus and Arthur A Stone (2016). "Understanding context effects for a measure of life evaluation: How responses matter". In: *Oxford economic papers* 68.4, pages 861–870.

Delle Fave, Antonella et al. (2011). "The eudaimonic and hedonic components of happiness: Qualitative and quantitative findings". In: *Social Indicators Research* 100.2, pages 185–207.

Demakakos, Panayotes, Anne McMunn, and Andrew Steptoe (2008). "Well-being in older age: a multidimensional perspective". In: edited by James Banks et al. London: United Kingdom: The Institute for Fiscal Studies, pages 115–177.

Deneulin, Séverine and Lila Shahani (2009). *An Introduction to the Human Development and Capability Approach: Freedom and Agency* London: United Kingdom: Earthscan.

DETR (1999). *Quality of Life Counts. Indicators for a Strategy for Sustainable Development for the United Kingdom: a Baseline Assessment* Technical report. London: United Kingdom.

Diener, Ed et al. (1999). "Subjective well-being: Three decades of progress". In: *Psychological bulletin* 125.2, pages 276–302.

Diener, Ed et al. (2010). "New well-being measures: Short scales to assess flourishing and positive and negative feelings". In: *Social Indicators Research* 97.2, pages 143–156.

DoH (2011). *The adult social care outcomes framework. Handbook of definitions. Version 2* Technical report. London: United Kingdom: Department of Health.

Dolan, Paul (1997). "Modeling valuations for EuroQol health states". In: *Medical Care* 35.11, pages 1095–1108.

Dolan, Paul and Daniel Kahneman (2008). "Interpretations of utility and their implications for the valuation of health". In: *The Economic Journal* 118.525, pages 215–234.

Dorji, Nidup et al. (2017). "Quality of Life Among Senior Citizens in Bhutan: Associations With Adverse Life Experiences, Chronic Diseases, Spirituality and Social Connectedness". In: *Asia Pacific Journal of Public Health* 29.1, pages 35–46.

DSS—The Welfare Reform Green Paper Consultation Team (1998). *New Ambitions for our Country: A New Contract for Welfare* Technical report. London: United Kingdom.

Easterlin, Richard A (2005b). "Building a Better Theory of Well-Being". In: *Economics and Happiness. Framing the Analysis* Edited by Luigino Bruni and Pier Luigi Porta. Oxford: United Kingdom: Oxford University Press, pages 29–64.

Farquhar, Morag (1995). "Elderly people's definitions of quality of life". In: *Social Science & Medicine* 41.10, pages 1439–1446.

Feldman, Fred (2010). *What Is This Thing Called Happiness?* Oxford: United Kingdom: Oxford University Press.

Fisher, Irving (1918). "Is "Utility" the Most Suitable Term for the Concept It is Used to Denote?". In: *The American Economic Review* 8.2, pages 335–337.

Forjaz, Maria Joao et al. (2011). "Measurement properties of the Community Wellbeing Index in older adults". In: *Quality of Life Research* 20.5, pages 733–743.

Frawley Ashley (2015). *Semiotics of Happiness. Rhetorical beginnings of a public problem* London: United Kingdom: Bloomsbury Academic.

Furlong, William J et al. (2001). "The Health Utilities Index (HUI) system for assessing health-related quality of life in clinical studies". In: *Annals of medicine* 33.5, pages 375–384.

Gallup (2018). *The State of American Well-Being. 2017 State well-being rankings* Technical report. https://wellbeingindex.sharecare.com/wp-content/uploads/2018/02/Gallup-Sharecare-State-of-American-Well-Being_2017-State-Rankings_FINAL.pdf.

Gasper, Des (2010). "Understanding the diversity of conceptions of well-being and quality of life". In: *The Journal of Socio-Economics* 39.3, pages 351–360.

George, Linda K (1981). "Subjective well-being: Conceptual and methodological issues". In: *Annual review of gerontology and geriatrics* 2.1, pages 345–382.

Gilhooly Mary Ken Gilhooly and Ann Bowling (2005). "Meaning and measurement". In: *Understanding quality of life in old age* Edited by Alan Walker, pages 14–26.

Gilson, Betty S et al. (1975). "The sickness impact profile. Development of an outcome measure of health care". In: *American Journal of Public Health* 65.12, pages 1304–1310.

Giraldez-Garcia, Carolina et al. (2013). "Individual's perspective of local community environment and health indicators in older adults". In: *Geriatrics & Gerontology International* 13.1, pages 130–138.

Goodrich, Kacey Billingsley Kaambwa, and Hareth Al-Janabi (2012). "The inclusion of informal care in applied economic evaluation: a review". In: *Value in Health* 15.6, pages 975–981.

Green, Marcus et al. (2017). *A summary of Age UK's Index of Wellbeing in Later Life* Report. London: United Kingdom: Age UK.

Guyatt, Gordon H, David H Feeny and Donald L Patrick (1993). "Measuring health-related quality of life". In: *Annals of internal medicine* 118.8, pages 622–629.

Hall, Sue et al. (2011). "Assessing quality-of-life in older people in care homes". In: *Age and ageing* 40.4, pages 507–512.

Haucap, Justus and Ulrich Heimeshoff (2014). "The happiness of economists: Estimating the causal effect of studying economics on subjective well-being". In: *International Review of Economics Education* 17, pages 85–97.

Hawthorne, Graeme, Jeff Richardson, and Richard Osborne (1999). "The assessment of quality of life (AQoL) instrument: a psychometric measure of health-related quality of life". In: *Quality of Life Research* 8.3, pages 209–224.

Hoefman, Renske J, Job van Exel, and Werner Brouwer (2013). "How to include informal care in economic evaluations". In: *Pharmacoeconomics* 31.12, pages 1105–1119.

Hoefman, Renske, Job van Exel, and Werner Brouwer (2017). "Measuring care-related quality of life of caregivers for use in economic evaluations: carerqol tariffs for Australia, Germany Sweden, UK, and US". In: *PharmacoEconomics* 35.4, pages 469–478.

Holstein, James A and Jaber F Gubrium (1995). *The Active Interview* Volume 37. Qualitative research methods. Thousand Oaks, CA: United States of America: SAGE Publications.

Horsman, John et al. (2003). "The Health Utilities Index (HUI): concepts, measurement properties and applications". In: *Health and quality of life outcomes* 1.1, page 54.

Huguet, Nathalie, Mark S Kaplan, and David Feeny (2008). "Socioeconomic status and health-related quality of life among elderly people: results from the Joint Canada/United States Survey of Health". In: *Social Science & Medicine* 66.4, pages 803–810.

Huppert, Felicia A and Timothy TC So (2013). "Flourishing across Europe: Application of a new conceptual framework for defining well-being". In: *Social indicators research* 110.3, pages 837–861.

Hyde, Martin et al. (2003). "A measure of quality of life in early old age: the theory development and properties of a needs satisfaction model (CASP-19)". In: *Aging & Mental Health* 7.3, pages 186–194.

Iparraguirre, José (1992). *The Quality of Life in Northern Ireland* ERINI Monograph 1. Belfast: United Kingdom.

Iparraguirre, José and Ruosi Ma (2015). "Efficiency in the provision of social care for older people. A three-stage Data Envelopment Analysis using self-reported quality of life". In: *Socio-Economic Planning Sciences* 49, pages 33–46.

Al-Janabi, Hareth, Terry N Flynn, and Joanna Coast (2011). "QALYs and carers". In: *Pharmacoeconomics* 29.12, pages 1015–1023.

Kahneman, Daniel, Peter P Wakker, and Rakesh Sarin (1997). "Back to Bentham? Explorations of experienced utility". In: *The Quarterly Journal of Economics* 112.2, pages 375–406.

Kahneman, Daniel et al. (2004b). "Toward National Well-being Accounts". In: *The American Economic Review* 94.2, pages 429–434.

Katz, Sidney and Barry J Gurland (1991). "Science of quality of life of elders: Challenge and opportunity". In: *The concept and measurement of quality of life in the frail elderly*

Edited by James E Birren et al. San Diego, CA: United States of America: Academic Press, pages 335–343.

Keyes, Corey LM (2002). "The mental health continuum: From languishing to flourishing in life". In: *Journal of health and social behavior* 43.2, pages 207–222.

Keyes, Corey LM, Dov Shmotkin, and Carol D Ryff (2002). "Optimizing well-being: The empirical encounter of two traditions". In: *Journal of Personality and Social Psychology* 82.6, pages 1007–1022.

Khan, Abdul Rashid and Ibrahim Tahir (2014). "Influence of Social Factors to the Quality of Life of the Elderly in Malaysia". In: *Open Medicine Journal* 1.1, pages 29–35.

Kilpatrick, Franklin Pierce and Albert Hadley Cantril (1960). "Self-anchoring scaling: a measure of individuals: unique reality worlds". In: *Journal of Individual Psychology* 16, pages 158–173.

Kim, Yunji and Seung Jong Lee (2014). "The development and application of a community wellbeing index in Korean metropolitan cities". In: *Social Indicators Research* 119.2, pages 533–558.

Kopec, Jacek A et al. (2000). "Measuring population health: correlates of the Health Utilities Index among English and French Canadians". In: *Canadian Journal of Public Health* 91.6, pages 465–70.

Krol, Marieke, Jocé Papenburg, and Job van Exel (2015). "Does including informal care in economic evaluations matter? A systematic review of inclusion and impact of informal care in cost-effectiveness studies". In: *Pharmacoeconomics* 33.2, pages 123–135.

Kumar, Ganesh, Anindo Majumdar, et al. (2014). "Quality of Life (QOL) and its associated factors using WHOQOL-BREF among elderly in urban Puducherry India". In: *Journal of clinical and diagnostic research: JCDR* 8.1, pages 54–57.

Kwong, Enid Wai-yung, Claudia Kam-yuk Lai, and Faith Liu (2014). "Quality of life in nursing home settings: Perspectives from elderly residents with frailty". In: *Clinical Nursing Studies* 2.1, pages 100–110.

Lam, Jonathan MC and Walter P Wodchis (2010). "The relationship of 60 disease diagnoses and 15 conditions to preference-based health-related quality of life in Ontario hospital-based long-term care residents". In: *Medical care* 48.4, pages 380–387.

Land, Kenneth C (2015). "The Human Development Index: Objective Approaches (2)". In: *Global Handbook of Quality of Life. Exploration of Well-Being of Nations and Continents* Edited by Wolfgang Glatzer et al. International Handbooks of Quality-of-Life. Dordrecht: The Netherlands: Springer, pages 133–157.

Larsen, Randy J, Ed Diener, and Robert A Emmons (1985). "An evaluation of subjective well-being measures". In: *Social Indicators Research* 17.1, pages 1–17.

Lawton, M Powell (1991). "A multidimensional view of quality of life in frail elders". In: *The concept and measurement of quality of life in the frail elderly* Edited by James E Birren et al. San Diego, CA: United States of America: Academic Press, pages 3–27.

Leeuwen, Karen M van et al. (2015). "Comparing measurement properties of the EQ-5D-3L, ICECAP-O, and ASCOT in frail older adults". In: *Value in Health* 18.1, pages 35–43.

Lera, Lydia et al. (2013). "Validity and reliability of the SF-36 in Chilean older adults: the ALEXANDROS study". In: *European Journal of Ageing* 10.2, pages 127–134.

Little, Ian Malcolm David (1950). *A Critique of Welfare Economics* Oxford: United Kingdom: Oxford University Press.

Ludwigs, Kai et al. (2017). "How Does More Attention to Subjective Well-Being Affect Subjective Well-Being?" In: *Applied Research in Quality of Life* pages 1–26. https://doi.org/10.1007/s11482-017-9575-y.

Machlup, Fritz (1943). *Forced or induced saving: an exploration into its synonyms and homonyms* Volume XXV. 1, pages 26–39.

MacKerron, George (2012). "Happiness economics from 35000 feet". In: *Journal of Economic Surveys* 26.4, pages 705–735.

McGee, Hannah M et al. (1991). "Assessing the quality of life of the individual: the SEIQoL with a healthy and a gastroenterology unit population". In: *Psychological medicine* 21.3, pages 749–759.

McGee, Hannah et al. (2011). "Quality Of Life And Beliefs About Ageing". In: *Fifty Plus in Ireland 2011. First results from the Irish Longitudinal Study on Ageing (TILDA)* Edited by Alan Barrett et al. Dublin: Republic of Ireland: The Irish Longitudinal Study on Ageing (Trinity College Dublin), pages 265–284.

McGrail, Kimberlyn M et al. (2009). "Income-related health inequalities in Canada and the United States: a decomposition analysis". In: *American Journal of Public Health* 99.10, pages 1856–1863.

McMahon, Darrin M (2006). *Happiness: A History* New York, NY: United States of America: Grove Press.

McNeil, J Kevin, Ml J Stones, and Albert Kozma (1986). "Subjective well-being in later life: Issues concerning measurement and prediction". In: *Social Indicators Research* 18.1, pages 35–70.

Miranda, Lívia Carvalho Viana, Sônia Maria Soares, and Patrícia Aparecida Barbosa Silva (2016). "Qualidade de vida e fatores associados em idosos de um Centro de Referência à Pessoa Idosa". In: *Ciência & Saúde Coletiva* 21.11, pages 3533–3544.

Mollenkopf, Heidrun and Alan Walker, editors (2007). *Quality of Life in Old Age. International and Multi-Disciplinary Perspectives* Volume 31. Social Indicators Research Series. Dordrecht, The Netherlands: Springer.

Mountain, LA et al. (2004). "Assessment of individual quality of life using the SEIQoL-DW in older medical patients". In: *QJM: An International Journal of Medicine* 97.8, pages 519–524.

Murphy, Kathy, Eamon O Shea, and Adeline Cooney (2007). "Quality of life for older people living in long-stay settings in Ireland". In: *Journal of Clinical Nursing* 16.11, pages 2167–2177.

Nations, United (1948). "Universal Declaration of Human Rights". In:.

Netten, Ann et al. (2009). *Outcomes of social care for adults (OSCA). Interim findings* PSSRU Discussion Paper 2648/2. Canterbury: United Kingdom: Personal Social Services Research Unit, University of Kent at Canterbury.

Netten, Ann et al. (2010). *Measuring the outcomes of care homes: final report* PSSRU Discussion Paper 2696/2. Canterbury: United Kingdom: Personal Social Services Research Unit, University of Kent at Canterbury.

Netuveli, Gopalakrishnan and David Blane (2008). "Quality of life in older ages". In: *British Medical Bulletin* 85.1, pages 113–126.

Netuveli, Gopalakrishnan et al. (2006). "Quality of life at older ages: evidence from the English longitudinal study of aging (wave 1)". In: *Journal of Epidemiology & Community Health* 60, pages 357–363.

Neugarten, Bernice L, Robert J Havighurst, and Sheldon S Tobin (1961). "The Measurement of Life Satisfaction". In: *Journal of Gerontology* 16.2, pages 134–143.

Ng, Yew-Kwang (1979). *Welfare Economics: Introduction and Development of Basic Concepts* Basingstoke: United Kingdom: The Macmillan Press.

Nguyen, Thang Tien et al. (2018). "Quality of life and its association among older people in rural Vietnam". In: *Quality & Quantity* https://doi.org/10.1007/s11135-018-0739-0.

OBoyle, Ciaran A (1994). "The schedule for the evaluation of individual quality of life (SEIQoL)". In: *International Journal of Mental Health* 23.3, pages 3–23.

OECD (2013a). *How's Life? 2013 Measuring Well-being: Measuring Well-being* How's Life? Paris: France: Organisation for Economic Co-operation and Development.

OECD Guidelines on Measuring Subjective Well-being (2013). Paris: France: Organisation for Economic Co-operation and Development.

Oishi, Shigehiro (2010). "Culture and well-being: Conceptual and methodological issues". In: *International differences in well-being* Edited by Ed Diener, John F Helliwell, and Daniel Kahneman. Oxford: United Kingdom: Oxford University Press, pages 34–69.

Oliver, Joseph P (1991). "The social care directive: Development of a quality of life profile for use in community services for the mentally ill." In: *Social Work and Social Sciences Review* 3.1, pages 5–45.

Oo, Min Yar, Sureeporn Punpuing, and Chalermpol Chamchan (2015). "Factors Affecting Quality of Life of Older People in Taungu Township, Bago Region, Myanmar". In: *Journal of Health Research* 29.4, pages 235–242.

Organization, World Health (1997). *WHOQOL. Measuring Quality of Life* Technical report. Geneva: Switzerland.

Ostermann, Thomas et al. (2005). "The Herdecke Questionnaire on Quality of Life (HLQ): validation of factorial structure and development of a short form within a naturopathy treated in-patient collective". In: *Health and quality of life outcomes* 3.1. doi: doi:10.1186/1477-7525-3-40.

Pareto, Vilfredo (1896). *Cours d'Économique Politique* Lausanne: Switzerland: F. Rouge.

Pham, Michel Tuan et al. (2015). "Affect as an ordinal system of utility assessment". In: *Organizational Behavior and Human Decision Processes* 131, pages 81–94.

Phillips, David (2006). *Quality of Life: Concept, Policy and Practice* London: United Kingdom: Routledge.

Pigou, A.C (1932). *The Economics of Welfare* 4th. London: United Kingdom: Macmillan and Co.

Poon, Leonard W and Jiska Cohen-Mansfield (2011). "Toward new directions in the study of well-being among the oldest old". In: *Understanding Well-Being in the Oldest Old* Edited by Leonard W Poon and Jiska Cohen-Mansfield. New York, NY: United States of America: Cambridge University Press, pages 3–10.

Power, Mick et al. (2005). "Development of the WHOQOL-old module". In: *Quality of Life Research* 14.10, pages 2197–2214.

Priebe, Stefan et al. (1999). "Application and results of the Manchester Short Assessment of Quality of Life (MANSA)". In: *International journal of social psychiatry* 45.1, pages 7–12.

Robichaud, Line et al. (2006). "Quality of life indicators in long term care: Opinions of elderly residents and their families". In: *Canadian Journal of Occupational Therapy* 73.4, pages 245–251.

Rodriguez-Blazquez, C et al. (2011). "Psychometric properties of the International Wellbeing Index in community-dwelling older adults". In: *International Psychogeriatrics* 23.1, pages 161–169.

Rojo-Perez, Fermina, Gloria Fernandez-Mayoralas, and Vicente Rodriguez (2014). "Spain, personal well-being index; application with people aged 50 years and older". In: *Encyclopedia of Quality of Life and Well-Being Research* Edited by Alex Michalos. Dordrecht: The Netherlands: Springer, pages 6236–6243.

Rudinger, Georg and Hans Thomae (1990). "The Bonn Longitudinal Study of Aging: Coping, life adjustment, and life satisfaction". In: *Successful aging: Perspectives from the behavioral sciences* Edited by Paul B Baltes and Margret M Baltes. New York, NY: United States of America: Cambridge University Press, pages 265–295.

Ryff, Carol D (1989). "Happiness is everything, or is it? Explorations on the meaning of psychological well-being". In: *Journal of Personality and Social Psychology* 57.6, pages 1069–1081.

Ryff, Carol D and Corey Lee M Keyes (1995). "The structure of psychological well-being revisited". In: *Journal of Personality and Social Psychology* 69.4, pages 719–727.

Ryff, Carol D and Burton Singer (1996). "Psychological well-being: Meaning, measurement, and implications for psychotherapy research". In: *Psychotherapy and Psychosomatics* 65.1, pages 14–23.

—— (1998). "The contours of positive human health". In: *Psychological inquiry* 9.1, pages 1–28.

Saint-Paul, Gilles (2011). *The Tyranny of Utility. Behavioral Social Science and the Rise of Paternalism* Princeton, NJ: United States of America: Princeton University Press.

Samuelson, Paul A (1938). "The numerical representation of ordered classifications and the concept of utility". In: *The Review of Economic Studies* 6.1, pages 65–70.

Schelling, Thomas C (1984). "Self-command in practice, in policy and in a theory of rational choice". In: *The American Economic Review* 74.2, pages 1–11.

Schumpeter, Joseph (1994). *History of Economic Analysis (with an introduction by Mark Perlman)* Oxford: United Kingdom: Oxford University Press.

Scitovsky Tibor (1976). *The joyless economy: the psychology of human satisfaction* New York, NY: United States of America: Oxford University Press.

Seligman, Martin (2002). *Authentic Happiness: Using the New Positive Psychology to Realize Your Potential for Lasting Fulfillment* New York, NY: United States of America: Free Press.

Sen, Amartya (2001). *Development as Freedom* Oxford: United Kingdom: Oxford University Press.

Seymour, David Gwyn et al. (2008). "Quality of life and its correlates in octogenarians. Use of the SEIQoL-DW in Wave 5 of the Aberdeen Birth Cohort 1921 Study (ABC1921)". In: *Quality of Life Research* 17.1, pages 11–20.

Shiner, Roger A and Lawrence J Jost (2003). *Eudaimonia and well-being: ancient and modern conceptions* Kelowna, BC: Canada: Academic Printing & Publishing.

Shmotkin, Dov (2011). "The pursuit of happiness: alternative conceptions of subjective well-being". In: *Understanding Well-Being in the Oldest Old* Edited by Leonard W Poon and Jiska Cohen-Mansfield. New York, NY: United States of America: Cambridge University Press, pages 27–45.

Sirgy M Joseph and Terri Cornwell (2001). "Further validation of the Sirgy et al's measure of community quality of life". In: *Social Indicators Research* 56.2, pages 125–143.

Sirgy M Joseph et al. (2000). "A method for assessing residents' satisfaction with community-based services: a quality-of-life perspective". In: *Social Indicators Research* 49.3, pages 279–316.

Srinivasan, Sylaja and Geoff Stewart (2004). "The quality of life in England and Wales". In: *Oxford bulletin of economics and statistics* 66.1, pages 1–22.

Strotz, Robert H (1953). "Cardinal utility". In: *The American Economic Review* 43.2, pages 384–397.

Szende, Agota, Mark Oppe, and Nancy J Devlin (2007). *EQ-5D value sets: inventory, comparative review and user guide* Dordrecht, The Netherlands: Springer.

Tellegen, Auke and Niels G Waller (2008). "Exploring personality through test construction: Development of the Multidimensional Personality Questionnaire". In: *The SAGE handbook of personality theory and assessment. Volume 2: Personality Measurement and Testing* Edited by Gregory J Boyle, Gerald Matthews, and Donald H Saklofske London: United Kingdom: SAGE, pages 261–292.

Thapa, Komilla and Lee A Rowland (1989). "Quality of life perspectives in long-term care: staff and patient perceptions". In: *Acta Psychiatrica Scandinavica* 80.3, pages 267–271.

Tonon, Graciela, editor (2017). *Quality of Life in Communities of Latin Countries* Community Quality-of-Life and Well-Being. Cham: Switzerland: Springer International Publishing.

Tourani, Sogand et al. (2018). "Health-related quality of life among healthy elderly Iranians: a systematic review and meta-analysis of the literature". In: *Health and Quality of Life Outcomes* 16.1. https://doi.org/10.1186/s12955-018-0845-7.

UNDP (1990). *Human Development Report 1990* New York, NY: United States of America: United Nations Development Programme.

—— (2016). *Human Development Report 2016. Human Development for Everyone* New York, NY: United States of America: United Nations Development Programme.

Van Praag, Bernard (1991). "Ordinal and cardinal utility: an integration of the two dimensions of the welfare concept". In: *Journal of Econometrics* 50.1-2, pages 69–89.

Veenhoven, Ruut (1984a). *Conditions of Happiness* Dordrecht: The Netherlands: Reidel Publishing Company.

—— (1991). "Is happiness relative?" In: *Social indicators research* 24.1, pages 1–34.

—— (2011). "Can We Get Happier Than We Are?" In: *The Human Pursuit of Well-Being. A Cultural Approach* Edited by Ingrid Brdar. Dordrecht: The Netherlands: Springer, pages 3–14.

Viner, Jacob (1925). "The utility concept in value theory and its critics". In: *Journal of Political Economy* 33.6, pages 638–659.

Vittersø, Joar (2016). "The most important idea in the world: An introduction". In: *Handbook of eudaimonic well-being* Edited by Joar Vittersø. Cham: Switzerland: Springer, pages 1–24.

Vlastos, Gregory (1991). *Socrates: Ironist and Moral Philosopher* Cambridge: United Kingdom: Cambridge University Press.

Walker, Alan (2005). *Understanding quality of life in old age* Maidenhead: United Kingdom: Open University Press.

Walker, Alan and Catherine Hagan Hennessy (2004). *Growing older. Quality of life in old age* Maidenhead: United Kingdom: Open University Press.

Walker, A. and Ariela Lowenstein. (2009). "European perspectives on quality of life in old age", European Journal of Ageing, 6(2): 61–66

Ware, J.E. et al. (1993). *SF-36 health survey: manual and interpretation guide* Boston, MA: United States of America: The Health Institute, New England Medical Center.

Watson, David, Lee Anna Clark, and Auke Tellegen (1988). "Development and validation of brief measures of positive and negative affect: the PANAS scales". In: *Journal of personality and social psychology* 54.6, pages 1063–1070.

WHO (1994). *Health Promotion Glossary (WHO/HPR/HEP/98.1)* Technical report. Geneva: Switzerland.

Wiggins, Richard Donovan et al. (2004). "Quality of life in the third age: key predictors of the CASP-19 measure". In: *Ageing & Society* 24.5, pages 693–708.

Wilhelmson, Katarina et al. (2005). "Elderly people's perspectives on quality of life". In: *Ageing & Society* 25.4, pages 585–600.

Wodchis, Walter P, John P Hirdes, and David H Feeny (2003). "Health-related quality of life measure based on the minimum data set". In: *International journal of technology assessment in health care* 19.3, pages 490–506.

Wood-Dauphinee, Sharon and J Ivan Williams (1987). "Reintegration to normal living as a proxy to quality of life". In: *Journal of Clinical Epidemiology* 40.6, pages 491–499.

Wouters, Sofie et al. (2015). "Do people desire to be healthier than other people? A short note on positional concerns for health". In: *The European Journal of Health Economics* 16.1, pages 47–54.

Wyss, K et al. (1999). "Validation of the Kiswahili version of the SF-36 Health Survey in a representative sample of an urban population in Tanzania". In: *Quality of Life Research* 8.1-2, pages 111–120.

Yeung, Polly and Vivien Rodgers (2017). "Quality of Long-Term Care for Older People in Residential Settings—Perceptions of Quality of Life and Care Satisfaction from Residents and Their Family Members/te Kounga O Te Taurima I Te Hunga Kaumatua I Nga Kainga Kaumatua-Nga Whakaaro O Te Tangata Mo Te Kounga O Te Noho Me Te Hari Mo Nga Mahi Taurima Mai I Nga Kaumatua Me O Ratou Whanau". In: *Nursing Praxis in New Zealand* 33.1, pages 28–43.

2

Theories and Empirical Findings

> **Overview**
>
> This chapter discusses the main theories and findings from empirical studies on happiness, quality of life, and related terms, with emphasis on later life. It presents alternative theories drawn from behavioural economics, social networks studies, personality traits, and genetics. Other topics include a description of the Easterlin paradox, the needs-based theories, and the links between inequality and happiness. Finally, there is a section on happiness along the life cycle.

The previous chapter reviewed conceptual and operational definitions. This chapter comments on the theoretical frameworks and the main empirical findings. Either explicitly or tacitly, there must be a theoretical viewpoint behind any scientific endeavour; otherwise, it would be a matter of blind measurement, 'measurement without theory' as Koopmans (1947) disapprovingly referred to a particular approach to the study of business cycles. A theory (even if half-baked) must be driving the process of, say, asking people how happy they are and then running regressions on their answers: at the very least, a theory would be informing the exact wording of the questions and the covariates to include in the statistical analyses. Yes, Parducci (1984, p. 5) opined that happiness is not an objectively observable event but a psychological or mentalist state, so he warned that '…it seems unlikely that any theory of happiness can ever be tested directly by the methods of science…The only test is subjective experience with all its shortcomings as a basis for inference'. However, the literature is not lacking in scientific rigour, both theoretical and empirical; Parducci himself has contributed to this. In fact, as Gilhooly et al. (2005) pointed out, more has been written in the academic literature about how to measure concepts such as happiness, quality of life, and alike, than about their definitions. Layard (2005, p. 145) cried out for 'a revolution in academia, with every social science attempting to understand the causes of happiness'. Let's see how they are doing.

'With the passing years, a man can feign many things, though not happiness' wrote the Argentine writer Jorge Luis Borges.[1] In fact, most scholars in the field of happiness studies would beg to disagree: men (and women!) seem to be feigning happiness when responding to surveys. This view is captured by Eckersley (2000, p. 17):

> Can people be telling the truth about their own levels of happiness? The present theory suggests that they are not, that their self-avowals exhibit a strong positive bias.

Several authors have identified a positive bias in the self-assessments of personal levels of happiness, life satisfaction, and similar constructs (as opposed to opinions about social indicators, which would exhibit a negative bias). This 'flawed self-assessment' (Dunning et al. 2004) (termed the human 'sense of relative superiority' by Headey and Wearing (1988)) leads to above-average or better-than-average effects, which whilst it is 'a view that violates the simple tenets of mathematics' (Dunning et al. 2004, p. 72), it is also 'one of the most robust of all self-enhancement phenomena' (Alicke and Govorun 2005, p. 85) and 'runs counter to a rational actor model' (Headey and Wearing 1988, p. 499). This above-average effect is present throughout the life course, but it does change with chronological age depending on the dimension under comparison: in domains in which older people tend to either underperform or are stereotyped as underperformers (e.g. athleticism), the opposite, 'worse-than-average' effect, has been observed (Zell and Alicke 2011).

Notwithstanding this bias, most scholars would also agree that respondents to surveys tend to answer in fairly predictable ways, so that, once this bias is corrected, the booming literature on economics and happiness and related concepts (as well as that emanating from psychology) is building persuasive stories about which factors determine and influence happiness, under which circumstances, and by how much (Ferrer-i-Carbonell 2013). These stories have been marred, though, by dissonant voices for various reasons—and to some extent they still are. Veenhoven (1984a) listed the following shortcomings in the earlier literature:

- Sloppy conceptualisations
- Uncritical acceptance of indicators
- Uncritical presentation of statistics
- Confusion between correlates and causes
- Incomplete coverage of studies
- Incomplete report of findings
- Little cultural relativism

In Veenhoven (1984b), this same author added two more items to the list of problems: confusion between moralising and the study of reality, and the preponderance of speculation over systematic observation. Veenhoven himself as well as many other researchers have helped overcome most of these problems, but some still persist. This is due, in part, to the very nature of scientific research in social sciences that makes advances in knowledge a rather tortuous and erratic process (Flyvbjerg and Sampson 2001; Law 2004; Rule 1997). Moreover, the plethora of terms used in the literature has not helped to convey consistent messages. Given the terminological and operational proliferation, it would be helpful for the task ahead in this chapter to use a classification of the approaches and theories to organise ideas. A number of typologies of 'what makes someone's life go best' have been proposed (see, for example, Parfit 1984, Appendix I, Dolan et al. 2006 or Navarro and Domínguez 2018).

In what follows, I am going to use a slightly modified version of the classification by Veenhoven and Ehrhardt (1995). Veenhoven and Ehrhardt identified comparison, folklore, and livability theories of happiness, and explained that these theories are focused, respectively, on social comparisons, traits, and satisfaction of needs (see also Dutt et al. 2009). In my version, I include individual comparisons (which is missing) in the first category, now relabelled 'comparison theories'. We get:

- Comparison theories, which contend that individuals evaluate their level of happiness in relative terms.
- Trait theories, which contend that happiness is ultimately a question of enduring personality dispositions (Costa et al. 1987; McCrae and Costa 1997).
- Needs-based theories, which contend that happiness refers to how effectively a society satisfy and gratify the needs of its individual members.

Let's see each type of theory in detail.

2.1 Comparison Theories

According to comparison theories, happiness is relative, or as Parducci (1984, p. 10) expressed: 'Nothing is valued in itself, only in its relationship to other things'. When expressed in terms of the relationship between income and happiness, the hypothesis that happiness is relative is known as the relative income theory, or the relative income theory of consumption (Duesenberry 1949).[2] This theory posits that the utility derived from consumption spending does not depend on the absolute level of consumption but on the relative level against the spending of other people.

Asserting that happiness is relative (either dependent on relative income or something else, such as relative health, education, and status, for example) begs the question: relative against whom? Here comes the economist's proverbial answer: it depends. The main distinction is that an economic agent may either compare her income, health, and so on against herself or against others. In other words, the happiness assessment may be based on an individual or a social context.

Individual comparisons are known as intrapersonal comparisons, because they consist of comparisons within one individual. Social comparisons, in turn, are known as interpersonal comparisons, as they involve more than one individual. Both types of comparisons can take a temporal dimension: comparing the present condition against past experiences or evaluations, or against expectations about the future (Redersdorff and Guimond 2006). Some studies suggest that this distinction has a life-cycle dimension: individuals would resort more to temporal comparisons in their childhood and teenage years, and to social comparisons between their working life; in later life, temporal comparisons would again take centre stage (Suls 1986).

It is important to take into account that, as Smith, Diener, and Wedell remarked, both individual and social comparisons are used to make happiness assessments:

> …a person may lag far behind in a race, ultimately finishing a distant last and yet still be happy with his or her performance because it reflects a personal best. Alternatively, a person who ranks first in a group may be unhappy with his or her performance if it represents a poor personal effort.
>
> (Smith et al. 1989, p. 324)

Self-comparisons

Self-comparisons are temporal comparisons through 'an internal and historical process' (Albert 1977, p. 490) that consist in comparing one's current situation with one's past or future situations. For example, older people tend to make negative or unfavourable assessments of their health when they compare their current situation with their past health status (Suls et al. 1991). This finding was reported by Angelini et al. (2012), who looked into data from a 2006 survey of individuals aged 50 or over from eleven European countries: life satisfaction increased with chronological age but the older the individual, she would compare with her younger self more negatively, due to deteriorating health.

Particularly when looking back, the salience of a situation or condition is more important than its frequency. As Gilbert (2007) commented (see also Powdthavee 2010), most people can remember what they were doing when they heard the news about (let alone witnessed) the terrorist attack on the World Trade Center in New York, the United States of America, on 11 September 2001, but hardly anyone remembers what they were doing on the morning of 10 September 2001, as it was for most of people 'just another day', unless for those it wasn't because it was the day they got married, their son was born, and so on, or they were born (although they can hardly 'remember' it!). Related to salience, the 'peak-end' rule states that people judge past experiences based on how they were at their highest intensity and their quality at the end (Do et al. 2008; Kahneman et al. 1997). Consequently, temporal comparisons are usually biased (either positively or negatively) as they consist of a comparison between a given, ordinary situation with a salient, extraordinary, and, therefore, 'unforgettable' moment. The comparisons are not made against, say, the average level over the life so far; the mist of time precludes such averaging process. The same effect operates when imagining the future. As Gilbert and Wilson (2007, p. 1353) illustrated:

> …when people who have missed trains in the past are asked to imagine missing a train in the future, they tend to remember their worst train-missing experience rather than their typical train-missing experience.

Another aspect related to temporal comparisons is that the mental picture used as a measure against which to compare a current situation or condition varies with the temporal distance or time frame. In psychology, the construal-level theory proposes that people tend to construe more distant events or situation more hazily, with less detail but more prototypical and extreme, whereas events or situations more immediate in time (as well as in terms of geography or space, cognition, affect, socio-economic status, visual perception, moral content, etc.) are remembered, imagined, or anticipated in more detail, less extreme (Liberman et al. 2002; Trope and Liberman 2003). The effects of salience, peak-end intensity, and temporal distance of the events which older people remember or imagine to make individual temporal comparisons have not been studied yet, as far as I know.

Time Orientation Finally, the time orientation is also important. Psychology studies in time perception distinguish three temporal frames or orientations and two time perspectives. The three time frames are the past, the present, and the future. Regarding each orientation individuals may have a negative or a positive perspective. Future, present, and past orientations influence different components of subjective well-being. Negative or positive perspectives, as can be predicted, are also significant. Six main perspectives are distinguished: past negative, past positive, future negative, future positive, and two related to the present: present hedonistic and present fatalistic. Present hedonistic is the perspective that seeks to 'seize the day' regardless of any future consequences. Present fatalistic is the perspective of hopelessness about any current course of action. Furthermore, researchers have also identified a balanced time perspective, present in individual who have the capacity to switch between temporal frames 'depending on situational demands, resource assessments, or personal and social appraisals' (Zimbardo and Boyd 1999, p. 1272).

Several studies have been carried out to understand the effect of orientation and perspective on happiness and subjective well-being in older adults (Gabrian et al. 2017). Individuals who are more past oriented tend to exhibit lower subjective well-being than future-oriented people, a finding that is independent of chronological age (Holman et al. 2016). Research on future orientation among older people found that a positive orientation towards the future is associated with more positive affect and a positive orientation towards the past is associated with more positive life satisfaction (Desmyter and Raedt 2012; Stolarski et al. 2014; Webster et al. 2014). Moreover, there is a negative association between balanced time perspective and chronological age: there is a loss of balance with advancing age and an increase in the present fatalistic perspective (Chen et al. 2016; Pethtel et al. 2018; Rönnlund et al. 2017).

Significant negative age relations were observed for past negative and future negative with a clear age-related increase in present fatalistic, while past positive, present hedonistic, and future positive were relatively stable across age. A significant age-related increase in deviation from balance was observed across methods (Cohen's ds 0.28–0.57), with the highest value for DBTP-E. Overall,

S-BTP and DBTP-E were more strongly associated with SWB than DBTP ($r = -0.40$), with the highest value for DBTP-E ($r = -0.53$). Analyses of separate age groups (60–65 vs 70–75 vs 80–90 years) revealed a trend of weakened association with balance in old-old age, for S-BTP and DBTP-E in particular. This seemed to reflect the fact that negative views of the future are strongly related to SWB in young-old adults but diminish in importance in late senescence (80–90 years).

Layard (1980) underlined the importance of expected income on utility and, hence, happiness. Utility would depend not only of current income but also of the expected income: you anticipate eagerly that pay increase; you feel happy already. Before I present these concepts, I want to go back to the utility function (in this I will follow). This author considered the following stylised facts, which by and large are consistent with most of the empirical evidence (see Layard 1980, p. 745):

- An individual's utility depends positively on her income but negatively on the income that in previous periods she expected to have in the present. This is in line with the adaptation theory: the expected bliss of that much anticipated increase in income fizzles out very quickly once it comes.
- The marginal utility of a given expected amount of income is positive. That is, it has a greater impact on utility today to anticipate that tomorrow there will be a big additional sum awaiting than a smaller one.
- The amount of income expected is positively associated with past income.
- The impact on marginal utility of an increase in income is lower than the impact of a reduction of the same amount. This corresponds to the 'endowment effect' seen in Chap. 8 in this volume.

Layard concludes that given the role of expected income on happiness (see MacLeod and Conway (2005) for the impact of positive anticipation on subjective well-being), redistribution may be unpopular in the short term and may reduce welfare, although it may be justified in the long run.

Social Comparisons

Social comparisons are comparisons of one's current situation against a reference group. They may be temporal or not. It has been suggested that social comparisons become more difficult and sparse in later life because of disengagement and increasing heterogeneity. These two processes would make it less likely or feasible that people compare with their peers in later life (Carstensen 1993).

There are various sources of social comparisons:

- They may be made relative to one or more prevalent social or cultural norms that set the standards that define what a satisfactory life or happiness are about. This approach is based on social construction theory: happiness would be, ultimately, a social construction (i.e. a typification that is meaningful in terms of a particular stock of knowledge shared by members of a society (Berger and Luckmann 1966)). Therefore, evaluations may be affected by cultural bias, that is by 'the systematic cross-national differences in quality of life perceptions which are not explained by objective measures of quality of life nor by demographic factors' (Ostroot and Snyder 1985, p. 243).
- Social comparisons may also be made against other people. Despite objectively 'good' living conditions, some individuals may still feel unhappy if their neighbour's grass looks greener than theirs. In this vein, Wills (1981) proposed the 'downward comparison' theory, which posits that instead of looking with envy at the immaculate lawn across the fence, individuals can increase their happiness and subjective well-being (the theory is usually presented in terms of subjective well-being) by comparing themselves with someone less fortunate (see below and Suls and Wheeler 2000).
- As already mentioned, Suls et al. (1991) found that negative self-reported health status by older people tend to reflect a temporal individual comparison they make against their past health condition. Furthermore, these authors reported that older individuals also make comparisons against a stereotype of a 'worse-off generalised other'. Suls et al. (1991, p. 1139), against a 'cognitively constructed stereotypical standard', which leads to a positive bias in subjective evaluations of health status.

Regarding their direction, individual and social comparisons can be lateral, downward, or upward (Ferring and Hoffmann 2007; Heckhausen and Krueger 1993). I mentioned downward comparisons above: assessments of one's current happiness or similar against oneself in the future or the past or against others in or perceived to be in a worse situation or condition than oneself. Similarly for upward comparisons against people (including ourselves) in or perceived to be in a better situation or condition than oneself. Finally, lateral comparisons are assessments consistent over time against our future or past situation or condition, or social comparisons with people in or perceived to be in a similar situation or condition to oneself. Upward comparisons, far from being a source of discouragement, are drivers for self-improvement. Downward comparisons are more frequently associated with the need to boost self-esteem. Ferring and Hoffmann (2007) carried out a study of temporal and social comparisons among over 2000 older people in Luxembourg. The study comprised comparisons along three different domains: physical fitness, mental fitness, and psychological resilience. Regarding social comparisons, lateral comparisons were the most frequent category (between 56 per cent and 68 per cent, depending on the domain), followed by roughly 25 per cent who applied upward comparisons. In other words, the majority of respondents opined to be in a similar condition than their peers, and a quarter said to be in a worse situation than their peers. The direction of temporal comparisons also showed a predominance of lateral comparisons, except for physical fitness, when respondents preferred to compare themselves upwardly, with a time they were in better physical shape. The authors also looked at whether the direction and category of comparison varied by age. They found that lateral comparisons decrease, and upward comparisons increase, with chronological age.

Sherrard (1994) interviewed among a small sample older people in Bradford, United Kingdom, to investigate both sources of comparison and their role on well-being. She reported significant class differences and also variations according to the dimension under comparison. For example, while both retired manual and professional workers resorted to downward comparisons, the former derived less well-being than the latter. Furthermore, the professional group was more likely to make upward comparisons with their former, younger selves, particularly regarding cognitive and physical condition.

Social Comparisons of Income That economic agents make social comparisons regarding their income is the subject of the relative income and conspicuous consumption theories referred to above. The link between relative income and happiness is that an increase in income changes the yardsticks used to measure happiness in relative terms. Furthermore, increasing attainment of goals also rises expectations (Galbraith 1958),[3] which in turn varies the subjective definition of fulfilment and success, and therefore moves the happiness goalposts. For example, according to Easterlin (1995), raising the incomes of all in a population would not increase their happiness, and this would happen as a result of two issues:

- Subjective well-being and happiness depend not only on the income of an individual but also on the incomes of others.
- The material norms on which the personal judgements about happiness are based change as the average income and standard of living raise.

The usual utility function in economics assumes the absence of social and individual comparisons. The utility derived from a given level of income or consumption by each agent does not depend on the level of income or consumption of any other agent. Furthermore, it is usually formalised mathematically as a cardinal function, which Layard (1980, p. 745) surmised it was the reason why it 'has never so far had any impact on practical affairs'.

The following mathematical model presents the situation of social comparisons (see Arrow and Dasgupta 2009). Assume time is infinite and continuous (see Chap. 7, Volume I). There is one consumption good. At any one period of time t, the consumption rate of each economic agent (say, each household) i is given by $c^i(t)$. For a population as a whole, the average consumption is given by the sum of the consumption of all the households. To simplify the algebra, the assumption is that there is a continuum of households, so we use integrals to sum the consumption of all the households[4]. Furthermore, the number of households is rescaled or normalised to equal 1. These assumptions render the following expression for the average population consumption in period t:

$$C(t) = \int_0^1 c^i(t) \cdot di \qquad (2.1)$$

The relative income hypothesis implies that a household's utility function at each period depends on the household income at that period and the average income for a reference group yet to be determined but in our simplified model is equivalent to the population as a whole. Therefore, in terms of consumption we get:

$$U^i = U[c^i, C] \qquad (2.2)$$

That is, the utility function of a household depends on the household's consumption and the average consumption of the population.

Two more assumptions. First, the usual assumption that a household's utility increases with household consumption. Second, and crucially, that a household's utility goes down as the average population consumption goes up:

$$\frac{\partial U}{\partial c^i} = U_c[c^i, C)] > 0;$$
$$\frac{\partial U}{\partial C} = U_C[c^i, C)] < 0 \qquad (2.3)$$

At $t = 0$, each household looks over time and considers their inter-temporal utility, W^i, which can be called 'household welfare'. We get:

$$W^i = \int_0^\infty e^{-\delta \cdot t} U[c^i, C(t)] \cdot dt \qquad (2.4)$$

where δ corresponds to the discount rate, which is assumed to be constant over time.

In this model, the social welfare is defined as the sum of all the household welfare functions, that is, the sum of all the households' inter-temporal utility functions. In symbols:

$$W = \int_0^1 W^i \cdot di = \int_0^1 \int_0^\infty \left\{ e^{-\delta \cdot t} U[c^i, C(t)] \cdot dt \right\} \cdot di \qquad (2.5)$$

Each household accumulates wealth over time as a result of the difference between their stock of capital (K) in each period and the consumption in the period:

$$\frac{\partial K}{\partial t} = F[K(t)] - c(t) \qquad (2.6)$$

Households maximise their utility over time subject to their stock of capital. Let's denote the price of capital by $p(t)$. Using the optimisation technique known as the Hamiltonian (see Volume I, Chap. 10), the maximisation of expression (2.5) subject to the constraint (2.6) requires that

$$\frac{dp(t)}{dt} = \delta - \frac{dF}{dK(t)} \qquad (2.7)$$

which gives the market equilibrium path. Equation (2.7) shows that, in equilibrium, the evolution of the price of the consumption good over time has to be equal to the marginal utility of consumption. Let's denote this price by $p^m(t)$. Because in equilibrium the optimal path for each household is the same, the equilibrium condition can be expressed thus:

$$U_c[C^m(t), C^m(t)] = p^m(t) \qquad (2.8)$$

Imagine, finally, that the government consists of a benevolent social planner that seeks to maximise social welfare over time, taking into account both arguments in each household utility equation: the consumption over time of each individual household and the average household consumption over time. Let's denote by $p^o(t)$ the equilibrium price at which social welfare over time is maximised. This will be satisfied when

$$U_c[C^o(t), C^o(t)] + U_C[C^o(t), C^o(t)] = p^o(t) \tag{2.9}$$

Remember that $U_C < 0$, therefore $U_c[C^o(t), C^o(t)] > p^o(t)$.

When will the equilibrium price at which each household unit maximises their utility over time be equal to the equilibrium price at which social welfare over time is maximised? Let's see:

$$\frac{\frac{dp^m(t)}{dt}}{p^m(t)} = \frac{\frac{dp^o(t)}{dt}}{p^o(t)} \tag{2.10}$$

From where we obtain:

$$\frac{p^o(t)}{p^m(t)} = \gamma \tag{2.11}$$

(where $\gamma > 0$).

We replace Eqs. (2.9) and (2.8) into Eq. (2.11) and obtain:

$$\frac{U_c[C^o(t), C^o(t)] + U_C[C^o(t), C^o(t)]}{U_c[C^m(t), C^m(t)]} = \gamma \tag{2.12}$$

If we define $\beta = \gamma - 1$, the relationship between the marginal social welfare function and the marginal household utility function can be expressed thus:

$$U_C(C, C) = \beta \cdot U_c(C, C), \tag{2.13}$$

from where the following corollary is obtained: if $c = C$, 'the absolute value of the negative effect of others' consumption is less than the gains from one's own consumption' (Arrow and Dasgupta 2009, p. F504).

Relative income is operationalised in at least four different ways:

- As the average income of the group of reference, as in Eq. (2.2)
- As the difference between the individual income and the average income of the group of reference

2 Theories and Empirical Findings   57

- With two potentially different effects depending on whether the individual income is above or below the average income of the group of reference (Ferrer-i-Carbonell and Ramos 2014)
- As the product of 'relative deprivation' following strictly upward comparisons (i.e. all each individual cares about is the distance to those who are better off) (D'Ambrosio and Frick 2012; Runciman 1972; Smith and Huo 2014).

Even if we accept that economic agents compare themselves to other people, we still need to know to whom and by how much.

- 'To whom' points to the reference group and 'by how much' to the intensity of the effect of the comparison. There are various candidates for a reference group of social comparisons. Imagine you are asked to say how much satisfied you are with your level of income, and let's assume you make a social comparison to come up with an answer. You may compare yourself with the richest person in the planet, or in your country—or with the poorest. Perhaps you do not compare with such extreme cases but you may place your income along a ranking of the estimated incomes of all the people you know, or perhaps only of your neighbours (judging by the cars on their driveways, for example). Or, perhaps you are not the mental ranking sort, and you compare yourself against the average income of your friends, neighbours, colleagues, and so on. Or maybe you make the comparison with someone much closer to home: a sibling or that in-law of yours who…(ok, I won't get you started).
- 'By how much' refers to both the distance between the individual making the comparison and the reference group and also about the effect the comparison has on the individual. Of course, lateral comparisons involve a much shorter distance than upward or downward comparisons. But 'how much' is also about the effect on the person making the comparison: you may live completely oblivious to and in total disregard of what others do or earn, or you may think of going to the extremes of changing your family name to Jones to stop trying to keep up with them to finally become one of them.

Clark and Senik (2010) presented evidence for eighteen European countries from 2006 showing that around 32 per cent of people aged between 16 and 65 years who were in paid employment compared their income mostly with work colleagues and with groups with whom they interacted more frequently, followed by almost a similar proportion who did not compare their income with anyone. Regarding intensity, even though the majority of people did not consider comparing their income with other people's income of importance, a sizeable 28 per cent of respondents declared such comparison to be important for them.

Similarly, Usui et al. (1985) studied social comparisons of financial situation and their association with self-reported life satisfaction among people aged 60 years or older in Kentucky, the United States of America. These authors found that life satisfaction was negatively affected by how much worse off a person felt she was compared to someone to whom she felt or was close—and vice versa (a 'paradoxical' finding, according to Usui et al., as it suggested that it improved a person's life satisfaction that someone to whom she felt close was in a worse financial situation than herself!). Comparisons with acquaintances, neighbours, or the 'neighbourhood' of residence were not significant (whether your neighbour's grass looks greener or not seems to be rather irrelevant after all).

In social comparison theory, the interpretation of the comparison refers to whether the individuals contrast or identify themselves with others. People may engage in four types of social comparison: upward identification, upward contrast, downward identification, and downward contrast (Buunk and Dijkstra 2017; Zee et al. 2001). Upward contrast comparisons, for instance, tend to evoke envy and other negative feelings and low satisfaction (see Frieswijk et al. 2004). Usui et al. (1985), from this point of view, reported that the older participants in their study who engaged in downward contrast comparisons on the financial domain with people they felt close to improved their life satisfaction as a result. Downward comparisons have been detected among people suffering from arthritis, spinal cord injury, or cancer, among other serious diseases. These comparisons helped them maintain similar levels of life satisfaction or well-being than healthy individuals of otherwise similar characteristics (Tennen et al. 2000). Lateral social comparisons may involve assessments whose yardsticks are social stereotypes. Age stereotypes, in particular, play a role in the evaluative processes of older people. Rothermund and Brandtstädter (2003) studied the relationship between age stereotypes and self-evaluations in almost 700 older people in Germany. The authors found evidence in support of the 'contamination hypothesis' proposed by Levy (1996), which posits that the age stereotypes prevalent in a society serve as a standard for positive, self-enhancing comparisons: by and large, the study participants '…held more positive views about themselves than about the "typical old person"' [p. 553].

Suls (1982) and Suls and Mullen (1984) indicated that the sources of social comparisons vary in relative importance along the life cycle, and that given the reduction in social contacts and social interaction in later life, older people would tend to rely more on temporal comparisons and less on comparisons against normative ideals, other people, or imagined others. However, Robinson-Whelen and Kiecolt-Glaser (1997) reported that social comparisons were determinant among older people in self-evaluations in three domains: personal health, income, and memory (with temporal comparisons being relevant only for the latter). Furthermore, older people would resort to downward comparisons as a self-enhancing mechanism to reduce negative affect and preserve happiness and life satisfaction despite onset or threat of frailty and other losses (see also Peck and Merighi 2007).

Neurophysiology has also produced evidence in support of the importance of social comparisons on subjective well-being. Fliessbach et al. (2007) found that social comparisons influenced the activation levels in the ventral striatum region in the brain—a region that is involved in the processing of reward and reward-related learning (Tremblay et al. 2009). Moreover, Dohmen (2011) showed that the activation in the ventral striatum increased with absolute income and decreased with lower relative income (see also Kedia et al. 2014).

Social comparisons have different effects depending on the dimension or aspect being compared. Two alternative conjectures in this regard are the 'mediation' and the 'moderator' hypotheses (Baron and Kenny 1986). In statistics, a moderator variable influences the strength and/or the direction of the relationship between two other variables. In turn, a mediator variable intervenes and accounts for such relationship. Figure 2.1 presents the distinction between moderator and mediator variables schematically:

Cheng et al. (2008) studied the relationship between living alone and low psychological well-being (measured by low levels of self-reported life satisfaction and depression) among older people in China. These authors found that downward social comparison was a moderator: those older people who lived alone reported lower psychological well-being than those who lived with someone, a difference that disappeared if the former applied downward social comparisons. Besides, in a study of older people suffering from depression in Hong Kong, Chou and Chi (2001) showed that social comparisons have either a moderating or a mediating role depending on the domain of life satisfaction the comparison is made. In particular, the authors reported that social comparisons over physical health moderated the relationship between functional disability and depression (see also Heidrich and Ryff 1993): when

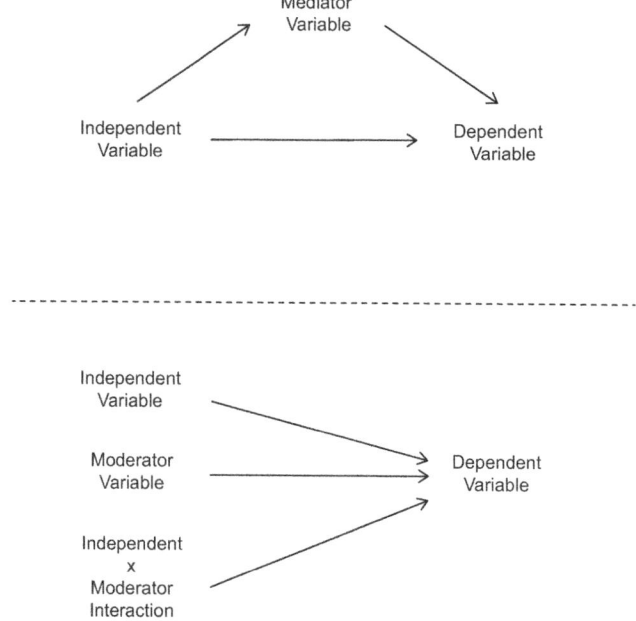

Fig. 2.1 Mediator and moderator variables

made over the personal financial situation, social comparisons mediated in the relationship between financial strain and depression.

The Rank-Income Hypothesis

In a study of data from a 2000 survey in Indonesia, Powdthavee (2009) found that a person's rank in the local community was determinant for where she would place herself on the ladder of life. This result is an instance of the rank-income hypothesis. This hypothesis, a variant of the relative income conjecture, posits that economic agents derive utility (or happiness, life satisfaction, etc.) from the ranked position of their income in a reference group. Brown et al. (2005) provide the best description of this hypothesis, which I repeat below because I am unable to improve on it and also because it was not included in the published version of the paper[5]:

> Consider Professor X, a relatively successful member of a small university department. Professor X earns $20,000 more than the average wage of professors in the department, and only $10,000 less than the most highly paid faculty

member in the department. In fact, Professor X is the third most highly paid member of the department. Compare the likely satisfaction of Professor X with that of Professor Y, a colleague in a different department and better paid discipline. Professor Y earns $10,000 more than Professor X, corresponding to $20,000 more than the average wage in Professor Y's department. Thus the salaries of Professor X and Professor Y are the same distance from the mean of their respective departments. Like her less well-paid colleague, Professor Y happens to earn just $10,000 less than the highest wage in her department. However, Professor Y is only the fifth most highly paid person in her department. Who will be more satisfied with their wage—Professor X or Professor Y?

Intuition and informal observation suggest that Professor Y may be less satisfied than Professor X, despite the fact that she is more highly paid and is identically located with respect with the mean and maximum departmental wages. To the extent this intuition is correct, it suggests that individuals care not just about their wage relative to some reference level, but also about the ranked position of their wage within their comparison set.

(Brown et al. 2005, p. 3)

Imagine a consumer with the option to buy two goods, x_1 and x_2, with prices p_1 and p_2, respectively, and an income equal to y. The optimal allocation of her income between x_1 and x_2 (remember there are only two goods, no possibility to save, etc.) corresponds to that one in which the ratio of the marginal utility derived from each good equals their relative prices. Algebraically, the agent's problem is to maximise her utility function given by:

$$U = U(x_1, x_2) \tag{2.14}$$

subject to

$$y = p_1 \cdot x_1 + p_2 \cdot x_2 \tag{2.15}$$

It is fairly straightforward[6] to find that the solution to this maximisation problem is given by

$$\frac{\frac{\partial U}{\partial x_1}}{\frac{\partial U}{\partial x_2}} = \frac{p_1}{p_2} \tag{2.16}$$

What would the consequences of assuming that rank income is important for utility be? Frank (1985) looked into this, and found that an agent's utility (that of every agent, remember) would be lower. One important difference between the rank-income hypothesis and the relative income hypothesis is that if an individual's income increases, it will not necessarily change his or her position in the reference group. For example, Boyce et al. (2010) reported rank-income effects on life satisfaction in the United Kingdom between 1997 and 2004: when rank income and absolute income were included together in a regression on life satisfaction, the latter indicator turned statistically not significant, and the same happened when the incomes of several alternative reference groups were factored in. Rank income dominated the explanation of variations in life satisfaction between individuals.

The rank-income hypothesis is an application of a more general theory with origin in psychology: the range-frequency theory (Parducci 1965). Its main conjecture is that category judgements (e.g. 'very' happy; 'somewhat' dissatisfied with life) depend on how the respondent mentally divides the range between the maximum and minimum categories (i.e. the range principle) and how often each category is presented (i.e. the frequency principle). Individuals use both principles: they compare themselves with extreme cases and position themselves along a ranking. In connection with this, Parducci (1984, p. 11) proposed the following theorem: 'The mean of the value judgements of all events in a context is directly proportional to their degree of skewing, with the relative weighting of position in the range being the constant of proportionality.'

This literature in psychology refers to the cognitive or mental representation used by individuals as a basis for their judgements and assessments as a 'stimulus' and the reference person or groups against whom the social comparisons are made as 'contexts'. With this terminology in mind, algebraically, the range-frequency theory can be expressed thus (Wedell and Parducci 2000):

$$SWB_{i,k} = w \cdot \left[\frac{(S_i - S_{min,k})}{(S_{max,k} - S_{min,k})} \right] + (1-w) \cdot \left[\frac{(rank_{i,k} - 1)}{N_k - 1} \right], \quad (2.17)$$

where SWB stands for subjective well-being; i and k are the stimulus and context, respectively; w is a weighting factor with values $0 < w < 1$; and S is a context invariant scale that defines the range values of a given stimulus. The second expression between square brackets expresses the frequency values, with *rank* denoting the rank in context k and N representing the maximum rank. Equation (2.17) captures the skewness (i.e. degree of asymmetry) of the distribution of the dimension under assessment (income, satisfaction, happiness, health, etc.) and renders each individual assessment proportional to the negative skewing of the distribution.

Using data from Germany for the period 1995–2011, Navarro and Domínguez (2018) showed that social comparisons of relative income are important for subjective well-being. First, the authors found that whilst a higher individual income increases individual subjective well-being scores, a higher average income of a reference group diminishes individual subjective well-being. Furthermore, they also found the existence of asymmetric effects: subjective well-being is negatively affected if individual income is below the average income of the reference group but is independent of income levels if it is above the average. In other words, the evidence supports the relative deprivation hypothesis. Also looking into data for Germany, Bartolini et al. (2013) reported that there would be a partial positive effect of increasing income on subjective well-being because around 75 per cent of the substantial direct effect of a rise in individual income would be wiped out by the negative effects of social comparisons and of hedonic adaptation.

Area-Level Studies A separate line of research applies a spatial approach, looking into area-level data. The assumption is that the reference group is made up of those people with greater physical closeness; as Luttmer (2005, p. 967) explained:

> Unless one assumes that neighbors' incomes affect an individual's *marginal* utility of other goods, the only behavior affected is the individual's choice of reference group implicit in the decision where to locate.

The results in the area-level literature have been fairly consistent. To name a few studies:

- Luttmer (2005) looked into a single-item measure of happiness from about 10,000 individuals in the United States of America in 1987–1988 and 1992–1994. This author found that although life satisfaction and own income were positively correlated, there was a negative correlation between own life satisfaction and the average income of the local area.
- Graham and Felton (2006) examined self-reported life satisfaction in Latin American countries from 2004 and, using a similar statistical approach to Luttmer (2005), reported significant negative ranking effects on life satisfaction, particularly in wealthier areas and cities
- Kingdon and Knight (2007) used data from a household survey carried out in South Africa in 1993 and found that based on a single-item measure,

life satisfaction increased with the average income of the neighbourhood but decreased with the average income of a wider geographical area. The authors found support for the conjecture that their findings would reflect a higher degree of altruism in very small geographical units.
- Clark et al. (2009) reported in a study of 9000 small neighbourhoods in Denmark from 1994 to 2001 that, conditional on the income of their own household, respondents reported higher satisfaction levels if their neighbours were richer. However, conditional on the income of their own household and the average income of their neighbourhood, respondents reported higher satisfaction levels as their own position in the income ranking in their neighbourhood increased. The authors concluded: 'Although richer neighbours appear to be welcome, being at the top of the pile still counts.' [p. 527]

Social Networks and Happiness

The literature on social comparisons and happiness has reported, as we have seen, disparate results regarding the link between levels of happiness and being in contact with others. Such contact may lead to upward comparisons that could constitute a source of envy or a springboard for self-enhancement, or downward comparisons that may be either a saddening and perturbing or a coping mechanism.

However, another strand of the literature, which focuses on the relationship between personal social networks and happiness, has found, practically unanimously, that the size of an individual's social network and the frequency of contacts have positive effects on levels of happiness. People with larger social networks and who interact more frequently with their friends and family members (although the type or composition of the networks is also relevant) tend to report higher levels of happiness, life satisfaction, well-being, and so on. This also applies to older people. Several studies, using different measures of happiness in different countries, concur that social networks and happiness are positively associated in later life.[7]

To illustrate with some recent studies[8]:

- Rafnsson et al. (2015) for people aged 50 years or older in England, using the SWLS and CASP-19 measures
- Kobayashi et al. (2015) for people aged 63 or older in Japan, using a single-item measure of life satisfaction

- Hsu and Chang (2015) for people aged 60 years of older in Taiwan, using a single-item measure of recent feelings of happiness
- Tomini et al. (2016) for people aged 50 years or older in sixteen European countries, using a single-item measure of life satisfaction
- Didino et al. (2017) for people aged 65 or older in Siberia, Russia, using single-item measures of happiness and of life satisfaction
- Rodriguez-Rodriguez et al. (2017) for people aged 50 or over in Spain, using satisfaction with relationships as a proxy for quality of life

Ateca-Amestoy et al. (2014) and Bárcena-Martn et al. (2017) are two of the few studies that focused on the link between social comparisons and social networks in connection with happiness and related concepts. They report differing findings. Using data for eighteen Latin American and Caribbean countries from 2007, Ateca-Amestoy, Aguilar, and Moro-Egido analysed the domain of wealth and its association with life satisfaction. They found that social comparison effects were only effectual for individuals with below-average wealth (in other words, social comparisons are mainly upwards) and, crucially, that social contacts enhance this effect. The upward social comparisons positively affect life satisfaction because the fact or perception that the average person is doing better financially than oneself sends a positive signal that acts as a source of hope overpowering any negative feelings of envy. This is an example of what Hirschman and Rothschild (1973) termed a 'tunnel effect':

> An individual's welfare depends on his present state of contentment (or, as a proxy, income), as well as on his expected future contentment (or income). Suppose that the individual has very little information about his future income, but at some point a few of his relatives, neighbors, or acquaintances improve their economic or social position. Now he has something to go on: expecting that his turn will come in due course, he will draw gratification from the advances of others for a while.
>
> (Hirschman and Rothschild 1973, pp. 545–546)

According to Hirschman and Rothschild, the 'tunnel effect' would be operative in developing countries, which is consistent with the findings by Ateca-Amestoy et al. (2014) in contrast to those reported by Bárcena-Martn et al. (2017). A bigger social network provides additional signals about the 'average person', thus reinforcing the comparison effect. A similar interaction was found in a study on data from Russia (Senik 2004).

Bárcena-Martn et al. analysed life satisfaction, income, and social networks using data from Germany for the period 1998–2011. These authors also found asymmetry in social comparisons but in this case for individuals on below-average income upward social comparisons negatively affected life satisfaction scores: envy overpowered any positive signals.

2.2 Easterlin Paradox

Easterlin (1974) looked into data on the Self-Anchoring Striving scale (see Chap. 1 in this volume) in fourteen countries and into responses to a single-item question about happiness ('how happy would you say you are—very, fairly, or not very?') in a survey carried out in the United States of America. And caused a perfect storm.

Within each country, the higher the income of individuals, the happier they report they are (see, for example, Herrera et al. 2012 for Colombia, and Castillo-Carniglia et al. 2012 for Chile). Poverty and happiness do not go hand in hand. So far, nothing earth shattering. However, Easterlin also looked into the macroeconomic data across countries. Here the data conveyed a different picture: there was hardly any difference at all in average happiness scores regardless national income levels per capita. In other words, within countries there was a positive association between income and happiness, but across countries this statistical association disappeared. Moreover, Easterlin also probed into national time series data for the United States of America between 1946 and 1970. These data suggested that happiness did not rise, despite increasing economic prosperity. The paradox that cross-sectional data render one conclusion and time series data a different one and also that the results from within a country differ from those between countries is known as the Easterlin paradox. According to Clark and Senik (2014a, p. 1), it is a 'double paradox', and, in the words of Stanca and Veenhoven (2015, p. 1), the paradox is the *starting point of the literature on economics and happiness*.

In a more recent study, Easterlin et al. (2010) examined the empirical evidence from available time series in developed and developing countries. The authors showed that, in the short term, happiness and national income go in tandem, but that over the long term (ten years, give or take) there is no association: happiness does not increase despite income rises. These authors restated the paradox thus:

…at a point in time both among and within nations, happiness varies directly with income, but over time, happiness does not increase when a country's income increases.

(Easterlin et al. 2010, p. 22463)

The same findings have been confirmed in several studies. To illustrate:

- Blanchflower and Oswald (2004) using data for Great Britain for the period 1973–1998 (question about life satisfaction) and the United States of America between 1972 and 1998 (question regarding happiness)
- Clark and Senik (2014b) studied happiness and economic growth in China, East Germany, and Russia and found that despite improvements in income per capita, since these countries transitioned away from a centralised economic system average happiness levels have remained fairly constant

A number of alternative explanations have been proposed to account for the Easterlin paradox (Clark et al. 2008). Easterlin himself provided an explanation based on the relative income theory of consumption. Happiness would depend on comparisons and changes rather than absolute levels. Richer people would report higher levels of happiness because they are richer, irrespective of their level of wealth or income. It's not how much you have or earn but that you have or earn more than others what matters. This line of reasoning echoes 'prospect theory' in behavioural economics; see Chap. 8 in this volume. Other explanations of the paradox are based on variations of set point theories (see next section).

Reflecting upon the Easterlin paradox, Frawley (2015, p. 21) concluded that the 'happiness index becomes an index of all the problems of modern society, a critique of modernity itself'.

2.3 Set Point, Adaptation Level, and the Hedonic Treadmill

Back in the 1940s, the US psychologist Harry Helson developed the theory of 'adaptation level' to explain the changes in the judgement of stimuli in terms of a frame of reference or 'adaptation level': when a stimulus coincides with that level, it evokes a neutral or indifferent response (Helson 1947, 1948). This level (which, despite its name, may consist of a range of levels rather than one single level[9]) is known, especially in psychology, as a 'set point', so these theories are also known as set point theories.

Adaptation level theory started in the field of visual perception but was later extended to motivation and affect.[10] The adaptation towards the set point is thought of a process of habituation whereby the more an individual is exposed to a given stimulus, the weaker the response is. This adaptation towards a set point is a cognitive process rather than a sensory process as, for example, the light and dark adaptation of the human eye Wilson and Gilbert (2008).

Brickman and Campbell (1971) applied this theory to a measure of 'subjective pleasure' as a proxy for happiness: life events may shift the happiness levels of an individual at any point in time to some degree but happiness eventually returns to its stable, baseline level. This notion, Brickman and Campbell realised, would lead to either a pessimist or an optimistic theme. The pessimistic theme was that the forces of adaptation would make individuals 'to seek new levels of stimulation merely to maintain old levels of subjective pleasure, to never achieve any kind of permanent happiness or satisfaction' [p. 289] as though they were on a hedonic treadmill. The optimistic position was that despite the chains of adaptation there would still be 'wise and foolish ways to pursue happiness, both for societies and for individuals'.

The hedonic treadmill concept, despite its pessimistic overtone, has stuck in the literature, especially as a result of the landmark paper by Brickman et al. (1978). In this paper, the authors compared the self-reported happiness levels of twenty-two lottery winners of significant amounts[11] in Illinois, United States of America, with a control group of randomly selected individuals. The measurement instrument was a set of questions about how happy they were at the moment, how happy they had been before winning (all the winners had won the lottery within eighteen months before the interviews), and how happy they expected to be in two years. Other questions included to rate how pleasurable each of a number of activities were, including talking with a friend, watching television, eating breakfast, buying clothes, and so on. The findings showed that winners were not happier, on average, and that they took less pleasure out of many activities than individuals in the control group. The same paper studied eleven paraplegic and eighteen quadraplegic individuals who had been victims of accidents in the previous twelve months. These respondents rated their current happiness levels lower than individuals in control groups, but not as low as expected.

The hedonic treadmill posits that individuals adapt to both pleasurable and negative events, experiences and activities, so that these events cease to have an effect on happiness. Take a person who believes that 'money can buy happiness' and pursues earning increasing amounts of money in her search for happiness. The hedonic treadmill conjecture predicts that this individual is doomed to fail because adaptation and habituation to new levels would lead to affective neutrality, rendering the search for happiness via higher incomes a futile endeavour: each improved financial position would yield no lasting effect on happiness. In connection with this, Frederick and Loewenstein (1999, p. 302) coined the term 'hedonic adaptation', which they defined as 'adaptation to stimuli that are affectively relevant'. Hedonic adaptation involves 'cognitive changes—in interests, values, goals, attention, or characterization of a situation' [p. 302], the conscious shift of the attention away from what may be perturbing, and a reinterpretation of those situations.

So, we have the notion of a 'set point' that developed into the 'adaptation level' theory that gave birth to the 'hedonic treadmill' concept, all encapsulated in the term 'hedonic adaptation'.[12]

The presence of set points in happiness, life satisfaction, and subjective well-being has been detected in several studies. The selection, optimisation, and compensation (SOC) theory introduced in Volume I, Chap. 4, focuses on transformations in the sensory, perceptual, cognitive, personal, and social domains that accompany the process of individual ageing. In this sense, it is a theory of adaptation in later life. There are three main models of adaptation in later life (Martin et al. 2011), which highlight the importance of life transitions at this stage in the life course:

- The Georgia adaptation model (Poon 1992; Poon et al. 1992), which proposes that the life satisfaction of older adults depends on how older people adapt to changes and life transitions. Adaptability in this model is a function of family longevity, environmental support, behavioural skills, nutrition, mental and physical health, and individual characteristics, and their inter-relationships.
- The developmental adaptation model (Martin and Martin 2002; Martin et al. 2011), which is an extension to the Georgia adaptation model. This model includes the influence of distal experiences such as childhood, economic status, trauma, and personality traits on the adaptation to life transitions and consequently on the life satisfaction of older adults.
- The resource change model, which is based on the 'conservation of resources' theory (Hobfoll 1988, 1989). This model predicts that the years from mid-life to later life are a stage of increased stress given they are characterised by transitions that involve net losses (i.e. more losses than gains) in social, health, cognitive, and financial resources.

These mechanisms would undergird the hedonic adaptation processes of older people. According to Lyubomirsky et al. (2005), the happiness set point determines around 50 per cent of the variation in happiness in a population, intentional activities contribute with another 40 per cent, and the remaining 10 per cent is due to circumstantial factors. The happiness set point varies from individual to individual and is determined genetically (given the importance that biological factors have on happiness through the set point, the next subsection looks into personality traits and genetics, the two aspects the literature has focused on). It is 'assumed to be fixed, stable over time, and immune to influence or control' (Lyubomirsky et al. 2005, p. 116). Chronological age belongs to a host of happiness-relevant circumstances along with gender, marital status, occupational status, health, religious affiliation, ethnicity, and so on, as well as national and regional geographical and cultural elements of the place of residence.

The previous paragraph makes it clear that chronological age does not seem to influence the set points by much. However, there are a number of questions that ought to be explored before reaching a conclusion in this regard:

- The set point levels, assumed to be fixed and stable, may not necessarily be so: preferences may shift.
- The adaptation may not be complete but partial: perhaps people do go back to a previous level of subjective well-being but merely tend towards it; the highs or lows may not disappear altogether but wane and taper off.
- Perhaps the adaptation process does not work the same from highs to lows as from low levels of happiness upwards: bouncing back after negative shocks may be a different process from habituation to positive stimuli.
- The adaptation mechanisms may not be same depending on the circumstance or stimulus (e.g. recovering from an unexpected loss of a loved one may not be one and the same process as recovering from losing one's job unexpectedly).

Let's consider each of these points, beginning with 'preference drift'.

Van Praag (1971) predicted that the concept of preference drift 'might lead to major changes in parts of the existing economy theory on consumer behavior and social welfare' [p. 338]. Preference drift is an application of adaptation theory to economics. Remember that according to adaptation theory, there is a set point that acts as an anchor towards which perception and other cognitive phenomena reverts or tends to after an exposure to a stimulus. In a study of individual welfare function in Belgium, Van Praag noted that when it came to evaluating income levels, the current income became the set point: how an individual evaluates monetary income (the level she considers 'sufficient') depends on the level of her current income. Preference drift takes place when the 'income aspirations shift with the income level actually attained' (Van

Praag 1971, p. 360). On receiving a marginal income, the drift implies a lower utility than what was expected before getting it. Importantly for the study of economics and ageing, given that older people are less in control of their income levels—especially if in retirement—than those of working age, van Praag added:

> The less consciously people can influence their own income level, the more fatalistically they tend to adapt to their income, since their income dos not adapt to them, and the more their actual income level sets their standard.
> (Van Praag 1971, p. 362)

Van de Stadt et al. (1985) estimated, using Belgium and Dutch data, that preference drift would erode around 60 per cent of the positive impact that an increase in income would be predicted to have on individual welfare. This effect only operates in the long term: in the short run, happiness and income do move together but as preferences change and people adapt to new levels of income, happiness moves back to a set point.

Di Tella et al. (2010) and Di Tella and MacCulloch (2006) presented evidence for Germany showing the same effect: 65 per cent of the immediate impact of an increase in income on happiness would be lost in four years. Di Tella et al. (2010), however, failed to find adaptation to the happiness effects of variations in occupational status. In addition, Easterlin (2005a) looked into data about things that respondents considered part of a good life in the United States of America between 1978 and 1994. This author reported that material aspirations rose once previously desired goods were acquired. To illustrate: a colour TV set, travelling abroad, a vacation home, and so on were included in the enumerations of goods that defined a good life by respondents who did not have these items, but were not included in their definitions once they owned them. In other words, adaptation was complete in the material domain (though this was not the case regarding marriage or number of children).

Think of the likely effect on your happiness before responding: if you had the possibility to buy a luxury, expensive car, would you settle for an inexpensive, frugal one instead? I am asking because if you think that the coveted masterpiece of vehicular performance will make you happier than a 'flivver' (all else equal), well, think again: Okulicz-Kozaryn et al. (2015) investigated the happiness impact of owning a luxury or an inexpensive car in the United States of America. His finding? Once he controlled for household income and house ownership, this author found that being the 'proud' owner of a luxury car will not add a jot to your happiness after a very short-term surge.

It seems that the Easterlin paradox is related to the futility of conspicuous consumption.

Adaptation may not be in full, but partial: the set point may not necessarily be reached. Partial adaptation would take place if the set point is not fully operative or if the set point itself changes over time. Lucas et al. (2004) analysed longitudinal data on life satisfaction and employment status for fifteen years from Germany and investigated adaptation to life satisfaction set points among individuals who had become unemployed. The authors found that life satisfaction levels were stable before unemployment but that even after years of finding another job, the experience of unemployment had left a permanent dent in the levels of life satisfaction: the new levels of reported life satisfaction were significantly lower (by about one standard deviation). A similar effect of unemployment was reported by Clark et al. (2008a) also with German data, particularly among men.

Oswald and Powdthavee (2008) studied hedonic adaptation (using a measure of psychological well-being) in a sample of individuals in the United Kingdom between 1996 and 2004 who were reported to have fallen into a level of disability that they were unable to at least climb stairs, dress themselves, walk at least for 10 minutes, or do housework activities on a normal day. The authors found that the onset of severe disability corresponded with a drop in one point in the life satisfaction score (along a 1 to 7 scale), and that the respondents showed between 30 per cent to 50 per cent of recovery (i.e. hedonic adaptation) within two years.

Cummins (2010) opined that this would be the case especially in high-income countries, with levels of income near their 'homeostatic ceiling' but that, at persistently low levels of income as in many developing countries, subjective well-being could drop and remain low.[13] The authors pointed out that this should not be confused with a change in the set point level. To explain the situation, they used the analogy of exposing a human body continually to low temperatures: it would not go back to its set point unless the thermal challenge ceased, but the core body temperature set point would remain at 37 °C.

Deaton (2012) reported that self-reported well-being dropped sharply in the aftermath of the 2008 financial crisis, and that by the end of 2010—and despite high unemployment—the levels of well-being had mostly recovered.

Vendrik (2013) applied an advanced time series econometric technique (integrated error-correction models) to data on life satisfaction for West Germany during the period 1984–2007. These types of models accommodate simultaneously short-term and long-term effects of different variables on life satisfaction and, given their dynamic specification, are able to test alternative hypotheses: social comparison, hedonic adaptation, and anticipation (i.e. that the expectation of a rise in income in period t increases life satisfaction in $t-1$). The findings showed support for the Easterlin paradox, as the models detected significant short-term effects but negligible long-term effects of income increase on life satisfaction levels. Furthermore, there was evidence of significant albeit incomplete adaptation to higher levels of income: the psychological set point theory of full adaptation was rejected, as well as the anticipation hypothesis.

Bottan and Truglia (2011) proposed another adaptation mechanism: the 'general habituation' channel, defined thus:

> …having experienced moments of happiness (unhappiness) in the present will directly make people prone to feelings of unhappiness (happiness) in the future, regardless of whether the original increase (decrease) in happiness was due to changes in income, health or love partners.
>
> (Bottan and Truglia 2011, p. 224)

In other words, general habituation assumes that people adopt two cognitive stances regarding future happiness levels considering their current state: either 'it's all downhill from up here' when things look bright or 'the good news is that it can't get any worse' when the chips are down. However, by applying time econometrics techniques to series for Germany, Japan, the United Kingdom, and Switzerland on satisfaction with life and also on satisfaction with various domains such as health, household income, and depending on the country, free or spare time, these authors *rejected* the hypothesis of general habituation. Instead, they found significant and positive *inertial* effects: the experience of high levels of happiness in the past makes it more likely to feel happy in the present.

Adaptation has been found to operate differently depending on the happiness construct under consideration and on the life event. Luhmann et al. (2012) carried out a meta-analysis of 188 different studies on the topic of subjective well-being adaptation to life events. The authors considered affective or psychological well-being and life satisfaction and several events (marriage, divorce, bereavement, childbirth, unemployment, reemployment, retirement, and relocation/migration), and found different effects depending the component of subjective well-being and the life event. For example, the 'honeymoon' positive effect of getting married only boosts life satisfaction and not psychological well-being, and only in the short term. In contrast, bereavement had a pronounced negative effect, especially on life satisfaction, but it was short-lived as people adapted and well-being increased gradually. Regarding unemployment, the meta-analysis showed persistent negative effects on subjective well-being even after three years following the loss of the job. In turn, retirement was shown as a neutral event: it had a negative impact on life satisfaction, but not on psychological well-being (which, the authors surmised, might reflect exaggerated positive expectations), with full adaptation in the following months as well-being returned to its previous level.

2.4 Personality Traits and Genetics

'A cheerful heart is good medicine', wrote the psalmist,[14] and he has been proved right: according to Boyce et al. (2013), personality is 'the strongest and most consistent cross-sectional predictor of high subjective well-being'. The literature suggests that higher measurements of happiness are associated with better physiological outcomes, physical functioning, and life expectancy, and also with health-related habits and practices such as diet, physical activity, sleep, alcohol consumption, and smoking. Besides these positive health effects of happiness, a large literature has looked into the biological processes that take place when happiness levels vary, and also into biological predictors of happiness—that is, to which extent happiness is predetermined by an individual's personality or genetic makeup (Ong and Patterson 2016).

According to trait theories, individuals would exhibit an underlying level of happiness around which they fluctuate as a result of life's ups and downs but to which they tend or are prone to remain. Some authors proposed that this temperamental predisposition towards a certain level of happiness would have a genetic origin. One of the most widely accepted models of personality is the 'five factor' or 'five trait' model (Digman 1990; McCrae and Costa 1997; Tupes and Christal 1992). The five factor personality model posits that the structure of personality can be described in terms of five factors or traits (the 'Big 5') known by the mnemonic OCEAN: openness (O), conscientiousness (C), extraversion (E), agreeableness (A), and neuroticism (N).

Using German data for the period 2000–2003, Budra and Ferrer-I-Carbonell (2018) investigated whether personality traits mediate between rank comparisons and life satisfaction. Apart from the Big 5, the authors considered locus of control and reciprocity. Locus of control refers to the degree people believe they have over life events. An individual's loci can be internal or external. Individuals with an internal locus of control believe that they have more control over the events in their lives than those with an external locus (Lefcourt 1991). Reciprocity is associated with trust, number of friends, empathy, and altruism (Perugini et al. 2003). Other than agreeableness and neuroticism, the other three personality traits as well as locus of control and reciprocity were found to be statistically significant mediators in the rank-life satisfaction relationship.

Mroczek and Spiro (2005) reported that of the OCEAN acronym, CEA were positively associated with life satisfaction, N was inversely associated, and O was not statistically significant (see also Lauriola and Iani 2017). A burgeoning literature has been examining the inter-relationships between life satisfaction and personality traits—that is, whether the causal relationship between happiness and personality traits is bi-directional. The evidence suggests it is. Magee et al. (2013) analysed the influence on life satisfaction of changes in the traits over time, rather than their levels. In a longitudinal data set for over 10,000 adults in Australia, they found that changes in personality traits predicted changes in life satisfaction four years later, and that the strength of the association diminished with chronological age as the traits are more stable with advancing years. Besides, these authors found that changes in life satisfaction influenced changes in personality traits. Similar findings were reported by Boyce et al. (2013), Soto (2015), and Specht et al. (2013). Tauber et al. (2016) reported, using data for Germany, that personality traits have a stronger predictive power of life satisfaction than life satisfaction of personality traits except in later life, when both would be inter-related.

Not only personality traits have been associated with happiness, but genetic factors as well. In a classic paper, Lykken and Tellegen (1996) examined genetic and socio-economic data and the responses to the well-being trait module of the MPQ scale from over 2000 twins born in Minnesota, United States of America, from 1936 to 1955. The results were indicative of a genetic basis for happiness/well-being: whilst income, education and socio-economic status accounted for 2 per cent of the variance in WB scores each, and marital status merely for 1 per cent, genetic variation explained between 44 per cent and 52 per cent of the variance. Hence, the authors' laconic conclusion: 'It may be that trying to be happier is as futile as trying to be taller and therefore is counterproductive' (Lykken and Tellegen 1996, p. 189).

Archontaki et al. (2013) looked into eudaimonic subjective well-being among 837 pairs of adult twins in the United States of America using the psychological well-being scale. The authors reported five distinct genetic factors on socio-psychological prosperity:

- self-control (a general substantial factor exerting influence over the other four factors)
- purpose (including purpose in life and personal growth, that is, seeking new experiences and having a sense of the future)

- social, positive relations with others
- growth linked to environmental mastery, and
- agency (management of goals and the necessary skills to achieve them)

Furthermore, De Neve et al. (2012) also explored differentials in subjective well-being among twins in the United States of America. The authors looked into the association between responses to satisfaction with life as a whole and a functional polymorphism on the serotonin transporter gene (SLC6A4 or 5-HTT or SERT) known as 5-HTTLPR (5-HTT-linked polymorphic region) and found that 33 per cent of the variation in life satisfaction was explained by genetic variation. Similarly, a 2015 meta-analysis of twenty-four different studies that explored the genetic influences on happiness, well-being, life satisfaction, and related concepts concluded that genetic factors explain around 35 per cent of the variance in the happiness-related constructs (Bartels 2015).

In a genome-wide association study (GWAS) of a combination of responses to questions about life satisfaction as a whole and happiness during the previous week in fifty-nine different birth cohorts, Okbay et al. (2016) reported a high genetic correlation between this combined measure of happiness and three single nucleotide polymorphisms (SNPs) (i.e. variations in a nucleotide in specific positions in the human genome).

Chen et al. (2013) reported that polymorphisms in the MAO-A (monoamine oxidase A) gene—in particular, low expression of this genotype—are associated with greater subjective happiness among women but not among men, after controlling for age, education, household income, marital status, employment status, mental and physical health, relationship quality, religiosity, abuse history, recent negative life events, and self-esteem.

These biological approaches, including the so-called agnostic (i.e. hypothesis-free) GWAS studies, provide a crucial piece of evidence for public health, economics, and other sciences: the (large) proportion of variance they cannot explain corresponds to *modifiable* determinants of happiness (Koellinger et al. 2016; Probst-Hensch 2017).

2.5 Needs-Based Theories

Needs-based theories focus on absolute measures of happiness.[15] These theories are based, directly or indirectly, on the hierarchy of needs model by Maslow (1943, 1954). Maslow proposed that human needs can be arranged along a hierarchy of 'pre-potency', that is that the satisfaction of one need presupposes or rests on the satisfaction of a more 'pre-potent' need. He identified five needs, from most basic to the least pre-potent: physiological safety, belonging and love, esteem, self-actualisation, and self-transcendence. Doyal and Gough (1991) rejected the proposition that human needs are hierarchically organised on the grounds (simplifying *in extremis* a rich line of thought) that needs as universalisable goals should not be confused with motivations or drives. Instead, the relative importance of needs is more fluid and depends on the situation. This situational hypothesis is supported by the evidence produced by Blane et al. (2004) in their study of the determinants of quality of life among people aged 65–75 years in Great Britain. They used the CASP-19 measure and found that participants' current situation was more influential on their level of quality of life, particularly material circumstances and serious health problems, thereby concluding that disadvantage earlier in the life course would not preclude good quality of life in later years. Two basic, necessary conditions to be able to participate in a society are the need for freedom from poverty and to act upon one's life chances. Calling to mind the distinction between 'freedom from' and 'freedom to' by Berlin (1969), Fromm (1942), and Taylor (1979), these two orientations are captured in the control (freedom from want and poverty) and autonomy (freedom to act) domains of the CASP-19 measure of quality of life (see previous chapter) (Blane et al. 2004).

Niedzwiedz et al. (2012) distinguishes between four theories that relate older people's quality of life with the socio-economic trajectories over their life courses: social mobility, accumulation, latent and pathway models.

- Social mobility models consider intra- and intergenerational mobility (see Part II in this volume) and predict that downward mobility negatively affects quality of life in later life, although other authors propose that any type of social mobility (including upwards) may cause strain with negative consequences on quality of life.
- The accumulation model—or cumulative advantage/disadvantage hypothesis (see Chap. 4 in Volume I)—posits that adverse circumstances earlier in the life course compound and negatively affect quality of life in later life.
- The latent model posits that adverse circumstances in childhood have detrimental effects later in life, given the key role of childhood in human development (see, in this regard, the concept of the 'life course' in that same chapter).
- Pathway models look at the whole trajectory over the life course, so that adverse situations earlier in life can be attenuated by later life.

In contrast to Blane et al. (2004), Wildman et al. (2018) did find that adverse childhood events would significantly reduce quality of life among people aged 62–64 more than proximal influences, supporting the latent model instead of a situational viewpoint more aligned to the pathway models.

Webb et al. (2010) looked into a number of baseline characteristics in a sample of people aged 50 or over in England and studied the changes in CASP-19 scores resulting from the levels of health, economic and social measures at baseline and the changes in health, economic and social measures over a period of six years. The authors found that whilst older people's circumstances decline over time, changes in these circumstances accentuate the rate of this decline. In addition, subjective or psychological socio-economic measures such as an improved *perception* of financial circumstances and of the quality of neighbourhoods can reduce the decline.

Drakopoulos and Grimani (2013) looked into data from thirty European countries and Turkey for 2007 and reported that the significance of the association between household income and happiness (measured with a single-item indicator) was greater among low-income families. He suggested that this was evidence of hierarchical behaviour *à la Maslow*.

Di Gessa et al. (2018) used the CASP-19 instrument to investigate whether working beyond state pensionable age affects the quality of life of older people in England. The authors observed that individuals who took the decision to be in paid employment beyond state pension age out of economic necessity showed lower quality of life and those who showed control over their decision reported the highest quality of life. Furthermore, among retirees, the greatest fall in quality of life was in those who reported they had retired involuntarily.

Clark et al. (2018) investigated the determinants of life satisfaction (single-item question) among people aged 50 or over between 2004 and 2012 in England. They reported that feelings of loneliness had the strongest association. Of course, it affected life satisfaction negatively, both within age groups and for men and women separately. A reduction in loneliness was also found to be the most effective change that had increased life satisfaction over the period. The second most important factor was the perception of having positive support. These results suggest that meeting the needs for social interaction and social activities is key to increase life satisfaction in later life, at least in developed countries where physiological and safety needs are by and large covered. This result, then, would be in accordance to Maslow's hierarchy of needs.

2.6 Inequality and Happiness

Could you be as happy, all else considered, if you lived in a country or region with high inequality as in an area with greater equality? This is the question that a number of studies investigated: whether happiness or life satisfaction are affected by the levels of inequality (usually, in income) in a given geographical area or group. The two main explanations for the inverse association between inequality and happiness are (Ferrer-i-Carbonell and Ramos 2014):

- Self-motives: greater inequality forebodes instability, shocks, criminality, and—especially for the well-off—a risk of falling down. However, not all agents would exhibit aversion to the risks behind greater inequality, as great disparity also brings great opportunities (although this optimism tend to be short-lived if the opportunities are not realised)—see also Binmore (2005).
- Care or regard for others: fairness and egalitarian concerns are a source of utility—see also Seabright (2006).

Other explanations stem from the evolutionary theory literature (Brosnan and Waal 2014; Henrich et al. 2010; Singer et al. 2006; Vogel 2004), which has found that even responses and the behaviour of babies (Schmidt and Sommerville 2011) and non-human primates (Brosnan 2013) are influenced by their perception of fairness.

Kahneman and Deaton (2010) presented some intriguing findings regarding income inequality, and the two components of subjective well-being for the United States of America. Life evaluation (measured with Cantril's ladder) is positively associated with income, whilst emotional well-being (an index of a number of emotions reported to have been felt or not the day before) ceases to be associated with income at fairly high levels (US$ 75,000 in 2010 prices). The author concludes that 'high income buys life satisfaction but not happiness, and that low income is associated both with low life evaluation and low emotional well-being' [p. 16489].

2.7 Happiness Along the Life Cycle

What does the evidence on happiness (and related measures) show along the life cycle? A literature review (Ulloa et al. 2013) summarises the main findings up to the start of the 2010s and discusses the drawbacks of some of those studies. The findings have been disparate: although most authors had found a

U-shaped relationship between happiness/SWB and chronological age, many of these studies were based on cross-sectional data, which means that they could not control for any cohort effects. Longitudinal studies, instead, showed anything from U-shaped, inverse U-shaped, flat, decreasing linear relations. For each type of finding, needless to say, the different authors suggested alternative explanations. In turn, the bulk of studies using longitudinal data and controlling for fixed effects had shown either a largely flat relationship or a combination of a flat the levels of happiness or SWB would not change along the life cycle.

Let's consider the predictions from the life-cycle hypothesis (see Volume I, Chap. 8). According to this hypothesis, income would exhibit a hump-shaped pattern by chronological age and consumption would be smoother. Consequently, if happiness were related to earnings, they would show a hump-shaped pattern by chronological age. Instead, if happiness were more dependent on consumption, it would follow a flatter shape over the life cycle. Which pattern is reflected in the data? *None of those.* Furnham (2014, p. 60) lists ten widely believed 'facts' about happiness, which are actually myths, among which he includes that young people are happier than older people. Most studies show a U-shaped relationship between happiness and age, not anticipated by the (overwhelmingly accepted) life-cycle hypothesis: happiness or subjective well-being (the two most common variables studied in this literature) would follow a U-shaped age pattern with the lowest point somewhere between the 30s and the 50s (see, for example, Oswald (1997), using data for twelve European countries between 1972 and 1990; Blanchflower and Oswald (2004) for both the United States of America for the period 1972–1998 and Great Britain between 1973 and 1998; Blanchflower and Oswald (2008) for the United States of America between 1972 and 2006 and sixteen European countries during the period 1976–2002; Van Landeghem (2012) for West Germany from 1984 to 2007 and East Germany between 1990 and 2007 and Schwandt (2016) for both West and East Germany between 1991 and 2004; or Cruz and Torres (2006), for Colombia in 2003).

Ulloa et al. (2013) concluded that cohort effects, the control variables (especially, whether health is included or not among the set of explanatory variables), the measure of interest (happiness, SWB, or satisfaction with one particular domain), and the statistical techniques (accounting for fixed effects, for example; see box) are of key importance. Some studies that used more flexible, non-parametric statistical approaches reported a U-shape pattern until the people are in their late 60s, and falling well-being thereafter (Gwozdz and Sousa-Poza 2010; Wunder et al. 2013). Besides, authors who focused on objective proxy variables, such as the use of anti-depressants (Blanchflower and Oswald 2011a) or diagnosed mental illness (Lang et al. 2011), also using longitudinal data sets, did find a U-shaped age pattern after controlling for fixed effects.

However, Bartolini et al. (2013), using data for Germany between 1996 and 2007, reported a negative association between subjective well-being and chronological age older than 65 years; for younger ages, the authors did find the U-shaped pattern. Population projections for Germany combined with the negative association between older life and subjective well-being led Bartolini, Bilancini, and Sarracino to predict that happiness will diminish over time.

Jivraj et al. (2014) looked into eudemonic, evaluative, and affective components of subjective well-being among people aged 50 or older in England during the period 2002–2003 to 2010–2012. The three constructs were considered separately. Eudemonic well-being was measured with the CASP-19 scale; the SWLS was used for evaluative well-being; and a depression scale was used as a proxy for affective well-being. The study found that older cohorts exhibited equal or higher SWB than younger cohorts in each of the measures of well-being. However, older cohorts also exhibited more pronounced declines in SWB as a result of individual ageing (once other age-related changes such as widowhood, retirement, and health decline were controlled for).

Recently, Beja (2018), using data from a worldwide survey, reported the presence of the U-shaped age pattern of happiness and the mid-life trough, but with the highest point in old age at a lower level than the highest point at younger ages. In other words, good econometric practice and the use of alternative measures seem to be conveying a fairly (though not totally) consistent message that the association between chronological age and happiness is non-linear and would reflect a U-shape.

As in other lines of enquiry within happiness studies, the researchers have applied different techniques to different measures of interest, so it is to be expected that this finding has not always been replicated. Mood, for example, was studied by Cameron (1975), who—using tests of statistical association such as the χ^2—reported that it was independent of chronological age. Or, take Mroczek and Kolarz (1998), who studied positive and negative on a sample of individuals in the United States of America. Using hierarchical regressions, the authors reported that negative affect diminished with chronological age among married men, though it was independent of age among women and single men. Concerning positive affect, the findings were that it increased linearly with age among men and at an accelerating rate among women.

> **Fixed Effects and Longitudinal Data Sets**
>
> Regarding econometric techniques, it is worth mentioning for the benefit of readers unfamiliar with rather ominous terms such as 'fixed effects' and their ilk that cross-sectional data sets contain data for units of observation (persons, countries, firms, etc.) for only one-time period, whereas longitudinal or panel data sets contain data for each unit of observation over time. For example, a data set on age, gender, marital status, and employment status of a sample of people aged 50 or over in Pennsylvania in October 2018 is cross-sectional. A data set on age, gender, marital status, and employment status of a sample of people aged 50 or over in Pennsylvania for each month between October 2010 and October 2018 is longitudinal. Cross-sectional data sets provide a snapshot; longitudinal data, a film. With cross-sectional data, all the variation in the data lies between units. Instead, with data for units over time, there are two sources of variation: differences between units as in cross-sections and differences within units, as individuals change with the passing of time (not just their chronological age changes but their marital status can—and even gender). Fixed effects control and capture the differences within units. (The appropriateness of using fixed effects against other econometric assumptions can be, and usually is, tested.)[16]
>
> In an interesting contribution, Li (2016) explained that studies using cross-sectional data can detect differences in average well-being or happiness across birth cohorts, whereas using longitudinal data and fixed effects they can detect changes over the life cycle of individuals. Moreover, this author pointed out that the literature has put forth several explanations for the opposing findings regarding the age pattern of happiness or well-being: unobserved heterogeneity, reverse causality, sample selectivity, parametrisation issues, participants' time-in-panel effects, and even interviewers' bias. He provided an alternative explanation: the multidimensionality of the data. Cross-sectional regressions provide information about whether in one particular point in time older people are more or less satisfied with their life or are happier than younger people. Fixed effects techniques, when applied to panel or longitudinal data from different cohorts, do control for individual fixed effects but not for cohort effects: they compare changes in happiness over time within each individual. The discrepant results derive, according to Li, from that hard nut to crack: the age-period-cohort effect (see Volume I, Chap. 2).

Since Ulloa et al. (2013), several more papers have been published on this topic. Despite that they have taken on board the recommendations by Ulloa et al. (2013), the results continue to be contradictory, although most of the evidence supports the U-shaped age-happiness relationship.

Cheng et al. (2017) reported a U-shaped relationship applying fixed effects in four different longitudinal data sets (two from Australia, and the other two from the United Kingdom and Germany, respectively). Instead, using longitudinal data for Great Britain between 1991 and 2008 and controlling for cohort effects, Bell (2014) concluded that the association between mental health and age would show improvement up to the mid-life years to worsen in old age, without the data providing evidence of a U-shape.

In contrast, Kolosnitsyna et al. (2017) investigated data for women aged 55 or over and men aged 60 or older in Russia between 2009 and 2012 and found a U-shaped relationship between chronological age and overall life satisfaction (men seemed to be increasingly happier than women as they became older and their nadir was met at a younger age than for women).

Moreover, Graham and Pozuelo (2017) confirmed the presence of the U-shaped pattern in data for forty-four out of the forty-six countries they studied. Interestingly, the turning points varied across countries.

A different approach looks into components of subjective well-being or into domains of life satisfaction, as they might show different dynamics with chronological age. Subjective well-being, as I explained in the previous chapter, comprises emotional or hedonic well-being and life satisfaction. Steptoe et al. (2015) found in English-speaking countries a U-shaped pattern between age and life satisfaction (measured with Cantril's ladder of life), but not in other regions of the world, and even in the former countries, hedonic well-being did not evidence a U-shape with age but a linear, positive association: '…the occurrence of a lot of stress or a lot of anger yesterday decreased throughout life, and more rapidly so after age 50 years. Worry remained high until age 50 years and reduced thereafter…' (Steptoe et al. 2015, p. 642).

In addition, Gana et al. (2015) studied positive affect among older people in Switzerland during the period 1992–2014 and found, using latent growth curve modelling, an inverse U-shape: positive affect (one of the components of subjective well-being) would reduce in later life.

Stone et al. (2017) reported an inverted U-shape in the age pattern of perceived psychological stress among older people in the United States of America using three different surveys. Graham and Pozuelo (2017) also reported this inverse U-shaped pattern between stress and subjective well-being in forty-six countries.

In an early contribution, Cutler (1979) investigated the multidimensionality of the life satisfaction construct and how its components and their relative importance changed with age. More recently, Bardo (2017) confirmed such age variations in reported satisfaction with family, friends, health, hobbies, and place of residence between 1973 and 1994 in the United States of America. Whereas satisfaction with health diminishes with chronological age, the older the people the more satisfied they are with their place of residence, and slightly more as well with their friends. In contrast, satisfaction with hobbies shows an inverted U-shaped age pattern. Bardo also presented the age pattern of life satisfaction net of cohort and period effects: life satisfaction increases with chronological age, although at a diminishing rate after age 79. Finally, life satisfaction by domain net of period and cohort effects exhibits similar patterns, which included the influences of period and cohort, especially with respect to health (inverse U-shaped) and family (U-shaped).

Using data for the United States of America between 1972 and 2004, Yang (2008) removed cohort and period effects to analyse the association between chronological age and a self-reported measure of overall happiness. The author reported five findings:

- The existence of significant net age, period, and cohort effects
- '…with age comes happiness' [p. 220]
- Convergence in happiness levels across social classes (measured by socio-economic status) over the life course—that is, rejecting the cumulative disadvantage hypothesis regarding happiness
- Significant period effects affecting happiness levels and trajectories of different sub-groups (e.g. by race or gender) differently
- Cohort effects were also significant, with baby boomers exhibiting less happiness on average than older and younger cohorts

Notes

1. 'Al cabo de los años, un hombre puede simular muchas cosas, pero no la felicidad' (Borges 2007, p. 474).
2. See also the related literature on conspicuous consumption theory (Arrow and Dasgupta 2009; Veblen 1912) and the bandwagon and snob effects introduced by Leibenstein (1950).
3. For the role of setting higher goals as motivating self-challenges towards action, see Bandura and Locke (2003).
4. See Chap. 3 in Volume II
5. Brown et al. (2008).
6. We form the Lagrangean

$$\mathscr{L} = U(x_1, x_2) + \lambda \cdot [y - p_1 \cdot x_1 - p_2 \cdot x_2] = 0 \qquad (2.18)$$

Maximising this expression with respect to x_1 and x_2 renders:

$$\frac{\partial U}{\partial x_1} - \lambda \cdot p_1 = 0 \qquad (2.19)$$

and

$$\frac{\partial U}{\partial x_2} - \lambda \cdot p_2 = 0 \qquad (2.20)$$

from where, solving for λ, we obtain Eq. (2.16) in the text.

7. Contacts with members of social networks in later life have also found to be positively associated with better mental health status, reduced risk of cardiovascular disease, cancer, and infectious diseases as well as overall mortality (Shor and Roelfs 2015).
8. See Larson (1978) for a survey of early studies in well-being in later life in the United States of America, which reviews the effects of social networks. See also Litwin and Shiovitz-Ezra (2010).
9. In gerontology, adaptation is also referred to as 'adjustment', although some authors opine that these two concepts are not exactly synonymous (Von Humboldt 2016).
10. For a short history of the theory of adaptation level and its relevance to economics, see Edwards (2018).
11. The three lowest amounts were US$50,000 in 1978 prices; around $200,000 in 2017 prices.
12. Typically, other authors use a different terminology: Cummins (2010, 2016) and Cummins and Wooden (2014) referred to adaptation level theory as 'subjective well-being homeostasis' and to the set point as the 'Homeostatically Protected Mood'.
13. See also Cummins et al. (2014).
14. Proverbs 17:22. The Holy Bible. New International Version (2011).
15. Need-based theories are also at the heart of the development economics and poverty studies that aim to understand the hierarchy of consumption (Doyal and Gough 1991; Gough 1994; Jackson and Marks 1999).
16. For an introduction, see Wooldridge (2013, ch. 14).

References

Albert, Stuart (1977). "Temporal comparison theory". In: *Psychological Review* 84.6, pages 485–503.

Alicke, Mark D and Olesya Govorun (2005). "The better-than-average effect". In: *The self in social judgment*. Edited by Mark D Alicke, David A Dunning, and Joachim Krueger. Studies in Self and Identity. New York, NY: United States of America: Psychology Press, pages 85–106.

Angelini, Viola et al. (2012). "Age, health and life satisfaction among older Europeans". In: *Social indicators research* 105.2, pages 293–308.

Archontaki, Despina, Gary J Lewis, and Timothy C Bates (2013). "Genetic influences on psychological well-being: A nationally representative twin study". In: *Journal of Personality* 81.2, pages 221–230.

Arrow, Kenneth J and Partha S Dasgupta (2009). "Conspicuous consumption, inconspicuous leisure". In: *The Economic Journal* 119.541, F497–F516.

Ateca-Amestoy, Victoria, Alexandra Cortés Aguilar, and Ana I Moro-Egido (2014). "Social interactions and life satisfaction: Evidence from Latin America". In: *Journal of Happiness Studies* 15.3, pages 527–554.

Bandura, Albert and Edwin A Locke (2003). "Negative self-efficacy and goal effects revisited". In: *Journal of applied psychology* 88.1, pages 87–99.

Bárcena-Martn, Elena, Alexandra Cortés-Aguilar, and Ana I Moro-Egido (2017). "Social comparisons on subjective well-being: The role of social and cultural capital". In: *Journal of Happiness Studies* 18.4, pages 1121–1145.

Bardo, Anthony R (2017). "A life course model for a domains-of-life approach to happiness: Evidence from the United States". In: *Advances in Life Course Research* 33, pages 11–22.

Baron, Reuben M and David A Kenny (1986). "The moderator-mediator variable distinction in social psychological research: Conceptual, strategic, and statistical considerations". In: *Journal of personality and social psychology* 51.6, pages 1173–1182.

Bartels, Meike (2015). "Genetics of wellbeing and its components satisfaction with life, happiness, and quality of life: A review and meta-analysis of heritability studies". In: *Behavior genetics* 45.2, pages 137–156.

Bartolini, Stefano, Ennio Bilancini, and Francesco Sarracino (2013). "Predicting the trend of well-being in Germany: How much do comparisons, adaptation and sociability matter?" In: *Social Indicators Research* 114.2, pages 169–191.

Beja, Edsel L (2018). "The U-shaped relationship between happiness and age: evidence using world values survey data". In: *Quality & Quantity* 52.4, pages 1817–1829.

Bell, Andrew (2014). "Life-course and cohort trajectories of mental health in the UK, 1991–2008 A multilevel age-period-cohort analysis". In: *Social science & medicine* 120, pages 21–30.

Berger, Peter L and Thomas Luckmann (1966). *The Social Construction of Reality. A Treatise in the Sociology of Knowledge.* London: United Kingdom: Penguin Books.

Berlin, Isaiah (1969). *Four Essays on Liberty.* Oxford: United Kingdom: Oxford University Press.

Binmore, Ken (2005). *Natural Justice.* Oxford: United Kingdom: Oxford University Press.

Blanchflower, David G and Andrew J Oswald (2004). "Well-being over time in Britain and the USA". In: *Journal of Public Economics* 88.7–8, pages 1359–1386.

—— (2008). "Is well-being U-shaped over the life cycle?" In: *Social science & medicine* 66.8, pages 1733–1749.

—— (2011a). *Antidepressants and age.* IZA Discussion Paper 5785. Bonn: Germany.

Blane, David B et al. (2004). "Life course influences on quality of life in early old age". In: *Social science & Medicine* 58.11, pages 2171–2179.

Borges, Jorge Luis (2007). "La Memoria de Shakespeare". In: *Obras Completas*. Buenos Aires, Argentina: Emecé.

Bottan, Nicolas Luis and Ricardo Perez Truglia (2011). "Deconstructing the hedonic treadmill: Is happiness autoregressive?" In: *The Journal of Socio-Economics* 40.3, pages 224–236.

Boyce, Christopher J, Gordon DA Brown, and Simon C Moore (2010). "Money and happiness: Rank of income, not income, affects life satisfaction". In: *Psychological Science* 21.4, pages 471–475.

Boyce, Christopher J, Alex M Wood, and Nattavudh Powdthavee (2013). "Is personality fixed? Personality changes as much as "variable" economic factors and more strongly predicts changes to life satisfaction". In: *Social indicators research* 111.1, pages 287–305.

Brickman, Philip and Donald Campbell (1971). "Hedonic Relativism and Planning the Good Society". In: *Adaptation-level theory: A symposium*. Edited by Mortimer Herbert Apley. New York, NY: United States of America: Academic Press, pages 287–302.

Brickman, Philip, Dan Coates, and Ronnie Janoff-Bulman (1978). "Lottery winners and accident victims: Is happiness relative?" In: *Journal of personality and social psychology* 36.8, pages 917–927.

Brosnan, Sarah F (2013). "Justice-and fairness-related behaviors in nonhuman primates". In: *Proceedings of the National Academy of Sciences* 110.Supplement 2, pages 10416–10423.

Brosnan, Sarah F and Frans BM de Waal (2014). "Evolution of responses to (un)fairness". In: *Science* 346.6207. https://doi.org/10.1126/science.1251776.

Brown, Gordon DA et al. (2005). *Does Wage Rank Affect Employees' Well-being?* IZA Discussion Paper 1505. Bonn: Germany.

—— (2008). "Does Wage Rank Affect Employees Well-being?" In: *Industrial Relations: A Journal of Economy and Society* 47.3, pages 355–389.

Budra, Santi and Ada Ferrer-I-Carbonell (2018). "Life Satisfaction, Income Comparisons and Individual Traits". In: *Review of Income and Wealth*. https://doi.org/10.1111/roiw.12353.

Buunk, Abraham P and Pieternel Dijkstra (2017). "Social Comparisons and Well-Being". In: *The Happy Mind: Cognitive Contributions to Well-Being*. Edited by Michael D Robinson and Michael Eid. Cham, Switzerland: Springer, pages 311–330.

Cameron, Paul (1975). "Mood as an indicant of happiness: Age, sex, social class, and situational differences". In: *Journal of Gerontology* 30.2, pages 216–224.

Carstensen, Laura L (1993). "Motivation for social contact across the life span: A theory of socioemotional selectivity". In: *Nebraska Symposium on Motivation*. Edited by Janis E Jacobs. Volume 40. Current Theory and Research in Motivation. Lincoln, NE: United States of America: University of Nebraska Press, pages 209–254.

Castillo-Carniglia, Álvaro et al. (2012). "Factores asociados a satisfacción vital en una cohorte de adultos mayores de Santiago, Chile". In: *Gaceta Sanitaria* 26.5, pages 414–420.

Chen, Tao et al. (2016). "Present-fatalistic time perspective and life satisfaction: The moderating role of age". In: *Personality and Individual Differences* 99, pages 161–165.

Chen, Henian et al. (2013). "The MAOA gene predicts happiness in women", *Progress in Neuro-Psychopharmacology & Biological Psychiatry*, 40: 122–125

Cheng, Sheung-Tak, Helene H Fung, and Alfred CM Chan (2008). "Living status and psychological well-being: social comparison as a moderator in later life". In: *Aging and Mental Health* 12.5, pages 654–661.

Cheng, Terence C, Nattavudh Powdthavee, and Andrew J Oswald (2017). "Longitudinal Evidence for a Midlife Nadir in Human Well-being: Results from Four Data Sets". In: *The Economic Journal* 127.599, pages 126–142.

Chou, K-L and I Chi (2001). "Social comparison in Chinese older adults". In: *Aging & Mental Health* 5.3, pages 242–252.

Clark, Andrew E et al. (2018). The Origins of Happiness: The Science of Well-Being over the Life Course. Princeton, NJ: United States of America. Princeton University Press.

Clark, Andrew E, Paul Frijters, and Michael A Shields (2008). "Relative income, happiness, and utility: An explanation for the Easterlin paradox and other puzzles". In: *Journal of Economic literature* 46.1, pages 95–144.

Clark, Andrew E and Claudia Senik (2010). "Who compares to whom? The anatomy of income comparisons in Europe". In: *The Economic Journal* 120.544, pages 573–594.

—— (2014a). *Happiness and Economic Growth: Lessons from Developing Countries*. Studies of Policy Reform. Oxford: United Kingdom: Oxford University Press.

—— (2014b). "Life Satisfaction in the Transition from Socialism to Capitalism: Europe and China". In: *Happiness and Economic Growth: Lessons from Developing Countries*. Studies of Policy Reform. Oxford: United Kingdom: Oxford University Press, pages 6–31.

Clark, Andrew E, Niels Westergård-Nielsen, and Nicolai Kristensen (2009). "Economic satisfaction and income rank in small neighbourhoods". In: *Journal of the European Economic Association* 7.2, pages 519–527.

Clark, Andrew E et al. (2008a). "Lags and leads in life satisfaction: A test of the baseline hypothesis". In: *The Economic Journal* 118.529, F222–F243.

Costa Jr, Paul T et al. (1987). "Longitudinal analyses of psychological well-being in a national sample: Stability of mean levels". In: *Journal of Gerontology* 42.1, pages 50–55.

Cruz, Jasson and Julián Torres (2006). "¿De qué depende la satisfacción subjetiva de los colombianos?" In: *Cuadernos de economa* 25.45, pages 131–154.

Cummins, Robert A (2010). "Subjective wellbeing, homeostatically protected mood and depression: A synthesis". In: *Journal of Happiness Studies* 11.1, pages 1–17.

Cummins, Robert A (2016). "The theory of subjective wellbeing homeostasis: A contribution to understanding life quality". In: *A Life Devoted to Quality of Life. Festschrift in Honor of Alex C. Michalos*. Edited by Filomena Maggino. Volume 60. Social Indicators Research Series. Cham: Switzerland: Springer International Publishing, pages 61–79.

Cummins, Robert A and Mark Wooden (2014). "Personal resilience in times of crisis: The implications of SWB homeostasis and set-points". In: *Journal of Happiness Studies* 15.1, pages 223–235.

Cummins, Robert A et al. (2014). "A demonstration of set-points for subjective wellbeing". In: *Journal of Happiness Studies* 15.1, pages 183–206.

Cutler, Neal E (1979). "Age variations in the dimensionality of life satisfaction". In: *Journal of Gerontology* 34.4, pages 573–578.

D'Ambrosio, Conchita and Joachim R Frick (2012). "Individual wellbeing in a dynamic perspective". In: *Economica* 79.314, pages 284–302.

De Neve, Jan-Emmanuel et al. (2012). "Genes, economics, and happiness". In: *Journal of Neuroscience, Psychology, and Economics* 5.4, pages 193–211.

Deaton, Angus (2012). "The financial crisis and the well-being of Americans (2011 OEP Hicks Lecture)". In: *Oxford economic papers* 64.1, pages 1–26.

Desmyter, Fien and Rudi De Raedt (2012). "The relationship between time perspective and subjective well-being of older adults". In: *Psychologica Belgica* 52.1, pages 19–38.

Di Gessa, Giorgio et al. (2018). "The decision to work after state pension age and how it affects quality of life: evidence from a 6-year English panel study". In: *Age and Ageing* 47.3, pages 450–457.

Di Tella, Rafael, John Haisken-De New, and Robert MacCulloch (2010). "Happiness adaptation to income and to status in an individual panel". In: *Journal of Economic Behavior & Organization* 76.3, pages 834–852.

Di Tella, Rafael and Robert MacCulloch (2006). "Some uses of happiness data in economics". In: *Journal of Economic Perspectives* 20.1, pages 25–46.

Didino, Daniele et al. (2017). "Exploring predictors of life satisfaction and happiness among Siberian older adults living in Tomsk Region". In: *European Journal of Ageing* 15.2, pages 1–13.

Digman, John M (1990). "Personality structure: Emergence of the five-factor model". In: *Annual Review of Psychology* 41.1, pages 417–440.

Do, Amy M, Alexander V Rupert, and George Wolford (2008). "Evaluations of pleasurable experiences: The peak-end rule". In: *Psychonomic Bulletin & Review* 15.1, pages 96–98.

Dohmen, Thomas et al. (2011). "Relative versus absolute income, joy of winning, and gender: Brain imaging evidence". In: *Journal of Public Economics* 95.3–4, pages 279–285.

Dolan, Paul, Tessa Peasgood, and Mathew White (2006). *Review of research on the influences on personal well-being and application to policy making*. Technical report. London: United Kingdom.

Doyal, Len and Ian Gough (1991). *A theory of human need*. Critical perspectives. Basingstoke: United Kingdom: Palgrave Macmillan.

Drakopoulos, Stavros A and Katerina Grimani (2013). *Maslow's needs hierarchy and the effect of income on happiness levels*. MPRA Paper 50987. Munich: Germany: University Library of Munich.

Duesenberry, James S. (1949). *Income, Saving, And The Theory of Consumer Behavior*. Cambridge, MA: United States of America: Harvard University Press.

Dunning, David, Chip Heath, and Jerry M Suls (2004). "Flawed self-assessment: Implications for health, education, and the workplace". In: *Psychological science in the public interest* 5.3, pages 69–106.

Dutt, Amitava Krishna and Benjamin Radcliff (2009). "Introduction: happiness, economics and politics". In: *Happiness, Economics and Politics: Towards a Multidisciplinary Approach*. Edited by Amitava Krishna Dutt and Benjamin Radcliff. Cheltenham: United Kingdom: Edward Elgar, pages 1–21.

Easterlin, Richard A (1974). "Does economic growth improve the human lot? Some empirical evidence". In: *Nations and Households in Economic Growth: Essays in Honor of Moses Abramovitz*. Edited by Paul A David and Melvin W Reder. New York, NY: United States of America: Academic Press, pages 89–125.

—— (1995). "Will raising the incomes of all increase the happiness of all?" In: *Journal of Economic Behavior & Organization* 27.1, pages 35–47.

—— (2005a). "A puzzle for adaptive theory". In: *Journal of Economic Behavior & Organization* 56.4, pages 513–521.

Easterlin, Richard A et al. (2010). "The happiness income paradox revisited". In: *Proceedings of the National Academy of Sciences* 107.52, pages 22463–22468.

Eckersley, Richard (2000). "The state and fate of nations: Implications of subjective measures of personal and social quality of life". In: *Social Indicators Research* 52.1, pages 3–27.

Edwards, José (2018). "Harry Helsons Adaptation-Level Theory, Happiness Treadmills, and Behavioral Economics". In: *Journal of the History of Economic Thought* 40.1, pages 1–22.

Ferrer-i-Carbonell, Ada (2013). "Happiness economics". In: *SERIEs* 4.1, pages 35–60.

Ferrer-i-Carbonell, Ada and Xavier Ramos (2014). "Inequality and happiness". In: *Journal of Economic Surveys* 28.5, pages 1016–1027.

Ferring, Dieter and Martine Hoffmann (2007). ""Still the same and better off than others?": social and temporal comparisons in old age". In: *European Journal of Ageing* 4.1, pages 23–34.

Fliessbach, Klaus et al. (2007). "Social comparison affects reward-related brain activity in the human ventral striatum". In: *Science* 318.5854, pages 1305–1308.

Flyvbjerg, B. and S Sampson (2001). *Making Social Science Matter: Why Social Inquiry Fails and How it Can Succeed Again*. Cambridge: United Kingdom: Cambridge University Press.

Frank, Robert H (1985). "The demand for unobservable and other nonpositional goods". In: *The American Economic Review* 75.1, pages 101–116.

Frawley, Ashley (2015). *Semiotics of Happiness. Rhetorical beginnings of a public problem.* London: United Kingdom: Bloomsbury Academic.

Frederick, Shane and George Loewenstein (1999). "Hedonic Adaptation". In: *Well-Being. The Foundations of Hedonic Psychology.* Edited by Daniel Kahneman, Ed Diener, and Norbert Schwarz. New York, NY: United States of America: Russell Sage Foundation, pages 302–329.

Frieswijk, Nynke et al. (2004). "The effect of social comparison information on the life satisfaction of frail older persons" In: *Psychology and aging* 19.1, pages 183–190.

Fromm, Erich (1942). *The Fear of Freedom.* International library of sociology and social reconstruction. London: United Kingdom: Paul Kegan.

Furnham, Adrian (2014). *The New Psychology of Money.* New York, NY: United States of America: Routledge.

Gabrian, Martina, Anne J Dutt, and Hans-Werner Wahl (2017). "Subjective time perceptions and aging well: A review of concepts and empirical research-a mini-review". In: *Gerontology* 63.4, pages 350–358.

Galbraith, John Kenneth (1958). *The Affluent Society.* London: United Kingdom: Hamish Hamilton.

Gana, Kamel, Yaël Saada, and Hélène Amieva (2015). "Does positive affect change in old age? Results from a 22-year longitudinal study" In: *Psychology and aging* 30.1, pages 172–179.

Gilbert, Daniel (2007). *Stumbling on Happiness.* New York, NY: United States: Vintage Books.

Gilbert, Daniel T and Timothy D Wilson (2007). "Prospection: Experiencing the future". In: *Science* 317.5843, pages 1351–1354.

Gilhooly, Mary Ken Gilhooly, and Ann Bowling (2005). "Meaning and measurement". In: *Understanding quality of life in old age.* Edited by Alan Walker, pages 14–26.

Gough, Ian (1994). "Economic Institutions and the Satisfaction of Human Needs". In: *Journal of Economic Issues* XXVIII.1, pages 25–68.

Graham, Carol and Andrew Felton (2006). "Inequality and happiness: insights from Latin America". In: *The Journal of Economic Inequality* 4.1, pages 107–122.

Graham, Carol and Julia Ruiz Pozuelo (2017). "Happiness, stress, and age: How the U curve varies across people and places". In: *Journal of Population Economics* 30.1, pages 225–264.

Gwozdz, Wencke and Alfonso Sousa-Poza (2010). "Ageing, health and life satisfaction of the oldest old: An analysis for Germany". In: *Social Indicators Research* 97.3, pages 397–417.

Headey, Bruce and Alex Wearing (1988). "The sense of relative superiority—central to well-being". In: *Social Indicators Research* 20.5, pages 497–516.

Heckhausen, Jutta and Joachim Krueger (1993). "Developmental expectations for the self and most other people: Age grading in three functions of social comparison". In: *Developmental Psychology* 29.3, pages 539–548.

Heidrich, Susan M and Carol D Ryff (1993). "The role of social comparisons processes in the psychological adaptation of elderly adults". In: *Journal of Gerontology* 48.3, P127–P136.

Helson, Harry (1947). "Adaptation-level as frame of reference for prediction of psychophysical data". In: *The American Journal of Psychology* 60.1, pages 1–29.

—— (1948). "Adaptation-level as a basis for a quantitative theory of frames of reference". In: *Psychological Review* 55.6, pages 297–313.

Henrich, Joseph et al. (2010). "Markets, religion, community size, and the evolution of fairness and punishment". In: *science* 327.5972, pages 1480–1484.

Herrera, Estela Melguizo, Ana Acosta López, and Brunilda Castellano Pérez (2012). "Factores asociados a la calidad de vida de adultos mayores. Cartagena (Colombia)". In: *Salud Uninorte* 28.2, pages 251–263.

Hirschman, Albert O and Michael Rothschild (1973). "The changing tolerance for income inequality in the course of economic development: with a mathematical appendix". In: *The Quarterly Journal of Economics* 87.4, pages 544–566.

Hobfoll, Stevan E (1988). *The Ecology of Stress*. New York, NY: United States of America: Hemisphere Publishing Corporation.

—— (1989). "Conservation of resources: A new attempt at conceptualizing stress". In: *American Psychologist* 44.3.

Holman, E Alison et al. (2016). "Adversity time, and well-being: A longitudinal analysis of time perspective in adulthood". In: *Psychology and Aging* 31.6, pages 640–651.

Hsu, H-C and W-C Chang (2015). "Social connections and happiness among the elder population of Taiwan". In: *Aging & mental health* 19.12, pages 1131–1137.

Jackson, Tim and Nic Marks (1999). "Consumption, sustainable welfare and human needs -with reference to UK expenditure patterns between 1954 and 1994". In: *Ecological Economics* 28.3, pages 421–441.

Jivraj, Stephen et al. (2014). "Aging and subjective well-being in later life". In: *Journals of Gerontology Series B: Psychological Sciences and Social Sciences* 69.6, pages 930–941.

Kahneman, Daniel and Angus Deaton (2010). "High income improves evaluation of life but not emotional well-being". In: *Proceedings of the National Academy of Sciences* 107.38, pages 16489–16493.

Kahneman, Daniel, Peter P Wakker, and Rakesh Sarin (1997). "Back to Bentham? Explorations of experienced utility". In: *The Quarterly Journal of Economics* 112.2, pages 375–406.

Kedia, Gayannée, Thomas Mussweiler, and David EJ Linden (2014). "Brain mechanisms of social comparison and their influence on the reward system". In: *NeuroReport* 25.16, pages 1255–1265.

Kingdon, Geeta Gandhi and John Knight (2007). "Community, comparisons and subjective well-being in a divided society". In: *Journal of Economic Behavior & Organization* 64.1, pages 69–90.

Kobayashi, Erika et al. (2015). "Associations between social networks and life satisfaction among older Japanese: Does birth cohort make a difference?" In: *Psychology and aging* 30.4, pages 952–966.

Koellinger, Philipp, Lars Bertram, and Gert G Wagner (2016). "Genes for Well-Being, Depression, and Neuroticism: No Need to Worry We Are Largely Responsible for our own Happiness". In: *LIFE* 10.3, pages 8–9.

Kolosnitsyna, Marina, Natalia Khorkina, and Hongor Dorzhiev (2017). "Determinants of Life Satisfaction in Older Russians". In: *Ageing International* 42.3, pages 354–373.

Koopmans, Tjalling C (1947). "Measurement without theory". In: *The Review of Economics and Statistics* 29.3, pages 161–172.

Lang, IA et al. (2011). "Income and the midlife peak in common mental disorder prevalence". In: *Psychological medicine* 41.7, pages 1365–1372.

Larson, Reed (1978). "Thirty years of research on the subjective well-being of older Americans". In: *Journal of Gerontology* 33.1, pages 109–125.

Lauriola, Marco and Luca Iani (2017). "Personality, positivity and happiness: A mediation analysis using a bifactor model". In: *Journal of Happiness Studies* 18.6, pages 1659–1682.

Law, John (2004). *After Method: Mess in Social Science Research*. International Library of Sociology, Abingdon: United Kingdom: Routledge.

Layard, Richard (2005). *Happiness. Lessons from a New Science*. London: United Kingdom: Penguin Books.

—— (1980). "Human satisfactions and public policy". In: *The Economic Journal* 90.360, pages 737–750.

Lefcourt, Herbert M (1991). "Locus of Control". In: *Measures of Personality and Social Psychological Attitudes. Volume 1 of Measures of social psychological attitudes*. Edited by John P. Robinson, Philip R. Shaver, and Lawrence S. Wrightsman. San Diego, CA: United States of America: Academic Press, pages 413–499.

Leibenstein, Harvey (1950). "Bandwagon, Snob and Veblen Effects in the Theory of Consumers' Demand". In: *Quarterly Journal of Economics* 64, pages 183–207.

Levy, Becca (1996). "Improving memory in old age through implicit self-stereotyping" In: *Journal of Personality and Social Psychology* 71.6, pages 1092–1107.

Li, Ning (2016). "Multidimensionality of longitudinal data: Unlocking the age-happiness puzzle". In: *Social Indicators Research* 128.1, pages 305–320.

Liberman, Nira, Michael D Sagristano, and Yaacov Trope (2002). "The effect of temporal distance on level of mental construal". In: *Journal of Experimental Social Psychology* 38.6, pages 523–534.

Litwin, Howard and Sharon Shiovitz-Ezra (2010). "Social network type and subjective well-being in a national sample of older Americans". In: *The Gerontologist* 51.3, pages 379–388.

Lucas, Richard E et al. (2004). "Unemployment alters the set point for life satisfaction". In: *Psychological science* 15.1, pages 8–13.

Luhmann, Maike et al. (2012). "Subjective well-being and adaptation to life events: a meta-analysis". In: *Journal of personality and social psychology* 102.3, pages 592–615.
Luttmer, Erzo FP (2005). "Neighbors as negatives: Relative earnings and well-being". In: *The Quarterly journal of economics* 120.3, pages 963–1002.
Lykken, David and Auke Tellegen (1996). "Happiness is a stochastic phenomenon". In: *Psychological science* 7.3, pages 186–189.
Lyubomirsky, Sonja, Kennon M Sheldon, and David Schkade (2005). "Pursuing happiness: The architecture of sustainable change". In: *Review of general psychology* 9.2, pages 111–131.
MacLeod, Andrew K and Clare Conway (2005). "Well-being and the anticipation of future positive experiences: The role of income, social networks, and planning ability". In: *Cognition & Emotion* 19.3, pages 357–374.
Magee, Christopher A, Leonie M Miller, and Patrick CL Heaven (2013). "Personality trait change and life satisfaction in adults: The roles of age and hedonic balance". In: *Personality and individual differences* 55.6, pages 694–698.
Martin, Peter and Mike Martin (2002). "Proximal and distal influences on development: the model of developmental adaptation". In: *Developmental Review* 22, pages 78–96.
Martin, Peter et al. (2011). "The model of developmental adaptation: implications for understanding well-being in old-old age". In: *Understanding Well-Being in the Oldest Old*. Edited by Leonard W Poon and Jiska Cohen-Mansfield. New York, NY: United States of America: Cambridge University Press, pages 65–78.
Maslow, Abraham H (1943). "A theory of human motivation". In: *Psychological Review* 50.4, pages 370–396.
— (1954). *Motivation and Personality*. New York, NY: United States of America: Harper & Row.
McCrae, Robert R and Paul T Costa Jr (1997). "Personality trait structure as a human universal". In: *American psychologist* 52.5, pages 509–516.
Mroczek, Daniel K and Christian M Kolarz (1998). "The effect of age on positive and negative affect: a developmental perspective on happiness." In: *Journal of Personality and Social Psychology* 75.5, pages 1333–1349.
Mroczek, Daniel K and Avron Spiro III (2005). "Change in life satisfaction during adulthood: findings from the veterans affairs normative aging study". In: *Journal of personality and social psychology* 88.1, pages 189–202.
Navarro, María and María Angeles Sánchez Domínguez (2018). "Ingreso y bienestar subjetivo: el efecto de las comparaciones sociales". In: *Revista de Economía Mundial* 48, pages 153–178.
Niedzwiedz, Claire L et al. (2012). "Life course socio-economic position and quality of life in adulthood: a systematic review of life course models". In: *BMC Public Health* 12:168. https://doi.org/10.1186/1471-2458-12-628.
Okbay, Aysu et al. (2016). "Genetic variants associated with subjective well-being, depressive symptoms, and neuroticism identified through genome-wide analyses". In: *Nature genetics* 48.6, pages 624–633.

Okulicz-Kozaryn, Adam, Tim Nash, and Natasha O Tursi (2015). "Luxury car owners are not happier than frugal car owners". In: *International Review of Economics* 62.2, pages 121–141.

Ong, Anthony D and Alicia Patterson (2016). "Eudaimonia, Aging, and Health: A Review of Underlying Mechanisms". In: *Handbook of Eudaimonic Well-Being*. Edited by Joar Vittersø. Cham: Switzerland: Springer, pages 371–378.

Ostroot, Nathalie M and Wayne W Snyder (1985). "Measuring cultural bias in a cross-national study". In: *Social Indicators Research* 17.3, pages 243–251.

Oswald, Andrew J (1997). "Happiness and economic performance". In: *The Economic Journal* 107.445, pages 1815–1831.

Oswald, Andrew J and Nattavudh Powdthavee (2008). "Does happiness adapt? A longitudinal study of disability with implications for economists and judges". In: *Journal of Public Economics* 92.5–6, pages 1061–1077.

Parducci, Allen (1965). "Category judgment: a range-frequency model". In: *Psychological review* 72.6, pages 407–418.

——— (1984). "Value judgments: Toward a relational theory of happiness". In: *Attitudinal Judgment*. Edited by J Richard Eiser. Springer Series In Social Psychology. New York, NY: United States of America: Springer-Verlag, pages 3–21.

Parfit, Derek (1984). *Reasons and Persons*. Oxford: United Kingdom: Oxford University Press.

Peck, Michael D and Joseph R Merighi (2007). "The relation of social comparison to subjective well-being and health status in older adults". In: *Journal of Human Behavior in the Social Environment* 16.3, pages 121–142.

Perugini, Marco et al. (2003). "The personal norm of reciprocity". In: *European Journal of Personality* 17.4, pages 251–283.

Pethtel, Olivia Lee, Marnie Moist, and Stephen Baker (2018). "Time perspective and psychological well-being in younger and older adults". In: *Journal of Positive Psychology and Wellbeing* 2.1, pages 45–63.

Poon, Leonard W (1992). *The Georgia Centenarian Study*. Amityville, NY: United States of America: Baywood Publishing Co.

Poon, Leonard W et al. (1992). "The Georgia Centenarian Study". In: *International Journal of Aging and Human Development* 34, pages 1–17.

Powdthavee, Nattavudh (2009). "How important is rank to individual perception of economic standing? A within-community analysis". In: *The Journal of Economic Inequality* 7.3, pages 225–248.

———(2010). *The Happiness Equation. The Surprising Economics of Our Most Valuable Asset*. London: United Kingdom: Icon Books.

Probst-Hensch, Nicole (2017). "Happiness and its molecular fingerprints". In: *International Review of Economics* 64.2, pages 197–211.

Rafnsson, Snorri Bjorn, Aparna Shankar, and Andrew Steptoe (2015). "Longitudinal influences of social network characteristics on subjective well-being of older adults: findings from the ELSA study". In: *Journal of aging and health* 27.5, pages 919–934.

Redersdorff, Sandrine and Serge Guimond (2006). "Comparing oneself over time: The temporal dimension in social comparison". In: *Social comparison and social psychology: Understanding cognition, intergroup relations and culture*. Edited by Serge Guimond. Cambridge: United Kingdom: Cambridge University Press, pages 76–96.

Robinson-Whelen, Susan and Janice Kiecolt-Glaser (1997). "The importance of social versus temporal comparison appraisals among older adults". In: *Journal of Applied Social Psychology* 27.11, pages 959–966.

Rodriguez-Rodriguez, Vicente, Fermina Rojo-Perez, and Gloria Fernandez-Mayoralas (2017). "Family and Social Networks and Quality of Life Among Community-Dwelling Older-Adults in Spain". In: *Quality of Life in Communities of Latin Countries*. Edited by Graciela Tonon. Community Quality-of-Life and Well-Being. Cham: Switzerland: Springer International Publishing, pages 227–253.

Rönnlund, Michael, Elisabeth Åström, and Maria Grazia Carelli (2017). "Time perspective in late adulthood: aging patterns in past, present and future dimensions, deviations from balance, and associations with subjective well-being". In: *Timing & Time Perception* 5.1, pages 77–98.

Rothermund, Klaus and Jochen Brandtstädter (2003). "Age stereotypes and self-views in later life: Evaluating rival assumptions". In: *International Journal of Behavioral Development* 27.6, pages 549–554.

Rule, James B (1997). *Theory and progress in social science*. Cambridge: United Kingdom: Cambridge University Press.

Runciman, Walter Garrison (1972). *Relative Deprivation and Social Justice. A study of attitudes to social inequality in twentieth-century England*. Harmondsworth: United Kingdom: Pelican Books.

Schmidt, Marco FH and Jessica A Sommerville (2011). "Fairness expectations and altruistic sharing in 15-month-old human infants". In: *PloS one* 6.10. https://doi.org/10.1371/journal.pone.0023223.

Schwandt, Hannes (2016). "Unmet Aspirations as an Explanation for the Age U-shape in Wellbeing". In: *Journal of Economic Behavior & Organization* 122, pages 75–87.

Seabright, Paul (2006). "The evolution of fairness norms: an essay on Ken Binmore's Natural Justice". In: *Politics, Philosophy & Economics* 5.1, pages 33–50.

Senik, Claudia (2004). "When information dominates comparison: Learning from Russian subjective panel data". In: *Journal of Public Economics* 88.9, pages 2099–2123.

Sherrard, Carol A (1994). "Elderly wellbeing and the psychology of social comparison". In: *Ageing & Society* 14.3, pages 341–356.

Shor, Eran and David J Roelfs (2015). "Social contact frequency and all-cause mortality: A meta-analysis and meta-regression". In: *Social Science & Medicine* 128, pages 76–86.

Singer, Tania et al. (2006). "Empathic neural responses are modulated by the perceived fairness of others". In: *Nature* 439.7075, pages 466–469.

Smith, Heather J and Yuen J Huo (2014). "Relative Deprivation: How subjective experiences of inequality influence social behavior and health". In: *Policy Insights from the Behavioral and Brain Sciences* 1.1, pages 231–238.

Smith, Richard H, Edward Diener, and Douglas H Wedell (1989). "Intrapersonal and social comparison determinants of happiness: A range-frequency analysis". In: *Journal of personality and social psychology* 56.3, pages 317–325.

Soto, Christopher J (2015). "Is happiness good for your personality? Concurrent and prospective relations of the big five with subjective well-being". In: *Journal of personality* 83.1, pages 45–55.

Specht, Jule, Boris Egloff, and Stefan C Schmukle (2013). "Examining mechanisms of personality maturation: The impact of life satisfaction on the development of the Big Five personality traits". In: *Social Psychological and Personality Science* 4.2, pages 181–189.

Stanca, Luca and Ruut Veenhoven (2015). "Consumption and happiness: An introduction". In: *International Review of Economics* 62, pages 91–99.

Steptoe, Andrew, Angus Deaton, and Arthur A Stone (2015). "Subjective wellbeing, health, and ageing". In: *The Lancet* 385.9968, pages 640–648.

Stolarski, Maciej et al. (2014). "How we feel is a matter of time: Relationships between time perspectives and mood". In: *Journal of Happiness Studies* 15.4, pages 809–827.

Stone, Arthur A, Stefan Schneider, and Joan E Broderick (2017). "Psychological stress declines rapidly from age 50 in the United States: Yet another well-being paradox". In: *Journal of Psychosomatic Research* 103, pages 22–28.

Suls, Jerry (1986). "Comparison processes in relative deprivation: A life-span analysis". In: *Relative deprivation and social comparison: The Ontario Symposium*. Edited by James M Olson, C Peter Herman, and Mark P Zanna. Volume 4. Hillside, NJ: United States of America: Lawrence Erlbaum Associates, pages 95–116.

Suls, Jerry M (1982). *Psychological perspectives on the self*. Volume 1. Hillside, NJ: United States of America: Lawrence Erlbaum Associates.

Suls, Jerry Christine A Marco, and Sheldon Tobin (1991). "The role of temporal comparison, social comparison, and direct appraisal in the elderly's self-evaluations of health". In: *Journal of Applied Social Psychology* 21.14, pages 1125–1144.

Suls, Jerry and Brian Mullen (1984). "Social and temporal bases of self-evaluation in the elderly: Theory and evidence". In: *The International Journal of Aging and Human Development* 18.2, pages 111–120.

Suls, Jerry and Ladd Wheeler (2000). *Handbook of Social Comparison: Theory and Research*. The Plenum Series in Social Clinical Psychology. New York, NY: United States of America: Springer Science+Business Media.

Tauber, Benjamin, Hans-Werner Wahl, and Johannes Schröder (2016). "Personality and life satisfaction over 12 years". In: *GeroPsych. The Journal of Gerontopsychology and Geriatric Psychiatry* 29, pages 37–48.

Taylor, Charles, editor (1979). *The Idea of Freedom: Essays in Honour of Isiah Berlin*. Oxford: United Kingdom: Alan Ryan.

Tennen, Howard, Tara Eberhardt Mckee, and Glenn Affleck (2000). "Social comparison processes in health and illness". In: *Handbook of Social Comparison*. Edited by Jerry Suls and Ladd Wheeler. The Plenum Series in Social Clinical Psychology. New York, NY: United States of America: Springer Science+Business Media, pages 443–483.

Tomini, Florian, Sonila M Tomini, and Wim Groot (2016). "Understanding the value of social networks in life satisfaction of elderly people: a comparative study of 16 European countries using SHARE data". In: *BMC Geriatrics* 16.1. https://doi.org/10.1186/s12877-016-0362-7.

Tremblay, Léon, Yulia Worbe, and Jeffrey R. Hollerman (2009). "Handbook of reward and decision making". In: edited by Jean-Claude Dreher and Léon Tremblay. Burlington, MA: United States of America: Academic Press, pages 51–77.

Trope, Yaacov and Nira Liberman (2003). "Temporal construal". In: *Psychological review* 110.3, pages 403–421.

Tupes, Ernest C and Raymond E Christal (1992). "Recurrent personality factors based on trait ratings". In: *Journal of personality* 60.2, pages 225–251.

Ulloa, Beatriz Fabiola López, Valerie Møller, and Alfonso Sousa-Poza (2013). "How does subjective well-being evolve with age? A literature review". In: *Journal of Population Ageing* 6.3, pages 227–246.

Usui, Wayne M, Thomas J Keil, and K Robert Durig (1985). "Socioeconomic comparisons and life satisfaction of elderly adults". In: *Journal of Gerontology* 40.1, pages 110–114.

Van de Stadt, Huib, Arie Kapteyn, and Sara Van de Geer (1985). "The relativity of utility: Evidence from panel data". In: *The Review of Economics and Statistics* 67, pages 179–187.

Van Landeghem, Bert (2012). "A test for the convexity of human well-being over the life cycle: Longitudinal evidence from a 20-year panel". In: *Journal of Economic Behavior & Organization* 81.2, pages 571–582.

Van Praag, Bernard (1971). "The welfare function of income in Belgium: An empirical investigation". In: *European Economic Review* 2.3, pages 337–369.

Veblen, Thorstein (1912). *The Theory of the Leisure Class: An Economic Study of Institutions*. Macmillan standard Library B. W. Huebsch.

Veenhoven, Ruut (1984a). *Conditions of Happiness*. Dordrecht: The Netherlands: Reidel Publishing Company.

——— (1984b). *Data-Book of Happiness. A Complementary Reference Work to 'Conditions of Happiness' by the same author*. Dordrecht: The Netherlands: Reidel Publishing Company.

Veenhoven, Ruut and Joop Ehrhardt (1995). "The cross-national pattern of happiness: Test of predictions implied in three theories of happiness". In: *Social Indicators Research* 34.1, pages 33–68.

Vendrik, Maarten CM (2013). "Adaptation, anticipation and social interaction in happiness: An integrated error-correction approach". In: *Journal of Public Economics* 105, pages 131–149.

Vogel, Gretchen (2004). "The evolution of the golden rule". In: *Science* 303.5661, pages 1128–1131.

Von Humboldt, Sofia, editor (2016). *Conceptual and Methodological Issues on the Adjustment to Aging*. Volume 15. International Perspectives on Aging. New York, NY: United States of America: Springer.

Webb, Elizabeth et al. (2010). "Proximal predictors of change in quality of life at older ages". In: *Journal of Epidemiology & Community Health* 65, pages 542–547.

Webster, Jeffrey Dean, Ernst T Bohlmeijer, and Gerben J Westerhof (2014). "Time to flourish: The relationship of temporal perspective to well-being and wisdom across adulthood". In: *Aging & Mental Health* 18.8, pages 1046–1056.

Wedell, Douglas H and Allen Parducci (2000). "Social Comparison: Lessons from Basic Research on Judgment". In: *Handbook of Social Comparison*. Edited by Jerry Suls and Ladd Wheeler. The Plenum Series in Social Clinical Psychology. New York, NY: United States of America: Springer Science+Business Media, pages 223–252.

Wildman, Josephine M, Suzanne Moffatt, and Mark Pearce (2018). "Quality of life at the retirement transition: Life course pathways in an early baby boombirth cohort". In: *Social Science & Medicine* 207, pages 11–18. https://doi.org/10.1016/j.socscimed.2018.04.011.

Wills, Thomas A (1981). "Downward comparison principles in social psychology". In: *Psychological bulletin* 90.2, pages 245–271.

Wilson, Timothy D and Daniel T Gilbert (2008). "Explaining away: A model of affective adaptation". In: *Perspectives on Psychological Science* 3.5, pages 370–386.

Wooldridge, Jeffrey (2013). *Introductory Econometrics: A Modern Approach (5th Ed)*. Mason, OH: United States of America: South-Western.

Wunder, Christoph et al. (2013). "Well-being over the life span: Semiparametric evidence from British and German longitudinal data". In: *Review of Economics and Statistics* 95.1, pages 154–167.

Yang, Yang (2008). "Social Inequalities in Happiness in the United States, 1972 to 2004: An Age-Period-Cohort Analysis". In: *American Sociological Review* 73.2, pages 204–226.

Zee, Karen I van der, Arnold B Bakker, and Bram P Buunk (2001). "Burnout and reactions to social comparison information among volunteer caregivers". In: *Anxiety, Stress and Coping* 14.4, pages 391–410.

Zell, Ethan and Mark D Alicke (2011). "Age and the better-than-average effect". In: *Journal of Applied Social Psychology* 41.5, pages 1175–1188.

Zimbardo, Philip G and John N Boyd (1999). "Putting time in perspective: a valid, reliable individual-differences metric". In: *Journal of Personality and Social Psychology* 77.6, pages 1271–1288.

3

Happiness and Policy

> **Overview**
> This chapter covers the attempts to incorporate happiness and related notions into policy making, from the Gross National Happiness to the National Accounts of Well-Being to the happy life expectancy measure, against the background of population ageing.

The sheer amount of work in international organisations and academic circles on quality of life and happiness has eventually influenced policy makers. As a result, recommendations and initiatives to incorporate its measurement on a par with the national accounts have multiplied. Many commentators have proposed that the growth of the national income should be replaced by the growth of happiness, quality of life, or well-being as the ultimate objective of economic policy. After all, as Clark et al. stated, echoing many other economists:

> Surely the ultimate aim of human endeavor must be to produce flourishing communities of people who are profoundly satisfied with their lives. It cannot be simply the creation of wealth.
>
> (Clark et al. 2018, p. 211)

This may sound far-fetched at first blush, but it is nothing new, as economists have been concerned about the maximisation of social welfare since the dawn of the profession (though not necessarily using this mathematical jargon). However, a 'social welfare' function is not the same as what most scholars working in happiness studies mean by 'happiness'.[1]

A problem with using happiness and the other widely measured metric, subjective well-being as a policy guide is that these constructs have components that vary fleetingly, depending on mood, which of course conspire against their applicability as policy-measuring rods. In turn, the components that tend to be more stable in the short run and only exhibit change in the longer term are subject to hedonic treadmill and adaptation and are influenced by social and self-comparisons. This involves other risks of its own.

Consider adaptation. If people adapt to even the worst-imaginable circumstances so much so that happiness levels over time return to at least a resemblance of what they were before calamity struck, would it be worth the effort to prevent ill from happening or to intervene to reduce its immediate deleterious impact? Instead, would a *laissez-faire* 'time heals' approach to policy not be justified? Similar considerations could be made on the basis of the role of social comparisons. This led Frey and Stutzer (two world experts on happiness) to assert the following, in the context of a discussion about the maximisation of happiness as a policy goal:

> What matters in our context is that the way to deal with hedonic adaptation and the aspiration treadmill are not part of social welfare maximization but must be decided on a more fundamental level. Thus one needs a social decision-making mechanism to indicate how adaptation and aspiration effects have to be dealt with in public policy. Obviously such decisions have grave consequences for economic policy to which the social welfare maximization approach does not contribute anything.
>
> (Frey and Stutzer 2009, p. 311)

Leaving the decisions to be taken 'on a more fundamental level' may not be the answer either. Think, for instance, of the various political uses of sports and other mass spectacles[2] and propaganda[3] to influence public opinion. In this regard, Popper (2013) warned that

> …of all political ideals, that of making the people happy is perhaps the most dangerous one. It leads invariably to the attempt to impose our scale of "higher" values upon others, in order to make them realize what seems to us of greatest importance for their happiness…
>
> (Popper 2013, pp. 441–442)

On another level, it is not straightforward how measures of self-reported life satisfaction or happiness can be used to guide policy making. For example, the World Values Survey (a set of representative surveys carried out in over 100 countries using a common questionnaire that started in 1981 and is based at Tilburg University in the Netherlands) reports feeling of happiness for over 100 countries. Figure 3.1 presents the results from the question about feeling of happiness by age group in four different countries for two periods, 2005–2009

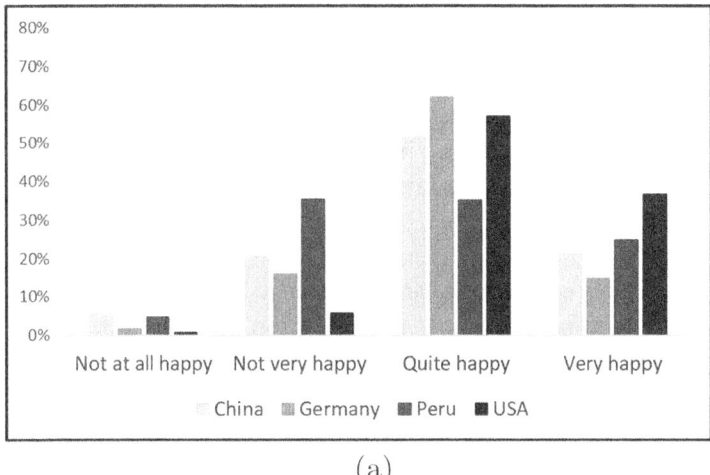

(a)

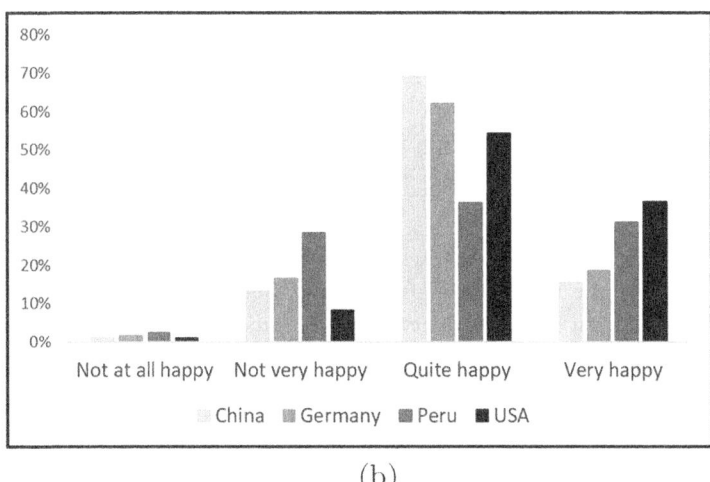

(b)

Fig. 3.1 Feeling of happiness among population aged 50 or over. (**a**) 2005–2009. (**b**) 2010–2014. *Source: World Values Survey*

and 2010–2014. The percentage of the population that reported to feel very happy increased between the periods in Germany and Peru, fell in China, and remained unchanged in the United States of America. It is remarkable how much the percentage of people who reported to feel quite happy has gone up in China.

Prycker (2010) reviewed the arguments in favour of and against the use of happiness in policy making, which summarised thus [Tables 1 and 2, p. 599]:

- Arguments in favour:
  - There is enough knowledge about the conditions and determinants of happiness.
  - Happiness can be measured quite accurately.
  - Happiness has intrinsic and instrumental value.
  - Increasing happiness is possible.

- Arguments against:
  - The measures of happiness are liable to be influenced.
  - There is no consensus about the definition of happiness.
  - There is not enough knowledge about which policies would increase happiness.
  - Promoting happiness might bring about adverse effects.
  - Happiness may be subject to an upper limit, so that increasing it beyond a certain level may not be possible.
  - If happiness is enshrined as the ultimate policy goal, it may create incentives towards paternalistic governmental interventions.

Frey, on reflecting about setting happiness or its related concepts as a policy objective to be maximised, asserted:

> Once the maximization of the aggregate happiness index of the population is taken to be the official goal of economic and social policy, one can no longer trust that survey respondents answer any questions about their subjective life satisfaction in an unbiased way. When citizens' happiness is taken as the measuring rod of politics, government politicians will make an effort to manipulate the aggregate happiness index in their favour. For both these reasons, the subjective well-being data are no longer a reliable measure of people's happiness. Governments should not be asked to maximize happiness.
>
> (Frey 2018, p. 29)

I share Frey's second concern in particular. This author proposed a different route and approach in which the role of policy is to set the institutional framework conducive to and facilitating of the individual search of happiness:

> A reasonable happiness policy should focus primarily on constructing and maintaining institutions that allow people to reach their own goals of happiness in the context of the society they live in.
>
> (Frey 2018, p. 34)

In this same line I classify a proposal by the Carnegie UK Trust that views well-being not only as a policy goal but also as a conversation (redolent of the philosophy of Martin Buber, Emmanuel Levinas and Jürgen Habermas), a policy framework, and an approach to service delivery (Trust 2018).

The movement to bring the constructs of happiness and quality of life to the political fore has been seen by many critical gerontologists as part of a novel image of later life in which notions such as active ageing and successful ageing are being promoted.

In a semiotic critique of the prominence of happiness in the public policy discourse in developed countries, Frawley opined that happiness in primordially a social phenomenon rather than a natural or empirical one. He went on to say that

> …'happiness' as a public problem is not a signifier of a desire to improve or optimize underlying mental states, nor is it even about these mental states at all. Rather, …it is implicitly a critique of change. It expresses a fear of the future articulated through a series of paradoxes that purport to describe the true nature of happiness and progress, but which really express a deep-seated uncertainty about the future and consequent desire to maintain the present.
>
> (Frawley 2015, p. 3)

Notwithstanding these qualms about the use of happiness indicators as policy objectives, a good number of initiatives have been proposed or implemented.

The United Nations have been producing its Human Development annual reports since 1990. Along with the United Nation's Human Development Index, economists have proposed other indicators either complementary or supplementary to gross national income—see Stiglitz et al. (2009) for an apology for such addition. Some of these metrics have been designed to gauge social welfare, whilst others—developed in ecological economics—aim to factor into the national accounts environmental costs and other negative externalities derived from productive activities, urbanisation, and so on. Among these initiatives we find the Index of Social Progress, the Social Development Index, the Measure of Economic Welfare, the Index of Economic Aspects of Welfare, the Index of Sustainable Economic Welfare, the Genuine Progress Indicator, the Index of Sustainable Economic Welfare, the Living Planet Index, the Ecological Footprint, the Environmental Sustainability Index, the Environmental Vulnerability Index, the Well-Being Index, and the Genuine Savings Index. I briefly touched upon the Human Development Index in Chap. 1 in this part and I am not going to concern with the other indicators as they have little relation with individual or population ageing. Instead, in this chapter I will focus on indicators that have been advanced with the specific aim to track the evolution of happiness in a society and discern how much it is impacted by policies and other national and international circumstances.

3.1 Gross National Happiness

The Constitution of the Kingdom of Bhutan, adopted on 18 July 2008, in its Article 9, point 2, sets out as one of the principles of state policy that 'the State shall strive to promote those conditions that will enable the pursuit of Gross National Happiness'. Despite the use of capital letters, Gross National Happiness (GNH) is not defined in the original constitutional text. It was left to economic and policy planners to elaborate on the concept, and by 2010 a GNH index was developed. The 2010 GNH index is comprised of nine domains and thirty-three indicators. The nine domains are (Ura et al. 2012a):

- Psychological well-being
- Health
- Time use
- Education
- Cultural diversity and resilience
- Good governance
- Community vitality
- Ecological diversity and resilience
- Living standards

Happiness in the GNH index is defined as having sufficiency in at least six out of nine domains, where 'sufficiency' is set as a minimum score for each of the indicators included in each domain. For example, the 'life satisfaction' indicator is an index of equally weighted subjective assessments of satisfaction with health, occupation, family relationships, standard of living, and work-life balance. Each question is structured on a 5-point scale, so that this indicator ranges from 5 (lowest satisfaction) to 25 (highest satisfaction). The sufficiency threshold for this indicator is set at 19.

The GNH declines almost linearly with advancing chronological age. With an average of 0.756 in 2015 for the population as a whole, it reaches almost 0.80 points among the age group under 30 years to fall below 0.65 for the 75-plus-year olds (although it was among the latter group that the index increased the most between 2010 and 2015) (GNH Research 2016). However, in some domains older people do better than the young: culture, governance, community, and psychological well-being (Ura et al. 2012b). In turn, a direct question about overall happiness—also included in the questionnaire though not in the GNH index—has the U-shaped age pattern for men (with its minimum point in the late 50s) but exhibits a negative trend for women.

3.2 National Accounts of Well-Being

The National Accounts of Well-being is a framework developed by the UK-based *new economics foundation* (NEF) in 2009. The framework comprises two constructs: personal well-being and social well-being. Personal well-being includes emotional well-being (defined as both the presence of positive feelings and the absence of negative feelings), satisfaction with life, vitality, resilience and self-esteem (including optimism), and positive functioning (which comprises competence, autonomy, engagement, and meaning and purpose in life). Social well-being is the combination of supportive relationships and trust and sense of belonging (nef 2009).

Using data for twenty-two European countries from 2006 to 2007, nef (2009) reported wide variations in the results by age group. Personal well-being was almost flat across the age groups in Scandinavian countries, the United Kingdom and the Republic of Ireland, but fell with chronological age in Western, Southern, Eastern, and Central European countries. Social well-being increased with chronological age in all the regions under study except in Central and Eastern countries. An analysis by component shows that whilst individuals aged 75 years or over score much less in vitality than those aged 25 or under, they score much higher in trust and sense of belonging, and slightly better also in absence of negative feelings, satisfaction with life, and positive functioning. The report suggested [p. 36]:

It is these types of insights which open the door for meaningful policy discussions. So, for instance, should a country like the UK target specific policies towards building feelings of trust and belonging among young people, or is it better to accept that these feelings change through the life course and young people are inevitably more likely to experience attachment and social interaction with peers and family members rather than through a wider sense of belonging? Conversely, with trust and belonging clearly high and important for many aged 75+, might we take better account of this in how future policy proposals are evaluated to ensure that those targeted towards older populations are 'proofed' to ensure this well-being component is safeguarded?

Still another approach was proposed by Diener and Tov (2012). Instead of aggregate indicators such as single-item questions about subjective well-being or happiness, these authors proposed the use of specific, ad hoc indicators of well-being with the view that they may be relevant to specific policy questions. Within the realm of retirement, the following examples were mentioned: life satisfaction of workers by age group, work satisfaction of retirees who return to work, and mood and emotions of older people at work and in retirement. Other examples closely related to ageing include the life satisfaction, mood, and emotions of caregivers when respite or day care is available and when it is not.

3.2.1 Time-Based National Well-Being Accounts

As I mentioned above in Chap. 1, Kahneman et al. expounded several reasons against the use of income as a proxy for utility based on the theories developed and the empirical evidence built up in behavioural sciences. Consequently, these authors recommend that a metric of experienced utility or subjective well-being should replace the canonical utility function dependent on income. However, not any measure will do, because of the various methodological pitfalls direct measures of happiness or subjective well-being are fraught with, as I have described in that chapter. Kahneman et al. set out the requisites that an appropriate metric of well-being should meet:

- It should capture actual hedonic and emotional well-being.
- It should assign appropriate weights to different activities and life stages.
- They should be as independent as possible of context and social and temporal comparisons.

In particular, they proposed the use of time-based national well-being accounts, populated with data from surveys that combine time use and satisfaction metrics (see also Kahneman et al. 2004a). National well-being would correspond to:

$$NWB = \sum_{i}^{N}\sum_{j}^{J} \frac{(h_{i,j} \cdot u_{i,j})}{N} \quad (3.1)$$

where $h_{i,j}$ is the amount of time that the individual i is engaged in activity or situation j, and $u_{i,j}$ stands for the net affective experience of that activity of situation (i.e. the net feeling or emotion experienced while engaged in j). For example, Eq. (3.1) can be also expressed thus:

$$NWB = \sum_{j=1}^{j=J} H_j \bar{u}_j + \sum_{i=1}^{i=N}\sum_{j=1}^{j=J} \frac{h_{i,j} \cdot (u_{i,j} - \bar{u}_j)}{N} \quad (3.2)$$

With advances in technology that allow to monitor changes in affect by means of mobile (cell) phones and other portable devices in real time (see Leon et al. 2011, 2010), there would be no need for time use and satisfaction surveys if the sample sizes were large enough and representative. Ishio and Abe (2017) presents an experiment in a small village in Japan that combines a day reconstruction method with wearable devices that track affect (the devices measured heart rate variability).

3.3 Happy Life Expectancy or Happy Life Years

Veenhoven (1996) introduced the notion of happy life expectancy (HLE) and its synonyms happy life years (HLY) and happiness-adjusted life years (HALYs) (see Veenhoven 2000, 2014a). HLY is similar to the quality-adjusted life year (QALY) indicator presented in Part II, Chap. 5. QALYs are calculated as the product of years of life and quality of life and used to carry out cost-effectiveness evaluations in healthcare. HLY are the product of life expectancy at birth and a happiness score normalised between 0 (lowest) and 10 (highest). Figure 3.2 shows the average life expectancy and happy life years for the period 2005–2014 in 158 countries. Countries that differ substantially in terms of life expectancy at birth (e.g. Sierra Leone and Tanzania, by 16.9 years) present similar happy life expectancy. This results from a higher life satisfaction score

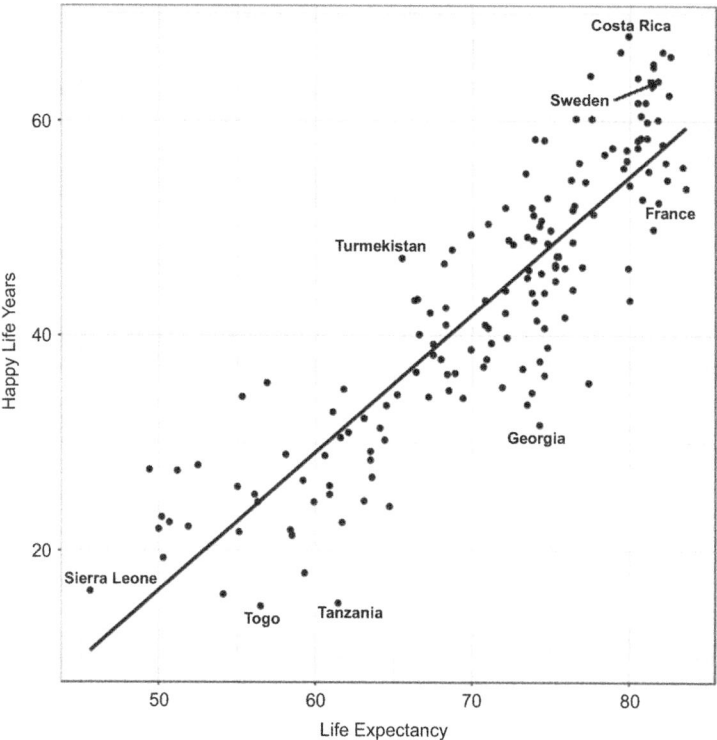

Fig. 3.2 Life expectancy and happy life years, 2005–2014. *Source: VeenhovenHLYdata*

in Sierra Leone (3.5) compared to Tanzania (2.5). In turn, France and Sweden present the same life expectancy at birth (81.8 years) but an 11.3 year gap in happy life expectancy as the life satisfaction score in France (6.4) lags behind that in Sweden (7.8).

According to Veenhoven (2014a, p. 2643), two-thirds of the variance in HLY across the world is explained by economic affluence, freedom, equality, solidarity, and justice.

Clark et al. (2018, ch. 15) proposed to evaluate policies using a single-item question of satisfaction with life as a whole. The idea was to use changes in the measure of life satisfaction to evaluate policy interventions as part of a cost-effectiveness exercise. The benefits of a policy would be measured in happiness-years, that is, in happiness-adjusted life years. By carrying out cost-effectiveness analyses, policy makers would be able to rank policy proposals according to 'happiness-benefits per unit of cost' Clark et al. (2018, p. 198).

3.3.1 Inequality of Happiness

Veenhoven and Kalmijn (2005) introduced the concept of inequality-adjusted happiness (IAH) (see also Kalmijn and Veenhoven 2014 and Veenhoven and Kalmijn 2005), which has been defined as *how well a nation combines a high level of happiness with an equitable distribution of happiness* (Veenhoven 2014b, p. 3253). This measure is a linear combination of the mean and the standard deviation of happiness scores. It is expressed by:

$$\text{IAH} = 8.28(m - s) + 17.2 \qquad (3.3)$$

where m is the average happiness score, and s is its standard deviation.

3.4 Closing Thought

Let me finish this chapter and part with a quotation and a question, which I preferred not to include in the Review and Comment section given that it makes for a good closing thought. In Kahneman and Deaton (2010, p. 16492), the authors opined:

> If measures of well-being are to be used to assess human welfare and to guide policy, the present findings raise the question of whether life evaluation or emotional well-being is better suited to these aims. The [former] is a serious contender for the best tool for measuring the degree to which individuals view themselves as achieving their goals, both material and other. But emotional well-being also is clearly important for individuals and for policy …. Not everyone will agree that enhancing the happiness experienced by those who are already quite happy is a legitimate policy objective. The policy goal of reducing suffering is likely to raise fewer objections, and measures of emotional pain may be useful for that purpose. This topic merits serious debate.

What do you think of making happy people happier as a policy objective? Emotional well-being refers to day-to-day feelings of joy, sadness, affection, or stress. Would you have any objections to reducing the negative emotions? And what about boosting the positive ones?

I agree with Kahneman and Deaton that the topic surely merits serious debate. I hope this part has provided you with enough elements to engage in and contribute to it.

Review and Reflect

1. Comment:

   > ...any want or satisfaction which exists at all exists in some amount and is therefore measurable, how exactly and how commensurably with other, we cannot tell until we have tried.
   >
   > (Thorndike 1940, p. 152)

2. Reflect on the following excerpts from a book by a philosopher referring to a book by two economists. I think they reflect the general approach by economists to, er, 'happiness'.

   > On my desk before me I have a copy of Happiness Quantified by Bernard van Praag and Ada Ferrer-i-Carbonell. The authors are economists associated with the "Leyden School" of economics. The Leyden School approach seeks to assess the well-being of groups of people not by looking at statistics concerning the amounts of money or consumer goods those people possess, but rather by trying to determine how happy they are. In the book in question, van Praag and Ferrer-i-Carbonell ...summarize results concerning different national groups, gender groups, age groups, income groups; they talk about satisfaction with work, marriage, leisure, health care and many other domains of life ...
   >
   > One surprising fact about this book concerns the connection between the title and the contents. The title strongly suggests that the book will be about the measurement of happiness. Yet the index contains only two references to "happiness"...Instead, the graphs and tables contain information about the extent to which people in various groups are satisfied with various aspects of their lives...levels of satisfaction with such domains as work, financial situation, health, housing, leisure, and the environment...
   >
   > I am not suggesting that the publishers put the wrong title on the book. Rather, I suspect that what has happened here is that the authors simply took it as obvious that if you want to measure the components of happiness, you just have to measure satisfaction with such domains as work, marriage, leisure, health care, and so on. And if you want to measure happiness in general, you just have to measure satisfaction with life as a whole. I suspect that van Praag and Ferrer-i-Carbonell took these assumptions to be beyond question. They don't discuss them; they don't attempt to defend them. Perhaps they are unquestioned presuppositions of their research.
   >
   > (Feldman 2010, pp. 70–71)

(continued)

3. Comment:

   Inequalities in quality of life should be assessed across people, socio-economic groups, gender and generations.
   (Stiglitz et al. 2009, p. 121)

4. The following quotation includes a definition of the so-called woodwork effect as well as a position I invite you to discuss within the context of long-term care related quality of life.

   *Perhaps most ubiquitous, contentious, and anxiety-provoking of all technical LTC topics is the woodwork effect: that is, the propensity for people who would shun nursing homes to come out of the woodwork to use more attractive forms of LTC, thus turning a potentially cost-effective alternative service into an expensive add-on...Indeed, it is high time to retire the woodwork effect as a concern. A confirmed woodwork effect could even be a sign of **success** if it meant that a state or community had effected wider access to the kind of LTC programs that people want to use. Rather than eliminating the spread of more user-friendly LTC, gerontologists are challenged to alter the essential nature of all LTC into more desirable forms while keeping down the price of services in all sectors (nursing homes, assisted living, and home care).*
   (Kane 2001, p. 294)

5. Consider the following excerpt from a critical examination of how sociologists have approached and engaged in happiness studies (by a sociologist specialised in happiness):

   a) most people are happy in modern nations, b) average happiness in nations is rising, c) inequality in happiness is going down, d) happiness depends heavily on the kind of society one lives in, but e) not very much on one's place in society. These remarkable findings are largely ignored in sociology, if not denied. This has several reasons. One reason is professional bias: most sociologists earn their living dealing with social problems are therefore not apt to see that people flourish. Another reason is ideological: many sociologists are 'critical' of modern society and can therefore hardly imagine that people thrive in these conditions. Lastly, some sociological theories play them false, in particular cognitive theories implying that happiness is relative.
   (Veenhoven 2014c, p. 537)

   Would a 'critical' approach to economics also dismiss the findings from happiness studies off-hand? And critical gerontology?

(continued)

Comment on the last point. How valid is the argument that the relativity of the happiness construct renders the five 'remarkable' findings listed in the beginning of the excerpt false?
6. Discuss the following explanation.

> ...if it is the case that persons occupying different positions in the life cycle -whether measured by chronological age, subjective age, or life cycle roles — possess differing characteristic psychological profiles of life satisfaction, then the very dependent variables employed to monitor the effects of life-course changes either may hide real changes or, perhaps more damaging, may point to changes which are not really there.
>
> (Cutler 1979, p. 574)

Is it the case? If so, how damaging is it and to what precisely?
7. Consider the following objection to using a measure of happiness as a policy objective, based on behavioural economics (see Part III in this volume).

> ...given that human beings suffer from hyperbolic discounting, it is not obvious that policies that are optimal from a public health standpoint would make people happier. Take, for example, a ban on junk food. Although it might have good health consequences, it might decrease the happiness of many individuals—some of whom are not overweight and enjoy junk food.
>
> (Graham 2008, p. 85)

Do you agree? Can you think of any mechanisms based on behavioural economics that could be implemented to nudge individuals who feel happy eating junk food away from that type of food without diminishing their levels of happiness?
8. Do you concur with Clark et al. (2008, p. 138) that 'taking relative income seriously is an important step towards greater behavioural realism in Economics, such that our models and empirical analysis move closer to how real people feel and behave'?

Should anything else be taken seriously towards greater realism in Economics?
9. Comment on what Layard (2005, pp. 224–231) calls the 'twelve truths of happiness':

- It is an objective dimension of the whole life experience.
- Human beings are programmes to seek it.
- The best society is the happiest.
- Societies will not become happier unless their members agree this is what they want.
- Human beings are deeply social being.
- Human beings want to trust each other.

(continued)

- Human beings are deeply attached to the status quo.
- Human beings are conscious of their relative status in their society.
- Human beings are adaptable.
- Additional income increases happiness less and less.
- Happiness depends on the inner life as much as the outer circumstances.
- Public policy can more easily remove misery than increase happiness.

10. Clark et al. (2018, p. 204), referring to a developed country, asserted:

    The basic inequality in our society is surely between people with different levels of happiness, not different levels of income.

    Do you agree with this assertion? How would this shift in perspective modify the evaluation of the distributional impacts of policies? Refer to Part II in this volume for distributive issues if needed.
11. Do you agree with the following assertion?

    In any ageing society, life satisfaction among older persons is the top priority of the public policy agenda because it is an important alternative assessment of group inequalities and public policy outcomes
    (Cheung and Chou 2017, p. 2)

    Discuss.
12. Berridge and Aldridge (2008, p. 2) remarked that experienced utility 'is what most people think of the term reward …the essence of what reward is all about'. In turn, writing in 1997 Kahneman et al. (1997) noted that 'With few exceptions, experienced utility is essentially ignored in modern economic discourse' and surmised that the economists' reluctance towards the concept of experienced utility stemmed from the (wrong) ideas that subjective hedonic experience could not be observed or measured and that choices provide all information needed to estimate a rational agent's utility. Comment.

Notes

1. Besides, to invoke famous economists and philosophers of the past to provide intellectual support for shifting policy away from economic growth towards a metric of happiness is akin to fall into the 'argumentum ad verecundiam' or appeal to authority fallacy (Copi et al. 2016).
2. There is a vast literature on this issue, including a specialised academic journal, *Sport in society. Culture, Commerce, Media, Politics*.
3. See O'Shaughnessy (2004).

References

Berridge, Kent C and J Wayne Aldridge (2008). "Decision utility, the brain, and pursuit of hedonic goals". In: *Social Cognition* 26.5, pages 621–646.

Cheung, Kelvin Chi-Kin and Kee-Lee Chou (2017). "Poverty, deprivation and life satisfaction among Hong Kong older persons". In: *Ageing & Society*, pages 1–19. URL: https://doi.org/10.1017/S0144686X17001143.

Clark, Andrew E, Paul Frijters, and Michael A Shields (2008). "Relative income, happiness, and utility: An explanation for the Easterlin paradox and other puzzles". In: *Journal of Economic literature* 46.1, pages 95–144.

Clark, Andrew E et al. (2018). *The Origins of Happiness: The Science of Well-Being over the Life Course*. Princeton, NJ: United States of America: Princeton University Press.

Copi, Irving M, Carl Cohen, and Kenneth McMahon (2016). *Introduction to Logic*. Abingdon: United Kingdom: Routledge.

Cutler, Neal E (1979). "Age variations in the dimensionality of life satisfaction". In: *Journal of Gerontology* 34.4, pages 573–578.

Diener, Ed and William Tov (2012). "National accounts of well-being" In: *Handbook of Social Indicators and Quality of Life Research*. Edited by Kenneth C Land, Alex C Michalos, and Joseph Sirgy. Dordrecht: The Netherlands: Springer, pages 137–157.

Feldman, Fred (2010). *What Is This Thing Called Happiness?* Oxford: United Kingdom: Oxford University Press.

Frawley, Ashley (2015). *Semiotics of Happiness. Rhetorical beginnings of a public problem*. London: United Kingdom: Bloomsbury Academic.

Frey, Bruno S (2018). *Economics of Happiness*. SpringerBriefs in Economics. Cham, Switzerland: Springer International Publishing AG.

Frey, Bruno S and Alois Stutzer (2009). "Should national happiness be maximized?" In: *Happiness, Economics and Politics. Towards a Multi-Disciplinary Approach*. Edited by Amitava Krishna Dutt and Benjamin Radcliff. Cheltenham: United Kingdom: Edward Elgar, pages 301–323.

GNH Research (2016). *A Compass Towards a Just and Harmonious Society. 2015 GNH Survey Report*. Technical report. Thimphu: Bhutan.

Graham, Carol (2008). "Happiness and Health: Lessons and questions for public policy". In: *Health Affairs* 27.1, pages 72–87.

Ishio, Junichirou and Naoya Abe (2017). "Measuring Affective Well-Being by the Combination of the Day Reconstruction Method and a Wearable Device: Case Study of an Aging and Depopulating Community in Japan". In: *Augmented Human Research* 2.2. URL: https://doi.org/10.1007/s41133-017-0006-2.

Kahneman, Daniel and Angus Deaton (2010). "High income improves evaluation of life but not emotional well-being" In: *Proceedings of the National Academy of Sciences* 107.38, pages 16489–16493.

Kahneman, Daniel, Peter P Wakker, and Rakesh Sarin (1997). "Back to Bentham? Explorations of experienced utility". In: *The Quarterly Journal of Economics* 112.2, pages 375–406.
Kahneman, Daniel et al. (2004a). "A survey method for characterizing daily life experience: The day reconstruction method" In: *Science* 306.5702, pages 1776–1780.
Kahneman, Daniel et al. (2004b). "Toward National Well-being Accounts". In: *The American Economic Review* 94.2, pages 429–434.
Kalmijn, Wim and Ruut Veenhoven (2014). "Index of inequality-adjusted happiness (IAH) improved: A research note". In: *Journal of Happiness Studies* 15.6, pages 1259–1265.
Kane, Rosalie A (2001). "Long-term care and a good quality of life: Bringing them closer together". In: *The Gerontologist* 41.3, pages 293–304.
Layard, Richard (2005). *Happiness. Lessons from a New Science*. London: United Kingdom: Penguin Books.
Leon, Enrique, Manuel Montejo, and Inigo Dorronsoro (2011). "Prospect of smart home-based detection of subclinical depressive disorders" In: *5th International ICST Conference on Pervasive Computing Technologies for Healthcare (Pervasive-Health)*. European Alliance for Innovation (EAI. Dublin: Republic of Ireland, pages 452–457.
Leon, Enrique et al. (2010). "Computer-mediated emotional regulation: detection of emotional changes using non-parametric cumulative sum". In: *Annual International Conference of the Engineering in Medicine and Biology Society (EMBC)*. IEEE Engineering in Medicine and Biology Society (EMBS). Buenos Aires: Argentina, pages 1109–1112.
nef (2009). *National Accounts of Well-Being: bringing real wealth onto the balance sheet*. Technical report. London: United Kingdom.
O'Shaughnessy, Nicholas Jackson (2004). *Politics and Propaganda: Weapons of Mass Seduction*. Manchester: United Kingdom: Manchester University Press.
Popper, Karl (2013). *The Open Society and Its Enemies. New One-Volume Edition*. Princeton, NJ: United States of America: Princeton University Press.
Prycker, Valérie (2010). "Happiness on the Political Agenda? PROS and CONS". In: *Journal of Happiness Studies* 11, pages 585–603.
Stiglitz, Joseph, Amartya Sen, and Jean Fitoussi (2009). *Report by the commission on the measurement of economic performance and social progress*. Technical report. Paris: France: Commission on the measurement of economic performance and social progress.
Thorndike, Edward Lee (1940). *Human nature and the social order*. New York: The Macmillan Company.
Trust, Carnegie UK (2018). *Wellbeing. What's in a name?* Dunfermline: United Kingdom.
Ura, Karma et al. (2012a). *A Short Guide to Gross National Happiness Index*. Technical report. Thimphu: Bhutan.

―― (2012b). *An Extensive Analysis of the GNH Index*. Technical report. Thimphu: Bhutan.

Veenhoven, Ruut (1996). "Happy life-expectancy". In: *Social indicators research* 39.1, pages 1–58.

―― (2000). "The four qualities of life ordering concepts and measures of the good life" In: *Journal of Happiness Studies* 1, pages 1–39.

―― (2014a). "Happiness Adjusted Life Years". In: *Encyclopedia of quality of life and well-being research*. Edited by Axel Michalos. Dordrecht: The Netherlands: Springer Netherlands, pages 2641–2643.

Veenhoven, Ruut (2014b). "Inequality-Adjusted Happiness" In: *Encyclopedia of quality of life and well-being research*. Edited by Axel Michalos. Dordrecht: The Netherlands: Springer Netherlands, pages 3253–3254.

―― (2014c). "Sociology's blind eye for happiness" In: *Comparative Sociology* 13.5, pages 537–555.

Veenhoven, Ruut and Wim Kalmijn (2005). "Inequality-adjusted happiness in nations egalitarianism and utilitarianism married in a new index of societal performance". In: *Journal of Happiness Studies* 6.4, pages 421–455.

Part II

Inequality and Poverty

4

Inequality

> **Overview**
>
> This chapter presents different theories of moral philosophy that are at the heart of economic analyses and policy discussions regarding inequality, poverty, and related concepts in later life. It also reviews various metrics designed to measure distribution and inequality. Other topics include the links between population ageing and income distribution and the intergenerational transmission of inequality.

4.1 Introduction

For the nineteenth-century British economist David Ricardo, the 'principal problem' in economics was the distribution of national income. For the contemporary US economist Robert Lucas, to focus on questions of distribution is the 'most poisonous' of the 'tendencies that are harmful to sound economics'.[1] It is between these two extremes that economists have approached (and, to some extent, are still approaching) the topics of inequality and distribution. In part, the reason is that more than in any other branch of economics, in the study of inequality and distribution there is a conspicuous meeting of normative and positive issues.

Over a century ago, echoing Ricardo's concern, Fisher rued that even though, in his opinion, no other problem was of such great interest as 'the problem of grading the population according to income', it so happened that 'the problem of discriminating the relatively rich and the relatively poor…ha[d] received so little scientific study' (Fisher 1912, p. 465). And, over half a century ago, Johnson stated that '…economic analysis provides a reasonably good theory of the functional distribution of income and nothing that can be called a theory of the personal distribution of income' (Johnson 1954, p. 175).

In turn, writing in 1996 and anticipating Lucas' disdain for the topic, the Dutch/British economic historian Mark Blaug was puzzled that inequality would remain a topic of research and debate: he considered it 'the great mystery of the modern theory of distribution…why anyone regards the share of wages and profits as an interesting problem' (Blaug 1996, p. 467).

However, the German economist Hans-Jürgen Krupp pointed out that the study of functional distribution would be useful if it could shed light on the understanding of personal distribution:

> There should be no doubt that the functional income distribution for society and politics in the narrower sense is only of secondary importance. Only if one could make conclusions from the functional distribution about the personal, the former would be socially relevant.[2]
>
> (Krupp 1967, p. 3)

Plenty of research has been carried out over the last decades on the distribution and inequality of income and wealth, but it was not until the work by French economist Thomas Piketty (Piketty 2014),[3] that these topics were brought back to the centre of the economic debate. Close to the interests of this textbook, Piketty found a strong association between demographic change and inequality. In particular, he postulated that population ageing would increase the level of capital intensity (or capital/labour ratio) as a consequence of the relative reduction in the labour force. Given that income from assets is more unequally distributed than income from employment, the rise in capital intensity would translate into a rise in income and wealth inequality (see also Goldstein and Lee 2014). Moreover, inequality and poverty have also been associated with negative outcomes in later life: from bereavement to poor housing, from social exclusion to mortality, from psychological distress to intensive caregiving demands, few aspects of later life are not related to income inequality and living on relative or absolute low income.

4.2 Distribution and Moral Theory

When it comes to inequality and distribution, the various moral philosophical assumptions usually made tacitly, and which underlie the economic analyses, have to be brought to light. Normative topics consist of policy and ethical aspects. Positive topics comprise descriptive, interpretative, explanatory, and predictive aspects. An understanding of what ought to be or what ought not to be done—a 'theory of justice'—underpins all policy considerations of course. However, discussions of justice become unavoidable when designing or analysing policies regarding inequality and distribution. Consider, for instance, the use of functions that assign different values to the elements in a set according to some characteristics in order to obtain a weighted magnitude (sum, mean, median, etc.). The 'problem' is that there is no such a thing as an 'unweighted' magnitude (e.g. an unweighted average). 'Unweighted' actually means 'equally weighted': all the elements or units in the set are given the same weight, that is, they are ascribed the same importance. In many branches of economics, an unweighted magnitude is adopted and the study goes on without much further ado. But in topics of inequality and distribution, the ethical significance or stance of such an apparently innocuous mathematical procedure is necessarily probed; the equality of treatment of the observational units hitherto underlying the choice of method is brought to the fore. In this sense, and more generally, the economics of inequality can be understood as a branch of applied ethics (i.e. of the 'attempt to apply general principles of normative ethics to particular difficult or complex cases' (Kagan 1998, p. 3)).

This is not the place to discuss the theories of moral philosophy in detail, but I want to start this part with a swift overview of some of the ethical views that inform current discussions around measurement, analyses, and policy developments in the economics of inequality and distribution: utilitarianism, libertarianism, contractualism, and the capabilities approach. These doctrines have been mentioned in other parts of this textbook, but I trust that a brief appearance here is in order.

Sen (2000) explained that a normative theory (or theory of justice) has a threefold informational basis:

- A basal space: the variables to which the claims about justice are made—that is, the objects of value.
- A focal combination: how the basal space is applied to assess the justice of a particular situation by means of aggregation—that is, how the alternative social states are ranked.

- A reference group: against whom it is claimed that an individual or a group is in an unequal relationship or is defined as poor in relative terms.

The first two informational contents require some unravelling. The basal (or relevant) space comprises the variable or variables which the different theories of justice centre on. For example, utilitarianism defines individual utility as its basal space whilst libertarianism puts individual rights and liberties in its stead; Rawls' version of contractualism sets out a basal space of 'primary goods' including basic rights and liberties, freedom of movement and choice, income and wealth, and so on, and the capabilities doctrine is based around the notions of capabilities and functionings. Capabilities refer to the set of all the things an individual is free to do or be, given the resources available. Functionings refer to the exercise of the capabilities. For example, going to the cinema is an activity that individuals in most countries are free to do—hence, it is part of their capabilities. If they are prevented from attending because of, say, income constraints, then they cannot exercise their freedom. The focal combination in utilitarianism depends on the assumptions regarding the social welfare function, but the most popular aggregation is the unweighted sum of individual utility functions. Libertarianism presents a binary focal combination according to which either all the basal space is applied and satisfied or no justice is achieved. Rawls' contractualism proposes the 'maximim' or difference principle as its focal combination: the utility of the worst-off individual or unit. The capability approach adopts the expansion of human capabilities, the freedom to access functionings, and the freedom from coercion and encroachment when choosing between functionings.

4.2.1 Value Claims

In economics, 'value' refers to the theoretical conjectures around the exchange value of goods and services. Basically, there are three main views: that the value of a commodity is:

- the result of an inherent or intrinsic characteristic of the commodity
- the result of the amount of labour socially necessary to produce the commodity
- the result of the demand for the commodity by consumers and how much each additional unit increases a consumer's marginal utility

Here I want to focus on the philosophical meaning of 'value', which refers to the goodness of the states of nature, as it is at the heart of the basal space of any theory of distribution. There is some overlap between the philosophical and the economic conceptualisations of the value of a commodity: moral philosophy and economics distinguish between intrinsic and instrumental value: either goodness is an intrinsic property of an action or state of nature or such action or state of nature causally generates value.[4] The important point I want to underline is what is claimed to be of value, or the predicate of value claims.

In moral philosophical theories of value, there is a separation between authors who assert that there is only one fundamental or utmost value (monism) and those who claim that there are many values of equal relevance and standing (pluralism). Again, this is reflected in the distinction in economics between individual utility or well-being as the only relevant value regardless of the utility or well-being of the rest of the agents in an economy, and the extension of individual and social 'welfare functions' to incorporate other considerations (other values) assumed to be of equal relevance.

In mainstream economics, individual utility is claimed to be the ultimate value. Income and wealth are usually assumed to have instrumental value, rather than intrinsic value, but income and wealth are also the most frequently used proxy variables for utility. Therefore, operationally, income or wealth is assumed to be a satisfactory approximation to utility, which means that the maximisation of utility would be akin to the maximisation of income or wealth. Discussions about the intrinsic or instrumental (positive or, more often, negative) value of an unequal distribution of utility in a population boil down to how income or wealth are distributed. However, the sub-discipline of economics of happiness promotes the adoption of direct measures of well-being, happiness, life satisfaction, and so on as predicates of value claims regarding states of nature derived from economic policies or interventions and repeatedly object that there is little association between income and happiness and well-being (see Part I in this volume).

Although disparities in income, and to a lesser extent in wealth, are possibly the most salient content in the inequality and distribution literature, they are by no means the only variables of interest: disparities in health, life expectancy, consumption, education, social participation, and opportunities have also been included in the basal space of alternative approaches to inequality and distribution. According to these views, it would not follow that the maximisation of income or wealth represents the best possible state of nature.

However, one can be concerned with income inequality without being utilitarian or even welfarist, given that there is plenty of evidence that inequalities in income affects functionings, health, life expectancy, and social participation (to name some of the alternative variables of interest in the inequality and distribution literature), so much so that income and wealth would constitute acceptable approximations to inequality in general. And this is evident in later life. Take oral health, for instance, and the capability approach. Mirroring the food example used above, we can say that toothpaste, brushes, dentures, dental care, and so on are the goods and services; that chewing is one of the related functionings; and that access to these goods and services is one of the capabilities conducive to oral health. In a series of studies, the German economist and dentist (an unusual combination!) Stefan Listl has shown that income-related inequalities in later life are associated with denture wearing (Listl 2012), chewing ability (Listl and Faggion 2012), and dental service utilisation (Listl 2011). Therefore, many economists would contend that a focus on income should not be equated with an uncritical embrace of utilitarianism.

4.2.2 Moral Desert

A study on implicit anti-fat bias reported strong discriminatory tendencies among a sample of health professionals specialising in obesity in Canada and the United States of America (Schwartz et al. 2003): obese people were mentally associated with laziness, stupidity, and worthlessness...by the professionals trained to treat their weight problem. From internalising this stigmatisation, it would not be a big step to be persuaded that obese people do not *deserve* any help from taxpayers. Note that this conclusion is unrelated to the utility gains obese people might experience from state-sponsored intervention, and that is also unrelated to the, say, potential net positive social welfare of the interventions. However, it is still utilitarian; the difference is that this theory modifies the definition of 'utility'. At the risk of sounding Manichaean, there would be a distinction between the utility of those deserving recipients of transfers (deserving poor, obese, etc.) and of that of the undeserving. This view is known as moral desert.[5] It has been defined succinctly (albeit rather crudely) thus: 'the moral value of achieving a one unit gain of well-being is greater, the greater the individual's level of deservingness' (Arneson 1997, p. 334).

Economic desert is generally tied with performance and contributions through one particular socially valued activity: paid employment. Hence, there is an element of paying back, of reciprocity. Need is not morally relevant in this view. Nor is merit. Closely linked to performance during paid employment is the notion of an agent's responsibility: to be deserving, the moral desert doctrine asserts that the poor should have tried and, dependent on age and other factors, should be still trying to find ways to get out of poverty by means of paid employment.

Related to the philosophical discussion of moral desert, it is worth mentioning the sociological approach to deservingness, which focuses on the criteria used by individuals across countries to mentally classify individuals in need between deserving and undeserving. Of the various sociological approaches to this issue, the most cited was developed by Oorschot (2000). It consists of five deservingness criteria:

- Control over neediness or locus of responsibility: individuals who appear personally responsible for their poverty would be seen as less deserving
- Attitude: gratefulness, pleasantness, and docility are characteristics that would make individuals in need appear more deserving
- Reciprocity: an essential aspect of the moral economy of the welfare state (Oorschot 2000, p. 14), individuals who exhibit higher reciprocation appear more deserving. Importantly, this criterion looks back not only at past contributions but also at the willingness to do something in return in the future.
- Identity: the more 'they' look like 'us', the more deserving they are. Closeness may be defined in terms of kinship, area of residence, ethnic background, migrant status, and so on,...including age.
- Need: irrespective of the level of control and personal responsibility, individuals who seem to be more in need appear more deserving

Oorschot opined that older people are usually the most deserving group (compared to unemployed, sick, migrant, or disabled adults):

> They cannot be blamed for their age, they are close to 'us' (they are our parents and grandparents, and we ourselves hope to live to an old age), they have extra age-related needs, they have earned their share in their productive life stage and they are not seen as an ungrateful and demanding group
>
> (Oorschot 2000)

This view was challenged by discourses portraying older people as affluent and 'greedy geezers' (Butler 1989; Cohen 1994; Street and Cossman 2006) or 'woopies'[6] (Falkingham and Victor 1991), and therefore less deserving. Whilst these views in some cases are 30 years old, they are still influential.

4.2.3 Utilitarianism

Consequentialism is a doctrine that holds that the consequences of acts ('act consequentialism') or rules ('rule consequentialism') are the only morally relevant factor. The reasons or motivations of an action or a regulation or norm do not matter, only its consequences. Welfarism is the ethical position that holds that the moral content of the states of nature are to be assessed exclusively by the individual well-being or utility they contain or create. Utilitarianism is a combination of consequentialism and welfarism. This school of moral philosophy holds that the greatest total well-being or utility is the only morally relevant factor. The combination or totality of individual utilities constitutes its basal space, so that claims about injustice regarding anything else become irrelevant. The assumption that a social welfare function is equal to the sum of the utility functions of all the individuals in society constitutes its focal combination.

4.2.4 Libertarianism

Libertarianism is the ethical doctrine according to which the preservation of personal autonomy ('self-ownership') and political and economic rights and freedom is the only morally relevant factor. The aim of libertarianism is not to maximise social welfare but the maximisation of 'empirical negative liberty': that is, the area within which individuals are or should be left to do or be what they are able to do or be, without interference by other persons (paraphrasing Berlin 1969).

4.2.5 Contractualism

Contractualism (as developed by US philosopher John Rawls[7]) emphasises reciprocity in social roles and is structured around two principles of justice (Rawls 1971): (a) that each individual is to have 'an equal right to the most extensive basic liberty compatible with a similar liberty for others' [p. 60] and (b) that inequalities are arranged so that they contribute to 'the greatest benefit to the least advantaged [members of society], attached to offices and positions open to all under conditions of fair equality of opportunity [p. 83]. The second principle is known as the 'maximin': the only morally relevant factor is the welfare of the worst-off member of society.

4.2.6 Capabilities

The capabilities approach is an egalitarian doctrine that holds that the only factors of moral relevance are an individual's functionings and capabilities. Functionings correspond to achievements of 'states of being and doing', which are made possible by commodities. Goods and services are not functionings, but means to functionings. Food is a commodity; being nourished is a functioning. Capabilities, in turn, are functionings an individual has access to. To continue with the example, having the means and access to food is a basic capability. The concept of capabilities is redolent of Berlin's notion of 'positive freedom'. Hence, the preservation of capabilities is the preservation of freedoms, and the development of capabilities is the development of freedoms.

4.2.7 Consequentialist and Deontological Approaches

Moral theories can be classified on the basis of their moral justification of acts into consequentialist and deontological. Consequentialist theories define the moral relevance of an act by its expected or probable outcomes of consequences. Deontological theories define the moral relevance of an act by its intrinsic elements: its probable or realised consequences are not morally relevant. Utilitarianism is a consequentialist theory: if an act increases social utility or welfare, it is morally acceptable. Rawlsian contractualism is deontological: it is based on two principles, the principle of equal liberty according to which each individual has an equal right to the most extensive liberties compatible with similar liberties and the difference principle, according to which inequalities resulting from an act are only acceptable if the worst-off individual is better off than if the act had not taken place. But the principle of equal liberty takes moral precedence over the difference principle (i.e. it is lexicographically superior).

4.3 Measurement of Distribution and Inequality

4.3.1 Inequality of Whom?

Apart from the basal space and the focal combination, the third informational component of a theory of justice is the reference group. Many studies of inequality and distribution in later life use younger populations as reference groups, whereas other studies concentrate on differences among older people.

In addition, apart from the reference group, it is important to define the observational unit: the 'inequality of whom?' question. The interest may be centred on the inequality among individuals, households, or families. From a different perspective, a study may concern the distribution between wage earners, agricultural labourers, capitalists, and so on. The first classification is known as the personal (or interpersonal or size) distribution. The second classification is known as the functional distribution. Personal or size distributions focus on income returns to individuals or households; functional distributions focus on the income returns to factors of production (Campano and Salvatore 2006). Personal distribution relates to the agents who receive income; functional distribution relates to the agents who produce it.

Personal or Size Distribution

Personal distribution of income or wealth is what most people have in mind when they refer to the income or wealth distribution. 'Distribution' has become a shorthand for 'personal/interpersonal/size distribution'. The 'personal' element is important: how income or wealth is distributed among individuals may not coincide with how they are distributed among households or families. Data on personal income and wealth tend to be collected at a household level, so most studies look into the distribution of household income or wealth. However, this choice implicitly adopts the strong assumption of equality in the intra-household relations of power to control household resources: that the resources within households are pooled among its members—that is, what Fritzell (1999, p. 65) termed a 'a nonmarket, collectivist and egalitarian distribution of resources within households'. If this assumption does not hold, measures based on individual data would differ from household-level measures. This is not a technicality to be brushed off lightly and as Ben-Porath (1982, p. 1) remarked: 'the family...is not merely a statistical nuisance that must somehow be suppressed in the analysis of income distribution'.

The most important dividing line in terms of intra-household power is gender, and there is abundant evidence that financial arrangements within households are less favourable to women than men (Cantillon and Nolan 2001; Corsi et al. 2016; Woolley 2003; Woolley and Marshall 1994). Attempts to incorporate intra-household inequality in the estimates of income distributions show that the assumptions around how resources are shared within households significantly affect the results. For example, in a study of income distribution in Sweden using data from 1990, Fritzell (1999) reported that under the equal sharing assumption, the 10 per cent richest households had, on average, incomes 2.37 times higher than the 10 per cent poorest households. This ratio went up to 2.42 times under a partial-sharing assumption, and to 2.66 times in a no-sharing scenario. Obviously, these considerations do not apply to single households. Of course, individual-based data circumvents the need to adjust or factor in intra-household inequalities.

One additional point to take into account is that households are not the same as families. 'Family' is a contested term in social sciences, but it would suffice to say that a family is a social institution and a locus of relationships and cultural practices, including socially expected practices.[8] Official statistics (e.g. the US Census) define a household as composed of one or more people living in the same housing unit whilst a family consists of two or more individuals related by birth, marriage, or adoption. People living alone form households but not families. However, these statistical definitions are not in sync with contemporary cultural understandings of a family in most developed countries. For example, the rise in same-sex couples with and without children and of co-residence arrangements among people not related by kin, marriage, or adoption shows changing definitions of and attitudes towards families not reflected in the official definitions. The structure of households in which older people live shows more heterogeneity than that of younger people. For example, there is an increasing number of older people's households with kin and non-kin members (Hinterlong and Ryan 2008), and of course with increasing age one-person households also become more common as well as households headed by women. On the other hand, family members not living in the same housing unit but in close vicinity may form 'local family circles' (Bonvalet and Andreyev 2003; Tillman and Nam 2008) or 'neo-extended families' (Tung et al. 2006). These family circles are relevant to older people and have economic repercussions in terms of the sharing of resources, exchange of services, and so on. As opined,

> Issues of social values and expectations lie behind what might otherwise appear to be a purely statistical question: whether we should measure inequality or poverty in terms of households or of families.
>
> (Atkinson 2015, p. 29)

Apart from intra-household gendered relations, another element of consideration is the inequality allocation of resources within intergenerational households consisting of adult children and their older parents (Pezzin and Schone 1997). However, no attempts have been made (as far as I can tell) to adjust size income distributions for unequal command within intergenerational households.

Functional Distribution

The other main approach to inequality and distribution is the functional partition of the different factors of production (hence, it is also known as 'factorial' distribution). This was the idea of income distribution that classical economists had in mind. Reflecting on the class divisions in Britain and in Western Europe in general in the late eighteenth and early nineteenth centuries, these authors focused on the share of national income earned by workers, landowners, and capitalists—that is, wages, rents, and interest (Sandmo 2000).

Contemporary studies focus on the wage and profit shares (i.e. the remuneration for labour and capital). There are links between the personal and functional distribution of income. These two dimensions of inequality are related, but are subject to different drivers. Inequality in access to education and consequently in the accumulation of human capital as well as in its remuneration (the skills premium), tied to changes in labour markets resulting from technological change and globalisation, can largely explain the increasing inequality in labour income evidenced in many developed countries since the 1980s. So can changes in organisational practices and taxation that favours high bonuses and compensations for top managers. In turn, saving behaviour, tax treatment of inheritances and inter vivos transfers, and dynamic developments in financial and real estate markets can explain inequality in wealth and ownership of capital goods, which has been increasing in developed countries since the 1950s. The accepted view is that inequality in factor shares is one driver of inequality in personal income, *but only one* (Atkinson 2015; Ryan 1996). In a long-term view of inequality in the United States of America, Lindert and Williamson (1976) concluded:

> The entire history of inequality [in] …highlights another important point: Inequality movements have not been the result of mere movements among demographic groups. Rather, they have followed trends in the basic occupational pay gaps as well as the level and dispersion in profit rates and rents.
> (Lindert and Williamson 1976, p. 8)

Moreover, according to Piketty (2014, p. 40), 'it is impossible to achieve a satisfactory understanding of the distributional problem without analyzing …' the functional and personal distribution of income. Regarding the functional distribution, he is interested in the share of national income between capital and labour, and in the inequality of labour income and of capital ownership.

4.3.2 Inequality of What?

It is not enough to define 'income' or 'wealth' as the topic of interest, for these concepts may comprise different things for different authors or be measured according to different methods. As Sen warned:

> While we can decide to close our eyes to this issue by simply assuming that there is something homogeneous called "the income" in terms of which everyone's overall advantage can be judged and interpersonally compared (and that variations of needs, personal circumstances, prices, etc. can be, correspondingly, assumed away), this does not resolve the problem—only evades it.
>
> (Sen 2000, p. 77)

In 1996, the Australian Bureau of Statistics set up a group of experts to address technical aspects of household income statistics. This group produced the *2001 Canberra Group Handbook*, reviewed in 2008 by the Conference of European Statisticians. In 2013, the Organisation for Economic Co-operation and Development (OECD) published its *Framework for Statistics on the Distribution of Household Income, Consumption and Wealth*, which complements the Canberra's handbook by providing a micro-data perspective on not only income, but household consumption and wealth as well, integrated within a framework that has variations in the stock of household wealth at its centre (OECD 2013b).

The 2011 edition of the Handbook (UNECE 2011) defines the household as the unit of observation, and defines household income as [p. 9]:

> …all receipts whether monetary or in kind (goods and services) that are received by the household or by individual members of the household at annual or more frequent intervals, but excludes windfall gains and other such irregular and typically one-time receipts

Disposable income is defined in a way that most people would not identify with their household income. Officially, it is composed of the sum of income from employment and property and the imputed income from the household production of services for own consumption (notably, the imputed net value of owner-occupied housing services and the imputed value of unpaid domestic services) plus any current transfers received, minus any current transfers paid (including taxes). Disposable income plus any social transfers in kind received form the household income. Most countries apply similar classifications, operational definitions, and compilation methods (based on household surveys).

Imputed rental income constitutes the largest non-cash component of household income in most developed countries. Furthermore, the inclusion of imputed rental income from owner-occupied housing is very relevant to studies of economics and ageing: in most countries, outright owners of housing stock on low incomes tend to be older people. For these households, the imputed rental income constitutes a large share of total disposable income, so that its inclusion increases the levels of inequality within, particularly older people on low incomes: renters are more disadvantaged compared to home owners. In addition, the inclusion of rental income increases the average income of poorer households proportionately more than that of better-off households with an overall impact of reducing poverty measures compared to poverty indicators that exclude these imputations (see Butrica et al. (2009) for a case study with data from the United States of America; Frick et al. (2010) for Belgium, Germany, Greece, Italy, and the United Kingdom; and Fessler et al. (2016) for Austria). Therefore, it is important to know what a study on income inequality and distribution is referring to when it talks about 'income'. I am not going to review any rental income imputation methods, but there are several methods to estimate the value of imputed rents (Garner and Short 2009) so even if the figures include the income rental imputations, it is worth investigating which method was used and its implications and impact on disposable income compared to other methods.

The definition of wealth, in studies of inequality and distribution, is generally restricted to financial (including pension savings) and material wealth (housing and non-housing). That is, other forms of capital such as human and social capital as well as collectively held assets are not included.

The latter type of capital is very relevant in later life, so this omission needs to be taken into account when carrying out studies on inequality of wealth before coming to conclusions and recommendations. For example, Bavel and Frankema (2013) looked at the distribution of private wealth in nine European countries[9] and found huge levels of inequality. One of the reasons for this, the author surmised, is that the computation of private wealth 'does not capture the collective and public arrangements that are put in place to guarantee lifetime income security' [p. 10]. Bavel and Frankema concludes that *a relative egalitarian distribution of the claims to collectively held assets* [p. 10] (e.g. state-guaranteed income security) compensates for the unequal distribution of private wealth in these countries and reduces the incentives to save to counter income shocks due to illness, unemployment, or old age.

The correlation between income and wealth is not as high as most people imagine: Keister (2014) estimated a correlation coefficient between 0.50 and 0.60 for the period 2001–2014 in the United States of America. Income and wealth are usually analysed separately, although Stiglitz et al. (2009) recommend that both variables should be considered jointly in studies of inequality and distribution. I would recommend that when considering other people's studies attention be paid to how the variables have been defined and measured. We should all (myself included!) be aware that 'All too often economists race ahead, drawing conclusions from figures that happen to be there, without asking whether the data are suitable.' (Atkinson 2015, p. 45)

After considering how income and wealth are usually defined and measured, let's see now how inequality is defined and measured.

4.3.3 Measures of Inequality

Several indices and statistical measures of inequality have been developed, alongside technical requisites responding to logical, welfare, statistical, and ethical considerations. The four most important properties that a measure or index of inequality must incorporate are as follows:

- A transfer of income or wealth from a wealthier unit (individual, household, family, etc.) to a poorer unit that does not reverse the relative ranks between these units (i.e. that after the transfer the poorer unit remains poorer than the wealthier unit) should reduce the value of the index. In other words, the statistic should reflect a decrease in inequality as a result of the transfer. This is known as the 'Pigou-Dalton condition' (Dalton 1920).
- A change in the income or wealth of all the units by a same scalar or factor (say, all the incomes are increased by 10 per cent) should not affect the measure of inequality. This is known as the 'mean independence' condition.
- A change in the number of units by the same proportion at each income or wealth level should not affect the measure of inequality. That is, if two or more identical groups are pooled, the index should not change. This is known as the 'population-size independence' condition.
- The measure or index of inequality should be additively decomposable. That is, it should be equal to the sum of the inequality within and between sub-groups in the population. In other words, if the population is broken down by any sub-groups (e.g. by gender, education, and chronological age),

the overall inequality measure or index must be equal to the weighted sum of the inequality between and within the sub-groups.[10] This is known as the 'decomposability' condition. If inequality in each sub-group has, for example, increased, the overall measure or index should equally show an increase in inequality.

Of the four properties listed above, the most relevant in terms of assessing the role of population ageing on changes in inequality is the decomposability condition.

Decomposition

According to Von Weizsäcker (1996), the question as to whether population ageing increases inequality is ill-defined, mostly due to the different levels of analysis at which the problem can be tackled and the fact that different aggregation levels tend to convey contrasting stories. Hence, the importance of considering the most popular methods in the literature to break down inequality measures into the contributions from different sub-groups.

Von Weizsäcker pointed out that in income inequality studies even a simple decomposition of the aggregate income between its intra- and intergenerational components render opposing results: with population ageing, the dispersion of the intragenerational component will increase, whereas the intergenerational component will decrease. This author presents the following framework to show the impact of population ageing on both components. Let's assume an economy is composed of two groups, $j = 1, 2, \ldots E$ workers and $i = 1, 2, \ldots R$ pensioners. Each worker earns a net income Y_j equal to the gross income A_j minus a fixed amount c deducted as pension contribution to a fund. In turn, each pensioner earns a retirement income P_i equal to the accumulated pension claims L_i times the pension rate p, which is assumed constant, times the average gross earnings of the working-age population (μ_A). In symbols,

$$Y_j = (1 - c) \cdot A_j$$
$$P_i = L_i \cdot p \cdot \mu_A \tag{4.1}$$

Let's aggregate each expression across all workers and pensioners, respectively, to obtain the distribution of individual income. The variance of a variable in a population can be partitioned into the groups forming that population. In this case, the decomposition allows us to distinguish between the variance within and between workers and pensioners. Total income (W) is equal to:

$$W = \sum_{j=1}^{j=E} Y_j + \sum_{i=1}^{i=R} P_i \qquad (4.2)$$

W can be decomposed into the average income for each group and the deviations of incomes for workers and pensioners against this average. For each individual in the population, we get:

$$W_{j,i} = \bar{W}_{j,i} + (Y_j - \mu_Y) + (P_i - \mu_P) \qquad (4.3)$$

In Eq. (4.3), $\bar{W}_{j,i}$ represents the average group income (i.e. the average income for workers and the average income for pensioners). From here, the variance of total income can then be decomposed into the variance across both groups and the variance within each of the groups, that is, into the variance by type of earner (which is not the same as the decomposition by type of income[11]). It is given by:

$$\text{var}\left(W_{j,i}\right) = \text{var}\left(\mu_{W_{j,i}}\right) + \text{var}(Y_j - \mu_Y) + \text{var}(P_i - \mu_P) \qquad (4.4)$$

The first term on the right-hand side, $\text{var}\left(\mu_{W_{j,i}}\right)$, represents the between-group variance, and the two other terms are the within-group variance for workers and pensioners, respectively. In other words, the total variance in the population is equal to the sum of the between- and within-group variances.

In our case, with the proportion of workers in the population equal to $E/(E+R)$ and therefore the proportion of pensioners in the population equal to $1 - E/(E+R)$, we can obtain this decomposition of the variance of total income in the economy:

$$\sigma^2 = \left(\frac{E}{E+R}\right) \cdot \sigma_Y^2 + \left[1 + \left(\frac{E}{E+R}\right)\right] \cdot \sigma_P^2 + \left(\frac{E}{E+R}\right)$$
$$\cdot \left(1 - \frac{E}{E+R}\right) \cdot (\mu_Y - \mu_P)^2 \qquad (4.5)$$

The expression $E/(E+R)$ is a function of population ageing. To simplify the notation, let's denote the proportion of workers in the total population by x (so $x = E/(E+R)$) and let's define the old-age dependency ratio (see Volume I, Chap. 5) as $\theta = R/E$. The important question is how much the dispersion in total income changes as a result of a change in the old-age dependency ratio. Of course, x is a function of R/E, and it is exclusively via x that the old-age dependency ratio affects the dispersion in total income in this model. Therefore we can estimate the partial first derivative of the dispersion in total income with respect to a change in x:

$$\frac{\partial \sigma^2}{\partial \theta} = \left(\sigma_Y^2 - \sigma_P^2\right) \cdot \frac{\partial x}{\partial \theta} + (1 - 2 \cdot x) \cdot (\mu_Y - \mu_P)^2 \cdot \frac{\partial x}{\partial \theta} \qquad (4.6)$$

The first term on the right-hand side represents the effects of demographic changes within the workers and the pensioners, that is, the intragroup effect. The second term represents the effects of demographic changes across or between groups, that is, the intergroup effect.

In all countries, the number of workers exceed the number of pensioners and the dispersion of income from employment is higher than the dispersion of pension income. Consequently, the intragroup effect is negative, and the inter-group effect is positive. Von Weizsäcker points out to the indefinite theoretical effect of population ageing on income inequality: income inequality (measured by its dispersion) could increase or decrease as a result of rising old-age dependency ratios (provided that the number of workers exceeds that of pensioners, and that the inequality in employment income exceeds the inequality in pension income).

The result also depends on the type of inequality indicator. Using the coefficient of variation, Von Weizsäcker demonstrated that under relatively weak assumptions population ageing would *increase* income inequality. The coefficient of variation (V^2) of a distribution is the ratio between the variance and the square of the mean:

$$V^2 = \frac{\sigma^2}{\mu^2} \qquad (4.7)$$

The decomposition into coefficient of variation of each group leads to (see Von Weizsäcker 1996, eq. 5):

$$V^2 = x \cdot \left(\frac{\mu_Y^2}{\mu^2}\right) \cdot V_Y^2 + (1-x) \cdot \left(\frac{\mu_P^2}{\mu^2}\right) \cdot V_P^2 + \left[\frac{x(1-x)}{\mu^2}\right] \cdot (\mu_Y - \mu_P)^2 \qquad (4.8)$$

With $E > R$ and $\mu_Y > (1 + \frac{1}{x}) \cdot \mu_P$, the derivative of V^2 with respect to θ is positive. In 'plain English': with the number of workers exceeding the number of pensioners and the dispersion of net earnings of the working population exceeding that of pensioners' income, a rise in the old-age dependency ratio increases income inequality. That the number of workers in a population exceeds the number of pensioners and that the inequality in income from employment is greater than that from pensions are, as mentioned, two stylised empirical findings. Therefore, using any of the measures of inequality I am going to present in this section, the conclusion is that via changes in the demographic structure, population ageing should increase income inequality, *ceteris paribus*. Population ageing has an impact on transfers of income between workers and pensioners (i.e. on the relative income between these two groups) as well as on the demographic structure, so that other things hardly 'paribus'. Von Weizsäcker showed that under pay-as-you-go pension systems (see Chap. 4 in Volume III) whichever way a government responds to the fiscal pressure of population ageing on the pension system (i.e. an increase in the contribution rate or a reduction in the pension benefits), the distributional impact is indeterminate and, ultimately, an empirical question. Moreover, Brewer and Wren-Lewis (2016) applied the decomposition in Eq. (4.13) to data for the United Kingdom between 1978/1979 and 2008/2009 and reported that an increase in the relative income of pensioners since 1991 and especially during the 2000s dampened the rise in inequality due to widening wage and self-employment income disparity over the period. In turn, Mookherjee and Shorrocks (1982) found no age effects on household income inequality in the United Kingdom between 1965 and 1980, also applying the decomposition in Eq. (4.13).

In the rest of this section, I briefly describe the most common measures of inequality in the literature of economics and ageing, and how they are decomposed by chronological age groups: the coefficient of variation, the Mean Logarithmic Deviation (MLD), the Gini coefficient including its age-adjusted variations, and the Theil indices. I also comment on the concentration index (CI) and the slope index of inequality (SII), principally used in health economics. Not all of these indices satisfy the four properties (Gini, for example, is not additively decomposable), but I do not tackle these discussions here; interested readers will benefit, among others, from Atkinson (2015), Campano and Salvatore (2006), Cowell (2011), and Sen (1997).

Variance of Logarithms

In statistics, the variance is a measure of the degree of dispersion in a frequency distribution. The average height of the pupils in two classrooms may be the same, say, 1.53 metres, but in one classroom there are three very tall pupils and three very short pupils, whereas in the other all the pupils are between 1.52 and 1.54 metres high. The average height of the first group of pupils has greater dispersion or variance than that the average height of the second group.

The variance of the distribution of a variable y ('height' in the example above) is equal to the sum of the squared differences between each of the i elements or observations (e.g. the pupils) and the mean for the population ($\bar{y}$), divided by the size of the population, n. In symbols,

$$\text{var} = \frac{\sum_{i=1}^{i=n} [y_i - \bar{y}]^2}{n} \tag{4.9}$$

This variance measures dispersion of inequality. However, as a measure of inequality it has one problem. Imagine the variable of interest is individual income, not height. An increase in the income of each individual in the population by a scalar (say, doubling everyone's income) should not affect the degree of inequality (the poorest is twice as rich now, but so is the richest individual; when it comes to how unequal the distribution is, nothing has changed). However, the variance would have increased. Instead, imagine that we transform the variable income by applying logarithms, so the focus now is not the distribution of income but the distribution of the logarithm of income in the population. Something remarkable happens: the variance of the logarithms is invariant to these changes; it remains the same as before the multiplication. The variance of the logarithm of a variable y is represented by:

$$\text{var}_l = \frac{\sum_{i=1}^{i=n} \left[\log\left(\frac{y_i}{\bar{y}}\right)\right]^2}{n} \tag{4.10}$$

where $\bar{y}$ now represents the mean of the logarithms of the variable.

Therefore, in empirical studies, income is transformed into the logarithm of income. The most common base[12] for logarithms used in economics is the mathematical constant $e = 2.718\ldots$, known as a natural logarithm. This transformation makes the most of some of the properties of natural logarithms, in particular,

- that for a variable that only adopts positive values (i.e. when a natural logarithm of a variable is a real-valued function), the natural logarithm is equal to the inverse of the exponential function of the variable. In symbols,

$$e^{\ln x} = x$$
$$\ln(e^x) = x \qquad (4.11)$$

- that the first derivative of the natural logarithm of a variable equals the inverse of the variable. In symbols,

$$\frac{d\ln(x)}{dx} = \frac{1}{x} \qquad (4.12)$$

Imagine that a population is broken down into $g = 1, 2, \ldots, G$ subgroups. The following approximation by Mookherjee and Shorrocks (1982) is used to decompose the changes in income inequality by sub-group over time (see also Shorrocks (1980) and Brewer and Wren-Lewis (2016, eq. 8)):

$$\Delta I_0 \approx \sum_{g=1}^{g=G} \left(\bar{v}_g \cdot \Delta I_{0g} \right) + \sum_{g=1}^{g=G} \bar{I}_{0g} \cdot \Delta v_g + \Delta_g \cdot \left[\bar{\lambda}_g - \overline{\ln(\lambda_g)} \right] \cdot \Delta v_g$$

$$+ \sum_{g=1}^{g=G} \left(\bar{v}_g \cdot \bar{\lambda}_g - \bar{v}_g \right) \cdot \Delta \ln(\mu_g) \qquad (4.13)$$

where Δ denotes changes between any two periods; λ_g is the ratio between the mean income for sub-group g and the population average income; and v_g is the ratio between the number of people in sub-group g and the size of total population (i.e. the share of g in total population). A bar (¯) over a variable indicates the average of its values between an initial period taken as the base and the current period.

Equation (4.13) reflects the three drivers behind the effects of demographic change on income distribution mentioned by Lindert and Williamson (1976) with which I started this section (Brewer and Wren-Lewis 2016, p. 296):

- 'pure' changes in inequality within groups: $\sum_{g=1}^{g=G} \left(\bar{v}_g \cdot \Delta I_{0g} \right)$
- changes due to variations in the size of the different groups: $\sum_{g=1}^{g=G} \bar{I}_{0g} \cdot \Delta v_g + \Delta_g \cdot \left[\bar{\lambda}_g - \overline{\ln(\lambda_g)} \right] \cdot \Delta v_g$

- changes due to variations in the relative income of the different groups: $\sum_{g=1}^{g=G} \left(\bar{v}_g \cdot \bar{\lambda}_g - \bar{v}_g \right) \cdot \Delta \ln(\mu_g)$

With x sub-groups in a population (e.g. different age groups), the variance of the distribution of a variable across the population (e.g. individual income) can be decomposed thus:

$$\text{var} = \sigma^2 = \sum_{g=1}^{g=x} c(g) \sigma_w^2(g) + \sum_{g=1}^{g=x} c(g) \left(\bar{\mu} - \mu_g \right)^2 \qquad (4.14)$$

where $\sigma_w^2(g)$ is the variance of the variable of interest (income, wealth, etc.) within group g—for example, the variance of income among people aged 65 or over.

Lam (1984) (see also Goldstein and Lee 2014) estimated the first derivative of the variance (of the logarithm) of income with respect to the size of a population:

$$\frac{\partial \log \sigma^2}{\partial n} = \alpha \cdot (\bar{g} - \bar{g}_b) + (1 - \alpha) \cdot (\bar{g} - \bar{g}_w) \qquad (4.15)$$

with $\alpha = \frac{\sum_{g=1}^{g=x} c(g)(\bar{\mu}-\mu_g)^2}{\sigma^2}$, that is, the between-group share in total variance, and the subscripts b and w stand for between- and within-group, respectively.

Kang (2009) investigated the population ageing effects on income and consumption inequality in South Korea from repeated cross-sectional data during the period 1982–2004 using the variance of the logarithms. Once controlling for cohort and period effects, the findings for consumption suggest that age effects on within-household inequality follow a U-shaped pattern with minimum at around chronological age 35: inequality decreases within younger households and increases among older households. In turn, the ages of the household have linear and positive effects on income inequality: income inequality is greater, the older the household head. Although, in an extension using data up to 2011, Kang and Rudolf (2016) reported that income inequality slightly rose and consumption inequality declined over the period. The authors predicted that population ageing would increase overall income and consumption inequality unless poorer households can get better access to financial markets.

Applying the variance of the logarithm of real per capita household income in Thailand over the years 1992–2011, Paweenawat and McNown (2014) failed to find any significant ageing effects.

Mean Logarithmic Deviation

Another measure of inequality frequently used based on the logarithm of income is the Mean Logarithmic Deviation (MLD):

$$\text{MLD} = \frac{1}{n} \cdot \sum_{i=1}^{i=n} \ln\left(\frac{\mu}{y_i}\right) = \ln(\mu) - \frac{1}{n} \cdot \sum_{i=1}^{i=n} [\ln(y_i)] \qquad (4.16)$$

with μ standing for the mean of the logarithms of income.

The MLD can be decomposed into between- and within-group inequality thus:

$$\text{MLD}_B = \ln(\mu) - \sum_{i=1}^{i=n} \left[\left(\frac{n_i}{n}\right) \cdot \ln(y_i)\right]$$

$$\text{MLD}_W = \sum_{i=1}^{i=n} \left(\frac{n_i}{n}\right) \cdot \left[\ln(y_i) - \frac{1}{n_i} \sum_{j=1}^{j=k} \ln(y_{i,j})\right] \qquad (4.17)$$

where $j = 1, 2, \ldots, k$ denotes the k sub-groups in the population.

Alimi et al. (2005) applied the MLD to data of household incomes in New Zealand between 1986 and 2013 and reported that the increase in inequality during the period was mostly due to change in age-specific distributions but that population ageing helped slow down inequality growth as the share of younger people declined.

The Lorenz Curve and the Gini Coefficient

In 1905, the US economist Max Otto Lorenz introduced a graphical device to represent the degree of inequality of income or wealth in a population (Lorenz (1905); see also Gastwirth (1972) and Yitzhaki and Schechtman (2013)). It plots the cumulated percentages of a population, from poorest to richest, on the horizontal axis and, in the vertical axis, the amount of income earned by each successive percentage of the population. An equal distribution would render a straight line (i.e. the bottom $x\%$ earn exactly $x\%$ of total income). Inequality translates into a convex curve below the straight line of perfect equality (also known as the equidistribution line).

If the income of i members of a population (with $i = 1, 2, \ldots, n$) is indexed in increasing order,[13] and the size of the population is normalised to 1 (or 100 per cent), the Lorenz curve is the representation of the following function:

$$L(p) = \frac{\int_{i=0}^{i=P} Q(q)dq}{\int_{i=0}^{i=1} Q(q)dq} \qquad (4.18)$$

In Eq. (4.18), $\int_0^P Q(q)dq$ represents the sum of the income (i.e. the cumulative income) of the poorest $p\%$ of the population. The denominator ($\int_0^1 Q(q)dq$) stands for the total income (the sum of all the incomes of the i individuals in the population). Because the size of the population is normalised to 1, the expression in the denominator is equal to the average income. Consequently, the Lorenz curve can be also expressed:

$$L(p) = \left(\frac{1}{\mu}\right) \cdot \int_{i=0}^{i=P} Q(q)dq \qquad (4.19)$$

Figure 4.1 presents the Lorenz curves for two hypothetical distributions:

In 1914, the Italian statistician Corrado Gini proposed a concentration index (now known as the Gini coefficient) to measure the degree of inequality in a distribution, based on the Lorenz curve (Gini 1913).[14] The Gini coefficient is 'probably the most common statistical index employed in social sciences for measuring concentration in the distribution of a positive random variable' (Giorgi and Gigliarano 2017, p. 1132).

The coefficient measures the area between the equidistribution line and the Lorenz curve. It ranges between 0 and 1, with 0 indicating equidistribution and 1, complete inequality (one person or household would earn 100 per cent of all income or hold 100 per cent of all wealth). It can be expressed thus:

$$\text{Gini} = 1 - 2 \cdot \int_{i=0}^{i=1} L(p)dp \qquad (4.20)$$

There are several other ways to express the Gini coefficient (see Giorgi and Gigliarano (2017, Table 1) and Yitzhaki (1998)[15]) Another frequent

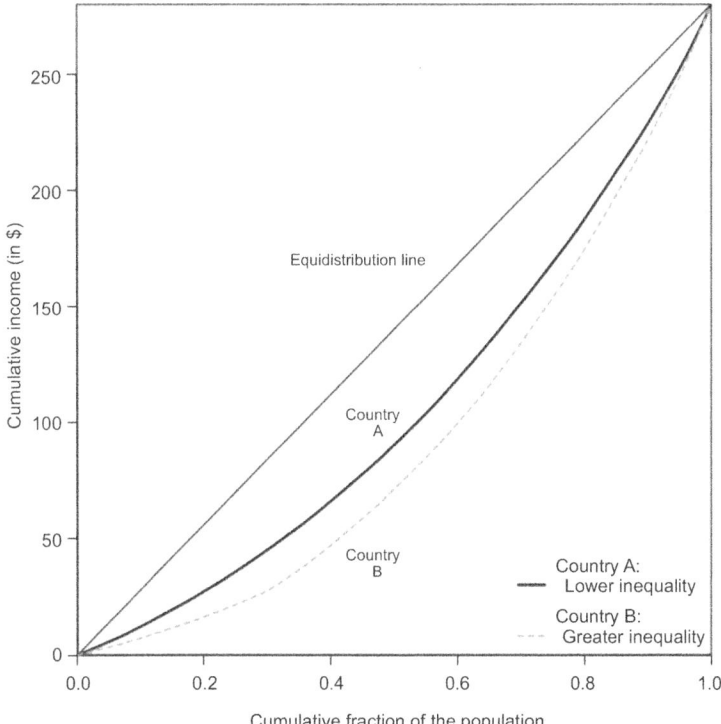

Fig. 4.1 Lorenz curve. *Source: Figure is illustrative, prepared with mock data*

specification, using income and population as discrete variables, is:

$$\text{Gini} = \frac{1}{2 \cdot n^2 \cdot \bar{x}} \cdot \sum_{i=1}^{i=n} \sum_{j=1}^{j=n} |x_i - x_j| \tag{4.21}$$

where $|x_i - x_j|$ indicates the absolute value of the difference between all pairs of incomes in the population.

Hong and Kim (2012) applied the Gini coefficient to data from repeated cross-sections of household income in South Korea between 1986 and 2006. The authors explored the relationship between household income inequality and the age of the head of the household. They found that income inequality among households increased with chronological age regardless of the birth cohort as a result of widening inequality in employment earnings and occupational retirement pension income.

Zhong (2011) used a regression-based decomposition of the Gini coefficient to investigate the effects of population ageing on income inequality in rural China in 1997, 2000, and 2006. The author found that a larger proportion of retired people in the population significantly increased income inequality, and suggested that the main underlying mechanism could be that a fall in the proportion of people of working age in the population. Coinciding with a process of rapid industrialisation, this would have increased the elasticity of income with respect to the labour ratio in households. In other words, the labour supply became more elastic, which dampened the effect of earnings from paid employment on inequality over the period.

Paglin-Gini and Age-Gini

The Gini coefficient is usually applied to cross-sectional data, that is, to data for one or more particular periods but not following the same group of individuals over time. For example, the Gini coefficient for the United States in 1979 was 34.6 and in 2016 it was 41.5.[16] This is based on two 'snapshots' at different points in time, but the individuals whose income was the variable of interest were not necessarily the same in both periods. One problem with this comparison is that, according to the life-cycle hypothesis, the age-income profile across a population in any one point in time follows an inverted U- or hump-shaped pattern. Therefore, as Paglin (1975) pointed out, the inequality across a population in one particular year could be the consequence of different proportions of people being at different stages in their life cycles. Hence, he argued that notion of equality of income underlying the Gini coefficient would make little economic sense: there would be equidistribution only if the age-income profiles were completely flat. Unless the population structure and the relationship between income and chronological age are taken on board, any estimate of inequality based on period income, such as the Gini coefficient, would lead to an erroneous conclusion and distorted distributive policies. For example, Chu and Jiang (1997) examined data from Taiwan between 1978 and 1993 to explore the effects of changes in the age structure on inequality in family income. The authors showed that the changes in the demographic structure of the population reduced the inequality in family income, and that this impact increased over the period. Therefore, they concluded that a Gini coefficient that did not incorporate these effects would hide the true magnitude of the reduction in income inequality.

The notion of equality of lifetime income across families, Paglin argued, means that each family would have the same income as any other family at each same stage in the life cycle: the equality of age-income profiles, not the equality of incomes between families of different chronological ages (see also Pyatt 1976). Therefore, this author proposed that instead of the distribution of period household income (i.e. the income distribution in one given period, say, one particular year), an estimate of lifetime family income should be used. The adjustment consists in obtaining the 'Age-Gini' coefficient, which represents the value of the Gini coefficient if each family of a given stage in the life cycle or chronological age had a lifetime income equal to the equitable income for their stage in the life cycle or chronological age. The Paglin-Gini coefficient is the difference between the Gini and the Age-Gini coefficients. The Paglin-Gini (PG) coefficient is obtained thus:

$$\text{PG} = \frac{\sum_{i=1}^{i=n} \sum_{j=1}^{j=n} |(x_i - x_j) - (\mu_i - \mu_j)|}{2 \cdot \mu \cdot n^2} \tag{4.22}$$

where $\mu_{i,j}$ represent the average incomes of age groups i, j, respectively.

Furthermore, Paglin proposed that inequality should be measured using families as observational units, instead of households or individuals. Using the income distribution for the United States of America from 1972, he estimated that this change alone would bring down the inequality measurement from a Gini coefficient of 0.359–0.239.

Paglin estimated the three coefficients for the United States of America for the period 1947–1972 using family-based data; Fig. 4.2 presents the results.

On the basis of the Gini coefficient, it seems that there was a gradual reduction in inequality over the period. In turn, the Paglin-Gini measure showed a drastic improvement towards greater equality in the family income distribution. The author explained that most of the increase in the Age-Gini coefficient behind the divergence between the Gini and the Paglin-Gini coefficients was due to two trends:

- The increasing accumulation of human capital by younger cohorts from the expansion in higher education. Compared to older cohorts at a same age or stage in the life cycle, cohorts of more recent vintage earned higher incomes on average as a result.
- The growing proportion of younger and older people in the population—age groups with higher within inequality—over the period. As low-income individuals are more likely to be older, population ageing would worsen inequality estimates.

Fig. 4.2 Gini, Paglin-Gini, and Age-Gini coefficients of Family Income Distributions United States of America, 1947–1972. *Source: Paglin (1975, Table 3)*

Beenstock (2004) applied these adjustments to data for Israel between 1983 and 1995 and found that changes in the age structure made a very small difference. However, applied to lifetime income, the Paglin-Gini coefficient conveyed a story of increasing equality over the period, as opposed to the findings based on the Gini coefficient.

Morley (1981) looked into personal income distribution in Brazil in 1960 and 1970, a period of increasing inequality. This author extended Paglin's approach by tracking the individuals who were classified as poor in the base period because:

> Anyone who compares incomes over time for the poor in Brazil is unwittingly comparing the incomes of two different groups with a high proportion of new entrants. He most assuredly is not tracking the base-period poor.
>
> (Morley 1981, pp. 288–289)

The findings suggested that almost all the rise in inequality as measured by the Gini coefficient was the result of increasing intracohort inequality. In contrast, variations in age-income profiles made no significant impact. Interestingly, Morley showed that a decomposition of the Gini coefficient that did not account for age effects would have resulted in opposite conclusions.

It is worth noting that lifetime distribution of can be measured either as inflow or an outflow of resources. As an inflow, the focus is on lifetime income from paid employment and other earnings plus any inheritance; a focus outflow centres on lifetime consumption plus any bequests (Lindert and Williamson 2016).

Wertz's Gini

Another proposal to adjust the Gini coefficient for age effects was suggested by Wertz (1979). This author indicated that the Paglin-Gini coefficient did not properly account for the life-cycle effects in cross-sectional income distributions. Following Almås et al. (2011), Wertz's Gini (WG) coefficient is defined by:

$$\text{WG} = \frac{\sum_{i=1}^{i=n} \sum_{j=1}^{j=n} \left| (x_i - \mu_i) - (x_j - \mu_j) \right|}{2 \cdot \mu \cdot n^2} \qquad (4.23)$$

with $\mu_{i,j}$ again denoting the average income of age groups i, j, respectively.

Compared to the Gini coefficient expressed in terms of absolute mean differences (Eq. (4.21)), the WG is based on the difference between each individual, household, or family income and the average income in their age group, whilst the Gini coefficient is based on the average income for the whole population. Compared to the PG coefficient (Eq. (4.22)), the measure of within-group distance in the WG coefficient is based on the difference between an individual's income and the mean for her age group, whilst the Paglin's indicator defines the distance as the difference between the incomes of any two individuals net of the absolute value of the difference in the average incomes for the age groups to which these individuals belong.

Almås and Mogstad (2012) considered that the PG and WG indicators failed to control for other age-related factors that could influence intragroup inequality. Therefore, they developed the Age-adjusted Gini (AG) coefficient.

Age-Adjusted Gini

Almås and Mogstad pointed out that while both the PG and the WG coefficients adjust the Gini coefficient for life-cycle effects, they do not control for other age-related factors that affect the level of income or wealth inequality within age groups, such as education, number of children, occupation, or marital status. Therefore, these authors proposed the AG coefficient, which is exactly like the WG indicator, except that the within-group mean income or wealth now captures other age-related drivers of income or wealth inequality.

The estimation of the AG coefficient starts from a linear regression of income or wealth on the age group and other age-related variables that could influence income or wealth inequality. Let's assume the variable of interest is wealth (w) inequality. The regression equation can be expressed thus:

$$\log w_i = \delta_i + X'_i \cdot B \qquad (4.24)$$

where δ represents the difference that being in age group i makes in terms of wealth inequality compared to a reference age group; X is a vector of individual characteristics apart from chronological age (education, number of children, etc.) and B represents the regression coefficients for each of the variables included in X.

The coefficient δ is the age effect net of any other age-related characteristics that could affect wealth inequality. This coefficient is the basis for the adjustment of the within-group average wealth estimates proposed by Almås and Mogstad. The formula for the age-adjusted mean is:

$$\tilde{\mu}_i = \frac{\mu \cdot n \cdot e^{\delta_i}}{\sum_{j=1}^{j=n} e^{\delta_j}} \qquad (4.25)$$

Once the age-adjusted means are obtained, the AG can be estimated by:

$$\text{AG} = \frac{\sum_{i=1}^{i=n} \sum_{j=1}^{j=n} \left| (x_i - \tilde{\mu}_i) - (x_j - \tilde{\mu}_j) \right|}{2 \cdot \mu \cdot n^2} \qquad (4.26)$$

Compared to the formula for the Gini coefficient (Eq. (4.21)) the means values in the AG indicator depend on the age of the individuals.

Almås and Mogstad (2012) compared the Gini, PG, WG, and AG coefficients of wealth inequality for seven developed countries in 2000. In all cases, the PG was much smaller than the other two indicators. WG estimates were remarkably similar to Gini, a consequence of being theoretically more consistent with the Gini coefficient than the PG. The AG coefficient resulted in an identical ranking of countries to the results from the Gini coefficient, and in similar levels of inequality. The authors concluded that age adjustments were not as important as previously reported for wealth inequality.

Almås et al. (2011) applied the AG to earnings data for Norway between 1967 and 2000, with intragroup means adjusted for education level, birth order, family size, immigration, parental education, and parental age at birth. The authors found significant effects of the age structure of the population on income inequality over the period. Using the Gini coefficient, the conclusion is that inequality decreased between 1967 and 1980, increased between 1980 and 1993, and decreased again during the period 1993–2000. Instead, using the AG coefficient, the conclusion is that inequality decreased less between 1967 and 1980 and increased also less between 1980 and 1993, compared to the Gini results, but dropped much more than the Gini coefficient suggested during the period 1993–2000.

Gini Re-centred Influence Function

Gini re-centred influence function (Gini-RIF) regressions are a statistical procedure to estimate the effect of a number of covariates on an inequality indicator such as the Gini coefficient (the RIF methodology has also been applied to other inequality indicators). RIF regressions proceed in two stages (Essama-Nssah and Lambert 2012; Firpo et al. 2009; Fortin et al. 2011). First, the influence of each individual on the Gini of the variable of interest (say, income) is modelled as a function of the structure and the distribution of the variable. For example, as a function of the wage structure and the distribution of wages, or the structure of pension wealth and its distribution, if the interest lies on inequality in pension wealth, and so on. An inequality measure should register a reduction in inequality if there is a greater concentration of units in the middle of the distribution compared to a smaller concentration. In turn, if the left tail gets longer, it means that there are proportionally more poorer units and this increase in inequality should be reflected in the indicator. Similarly, if both tails are extended, there are more poorer and more richer individuals (the so-called middle-class squeeze). The first stage in the RIF approach estimates this individual influence. The second stage consists of a regression of individual influences on the Gini index on covariates such as age, gender, unionisation, education, and so on.

RIF regressions, then, are useful to assess the relative importance of a number of covariates on inequality.

- Olivera (2018) used this method to study the distribution of pension wealth in twenty-six European countries in 2006 and 2014. Pension wealth inequality grew over the period. RIF regressions showed that attainment of tertiary education had a diminishing influence on pension wealth inequality.
- Heger (2018) constructed indicators of health inequality for the United States of America and Canada using data from 2002 to 2003. After applying RIF regressions, the author concluded that there is a higher proportion of individuals in the United States of America reporting low health status, which is not driven by ethnic background or health insurance coverage, but principally by a large effect of socio-economic variables at the bottom of the health distribution that the much higher spending per person in the United States of America cannot counterbalance.

Other decomposition methods have been developed, including the variance accounting approach (Juhn et al. 1993), the kernel re-weighting method (DiNardo et al. 1996), the methods based on counterfactuals (Machado and Mata 2005), the partial distributional policy approach (Rothe 2012), and the distributional regression technique (Chernozhukov et al. 2013).

Theil Index

Say we have information on income from older and younger people, or from older men and women in a population, and we want to know whether older and younger people, or older men and women, differ in some respect—for example, average income or inequality or trends over time, and so on. A measure of the information is the distance between the available information and the information needed to discern or distinguish between the groups. In information theory, 'entropy' refers to the amount of information missing before a message is received, and the reduction in entropy attributable to a message represents how much more it is known and how much less information is missing.

Benish (1999) provides the following example. A doctor tries to figure out what could be going wrong with her patient. From the information already gathered, the doctor has discarded several conditions, but still there are a number of equally possible conditions her patient may have. More information is needed for an accurate diagnosis. Let's denote with p_i the probability that diagnosis i is correct (and let's assume that the diagnoses are mutually exclusive: the patient either has one condition or another). If the information already in the hands of the doctor points to a specific condition and all she needs to confirm it is a white blood cell count, when the additional information arrives confirming the condition, it will not take her by surprise: there was a high probability that her patient had that condition in particular. However, if it turns out that the additional information confirms the patient has something else, a condition the doctor had not thought of because the available information she had prior to the test had not suggested its occurrence, then she will be taken by surprise: low probability of occurrence equals high surprise.

An important indicator is the measure of how much additional information is necessary to be able to discern or identify the patient as having one condition or another. More generally, we can think of the additional information needed to distinguish to which sub-group of a population each observational unit belongs. This is known as the entropy. For a distribution of a discrete variable, Kullback and Leibler (1951) demonstrated that the entropy can be represented by the following formula[17]:

$$I = \sum_{i=1}^{i=k} p_i \log\left(\frac{1}{p_i}\right) \quad (4.27)$$

where k represents the possible outcomes, and p is the probability of occurrence of i. The element $\frac{1}{p_i}$ is known as the 'surprisal' (M 1961).

Using the formula of entropy, Theil (1967) defined two indices of inequality. The Theil's entropy indices T and L. For a population of $n_{i,j}$ individuals in i groups, the Theil-T inequality index of the distribution of a variable Y, say income, divided into j classes is:

$$T_T = \sum_{i=1}^{i=n} \sum_{j=1}^{j=k} \left[\left(\frac{y_{i,j}}{Y}\right) \cdot \ln\left(\frac{y_{i,j}/Y}{n_{i,j}/n}\right)\right]$$

$$T_T = \sum_{i=1}^{i=n} \left(\frac{Y_i}{Y}\right) \cdot \sum_{j=1}^{j=k} \left[\frac{y_{i,j}}{Y_i} \cdot \ln\left(\frac{y_{i,j}/Y_i}{n_{i,j}/n_i}\right)\right] + \sum_{i=1}^{i=n} \left[\left(\frac{Y_i}{Y}\right) \cdot \ln\left(\frac{Y_i/Y}{n_i/n}\right)\right]$$

(4.28)

This index can be decomposed into within-group inequality (T_w)—the first term on the right-hand side of Eq. (4.28) and between-group inequality (T_b), the second term:

$$T_T^w = \sum_{i=1}^{i=n} \left(\frac{Y_i}{Y}\right) \cdot \sum_{j=1}^{j=k} \left[\frac{y_{i,j}}{Y_i} \cdot \ln\left(\frac{y_{i,j}/Y_i}{n_{i,j}/n_i}\right)\right]$$

$$T_T^b = \sum_{i=1}^{i=n} \left[\left(\frac{Y_i}{Y}\right) \cdot \ln\left(\frac{Y_i/Y}{n_i/n}\right)\right] \qquad (4.29)$$

The within-group contribution to Theil-T inequality is equal to T_T^w/T_T; the between-group contribution corresponds to T_T^b/T_T.

The Theil-L index inverts the shares of the population and income compared to the Theil-T index:

$$T_L = \sum_{i=1}^{i=n}\sum_{j=1}^{j=k}\left[\left(\frac{n_{i,j}}{n}\right) \cdot \ln\left(\frac{n_{i,j}/n}{y_{i,j}/Y}\right)\right]$$

$$T_L = \sum_{i=1}^{i=n} \ln\left(\frac{Y}{n}\right) - \sum_{i=1}^{i=n}\sum_{j=1}^{j=k}\left[\left(\frac{n_{i,j}}{n}\right) \cdot \ln y_j\right] \qquad (4.30)$$

This index can also be decomposed into within- and between-group inequality:

$$T_L^w = \sum_{i=1}^{i=n} \left(\frac{n_i}{n}\right) \cdot \sum_{j=1}^{j=k}\left[\left(\frac{n_{i,j}}{n_i}\right) \cdot \ln\left(\frac{n_{i,j}/n_i}{y_{i,j}/Y_i}\right)\right] \qquad (4.31)$$

$$T_L^b = \sum_{i=1}^{i=n}\left[\frac{n_i}{n} \cdot \ln\left(\frac{n_i/n}{Y_i/Y}\right)\right] \qquad (4.32)$$

Again, we can obtain the expressions for the within- and between-group contributions to total inequality. The within-group contribution to Theil-L inequality is equal to T_L^w/T_L; the between-group contribution corresponds to T_L^b/T_L.

Using the Theil-L index on data for Sri Lanka from 1963 to 1987, Karunaratne (2000) estimated that the projected population ageing would not affect the within-group contribution to total inequality, but would increase the contribution of between-group inequality, thus deepening total inequality in that country.

Concentration Index

The Concentration Index (CI) (Kakwani 1977a,b) is a measure of inequality mostly used in health economics (Kakwani et al. 1997). The index is constructed first by plotting a curve with individuals ranked according to socio-economic status starting from the most disadvantaged on the horizontal axis and the cumulative proportion of health on the vertical axis (it is usually presented in terms of ill health). The curve below the equidistribution line is known as the health 'concentration curve'. The CI is defined as twice the area between the concentration curve and the equidistribution diagonal. For a population of n individuals, the CI is calculated with this formula:

$$\text{CI} = \frac{2}{n \cdot \bar{\mu}} \cdot \sum_{i=1}^{i=n} \mu_i \cdot R_i - \left(1 + \frac{1}{n}\right) \qquad (4.33)$$

where μ_i is the ill health of individual i and $\bar{\mu}$ denotes the average level of ill health in the population; R_i corresponds to the fractional rank of individual i in the population according to her relative socio-economic status. CI ranges between minus 1 and plus 1. The closer CI is to minus 1, the more disproportionate is the concentration of ill health among the poorest or most disadvantaged individuals. Let's change the variable of interest to gain a better understanding of how to interpret the CI score. Imagine that instead of health status, the focus is on utilisation of healthcare services. A CI > 0 would indicate that there is a greater utilisation of healthcare services among better-off individuals; a 'pro-poor' policy would seek to turn the concentration closer to minus 1. If the index is equal to zero, income or socio-economic status are not related to healthcare utilisation inequality.

The average and the distribution of health status in a population depends on many factors other than socio-economic status. One of these confounding influences comes from the demographic structure: other things equal, an older population would have not only a worse level of health on average, but also a higher concentration of people of ill health. One method to control for the confounding effects of ageing and other factors is to run a regression of these variables on the health status variable. The predicted individual value of the ill health coefficient from the regression, say μ_i^*, represents the estimate of ill health once the effects of age, gender, and other covariates on the health of each individual have been controlled for. With this estimate, Kakwani et al. (1997) developed the following adjustment to CI to account for the effect of demographic factors in health inequality:

$$I^* = \mathrm{CI} - \mathrm{CI}^* = \frac{2}{n} \cdot \sum_{i=1}^{i=n} \left[\left(\frac{\mu_i}{\bar{\mu}} - \frac{\mu_i^*}{\bar{\mu}^*} \right) \cdot R_i \right] \qquad (4.34)$$

where $\bar{\mu}^*$ is the average predicted or adjusted health estimate.

As mentioned above, this index is widely used in health economics. To illustrate:

- Using this adjusted index, Ourti (2003) found that inequality in ill health among people aged 65 or over in Belgium between 1994 and 1998 presented a significant socio-economic gradient. The gradient is mainly associated with differential mortality by income group, which favours individuals on higher incomes.
- With another econometric approach to adjust for the effect of confounding covariates, Siegel et al. (2013) used the CI to study income-related inequalities in diabetes, hypertension, and obesity in Germany in 2002 and 2006. The authors reported the presence of socio-economic gradients in the three conditions, and that these gradients vary across age groups among women but not men: inequalities in hypertension and diabetes are most detrimental among middle-aged women, whereas obesity affects lower-income women of all ages.
- Listl (2011) used this index (and the slope index described below) to study income-related inequalities in the utilisation of dental services by people aged 50 or over in fourteen European countries. This author found a disproportionate concentration of utilisation among richer individuals in all countries (i.e. CI > 0).

Slope Index

Imagine a variable of interest h measured over a population of i individuals—say, flu vaccination or access to oral health services. We rank the individuals according to a socio-economic variable (e.g. income) and obtain the fractional rank of each individual: R_i. Then we apply a linear regression of h onto R:

$$h_i = \alpha + \beta \cdot R_i + \epsilon_i \qquad (4.35)$$

where α is the constant, β the regression coefficient of interest, and ϵ an error term.

The regression coefficient β is the Slope Index of Inequality (SII) (Pamuk 1985) (see also Regidor 2004). The SII measures the average change in the variable of interest as one moves from the lowest ranked individual to the highest. In this sense, it measures the absolute advantage of earning higher incomes. The relative advantage can be measured by dividing the coefficient β by the variable of interest.

4.4 Population Ageing and Distributional Issues

The interaction between demographic changes and changes in income distribution responds to three different drivers (Lindert and Williamson 1976):

- Changes in the relative income of the different age groups (e.g. the effects on inequality if pensioner income grows over time relative to other groups)
- Changes in the income inequality within the different groups (e.g. if the dispersion in pensioner income grows more than the inequality in income from wages or self-employment)
- Changes due to variations in the size of the different groups in relation to total population (the 'pure' effect of the process of population ageing)

The 'pure' effect of population ageing, that is, an increase in the proportion of older people in a population, has been found to accelerate the growth in inequality:

- Deaton and Paxson (1994) reported increasing inequality in consumption and income with *individual ageing* in Taiwan, the United States of America, and the United Kingdom for the periods 1976–1990, 1980–1990, and 1969–1990, respectively.
- Tsakloglou (1993) decomposed income inequality in Greece between 1974 and 1982 according to different groupings. Regarding chronological age of heads of households, this author found that the older the household head, the greater the inequality within age groups.
- Similarly, Cameron (2000) found that the process of *population ageing* in Java, Indonesia, between 1984 and 1990 contributed in a small but statistically significant effect on the widening distribution of national income per person during the period

- Looking into data for Japan between 1979 and 1989, Ohtake and Saito (1998) found that as the first cohorts of baby boomers started reaching 40 years of age, consumption inequality began to rise (half of the increase in consumption inequality during the period could be attributed to population ageing)
- Zhang and Xiang (2014) reported that 60 per cent of aggregate consumption inequality in China during the period 2003–2009 was caused by the cohort effects of population ageing. Dong et al. (2018) confirmed this finding using a Chinese panel data for the period 1996–2011.
- Estudillo (1997) reported a positive relationship between the age of the household head and income inequality within the age groups in the Philippines between 1961 and 1991.

As it is almost inevitably the case in economics, other studies failed to confirm these findings. For example:

- Jäntti (1993) researched household income inequality in Canada, the Netherlands, Sweden, the United Kingdom, and the United States of America between 1979/1983 and 1986/1987, depending on the country. This author concluded that none of the demographic factors investigated (namely, changes in the share of single-parent or single-person households, earnings, or sizes of younger cohorts or age structures) played 'any major role in the increase in inequality' [p. 2] in those countries during that period.
- Biewen and Juhasz (2012) concluded that population ageing only played a minor role in rising income inequality in Germany between 1999/2000 and 2005/2006, compared to the increasing wage inequality and the changes in the tax system. Nevertheless, the bulk of the evidence suggests that the pure effect of population ageing increases income inequality.

The theoretical explanation for the positive association between ageing and inequality is found in the life-cycle theory. The smoothing of consumption over the life cycle implies that the optimal inter-temporal consumption level in any one period is equal to that of the previous period plus or minus an error term that would account for minor temporal deviations around the lifetime consumption average. Deaton and Paxson (1994) shows that if we take this assumption, combined with the additional assumption that those temporal deviations are independent from the level of consumption in the previous period, then the variance of the consumption levels of an individual (or household) increases over time, that is, with chronological age. In symbols, the first assumption can be expressed thus:

$$c_{i,t} = c_{i,t-1} + u_{i,t} \tag{4.36}$$

where $c_{i,t}$ represents the consumption level of individual or household i in period t, and u stands for the error term. In statistics, the covariance between two variables is equal to their joint variability. Consequently, the second assumption can be expressed as the requisite that the covariance between the consumption in a previous period and the deviation in consumption around its lifetime average level is equal to zero:

$$cov(c_{i,t-1}, u_{i,t}) = 0 \tag{4.37}$$

The variance of a variable is a measure of its variability around its mean. The variance of a sum of two independent variables is equal to the sum of the variances of each variable. Therefore, from Eq. (4.37), we obtain:

$$\text{var}\left[c_{i,t}\right] = \text{var}\left[c_{i,t-1}\right] + var\left[u_{i,t}\right] \tag{4.38}$$

Equation (4.38) shows that the variance of consumption for each individual increases over time. Aggregating individuals by cohort, we obtain that the dispersion in consumption levels of each cohort increases over time, that is, as each birth cohort passes through their life cycle. In particular, older cohorts exhibit more variance in their consumption levels than younger cohorts.

I mentioned above a distinction between inequality within and between groups or sub-groups in a population. When considering different sub-groups, it is important to disentangle the contribution to total inequality of the inequality within each sub-group from the effects of the inequality between them. We can think of the inequality within sub-groups as the inequality that would exist in a population if all the sub-groups had no differences in terms of the variable of interest (say, if there were no differences across sub-groups in terms of income). Likewise, the inequality between sub-groups can be defined as the inequality that would exist in a population if there were no differences within sub-groups, if every individual in each sub-group had the same income as the average income for her sub-group (Anand 1983). For example, in the paper already cited, Estudillo (1997) found that within-group inequality was the main driver of total inequality in the Philippines between 1961 and 1991, with the between-group inequality having remained stable and only marginally contributed to overall inequality.

In this regard, the relevant question is whether the ageing of the population influences the variance of income or wealth in the population. In other words, whether population ageing increases or reduces income or wealth inequality. Younger and older people tend to have relatively lower incomes than middle-aged individuals; savings tend to peak at around retirement age. Population ageing may affect the distributions of income or wealth differently depending the stage of the demographic transition the country in question is going through: a greater concentration of individuals at the peak of their earning and accumulation stages would have a different effect to an increase in the population share of the oldest old. For example, Motonishi (2006) failed to find any significant age effects on household income inequality in Thailand between 1975 and 1998. The explanation this author offered was that the process of population ageing was only starting in Thailand back then, which manifested in a reduction in the ratio of very young people most of whom were not in the labour market yet, compared to individuals in later life.

It should be apparent that there is more to income or wealth inequality in any country or region than the effects from demographic drivers. Population ageing is but one of the many economic processes that affect income or wealth inequality. Jenkins (1995) mentioned the following additional economic changes as potential influences:

- Household composition
- Employment structure
- Industrial structure
- Unemployment
- Business cycles
- Income tax and benefit
- Wage dispersion
- Income from capital and assets

In a scoping review of inequalities in later life, Walsh et al. (2017) focused on the following:

- Subjective well-being
- Physical and mental health
- Life expectancy and healthy life expectancy
- Financial security
- Social connections
- Living environment

Whether increasing within-cohort inequality over time implies that the population as a whole also exhibits increasing variability depends on its demographic structure: an ageing population has increasing proportions of older people over time, which means that it will exhibit greater variance in its various domains; in other words, inequality will increase. This will be the case even if we assume that there are no links between successive cohorts (e.g. bequests). Within cohorts, ageing increases inequality over time. In a population as a whole, ageing also increases inequality but less so, because new cohorts partially replace older cohorts, and the new cohorts are of smaller size if the population is undergoing an ageing process.

In econometric applications, the decomposition proposed by Ohtake and Saito (1998) is often used:

$$\text{var}\left[\ln\left(C_{j+k}\right)\right] = \sum_{m=0}^{m=j} (a_m \cdot \text{cohort}_m) + \sum_{n=0}^{n=k} (\beta_n \cdot \text{age}_n) + \varepsilon \quad (4.39)$$

where j corresponds to the year of birth and k the age (so that in each period t, for each individual $t = j + k$). In turn, a_m is the effect of cohort m and b_n the age effect. Finally, *cohort$_m$* stands for cohort dummy variables, and *age$_n$* represents age dummy variables.

A different approach that links population ageing and inequality, used in income inequality studies, focuses on the factorial distribution of income. According to this view, population ageing widens the disparity across the income distribution because it is negatively related to labour income and positively related to capital income. With fewer people of advancing age in paid employment or self-employment, the ageing of a population means that the share of labour income on total income decreases and that of capital income increases. Labour income is generally more equally distributed than capital income, which leads to a deepening of inequality with the ageing of the population (Wang et al. 2018).

4.5 Intergenerational Transmission of Inequality

Yellen (2016) listed four sources of transmission of income and wealth across generations:

- Parental and subsidised investment in children's human capital
- Affordable higher education

- Private business ownership
- Inheritance (and inter vivos transfers)

By means of these sources, not only does each generation seek to pass onto the next generation their social standing but also to provide them with a platform to climb the social ladder as well, to move up the social scale.

Downward (i.e. from parents to children) inter vivos transfers and bequests can reproduce (i.e. transmit) inequality between generations. There is ample evidence of transmission of inequality at early ages, with different investments in the quantity and quality of children's human and social capital by social class or income level. Less researched is the link between older parents and their adult children, and even less still that between grandparents and their grandchildren.

Albertini and Radl (2012) investigated how parental class influences the amount and frequency of inter vivos financial support by older parents to their adult children. Using data from eleven European countries for 2004/2005, these authors found that downward financial transfers are substantial and primordially driven by the parents' fear that their adult children may fall economically behind. The financial support is a form of investment in the adult children's social status (defined as 'all social and economic resources defining an individual's position in the social hierarchy' [p. 112]). However, instead of the intergenerational transmission of resources and opportunities to move upwards socially and financially, Albertini and Radl described a situation of downward financial transfers to avert downward social mobility. These transfers seek to assure the reproduction of social status in the next generation. This motivation can explain why, as a proportion of their resources, poorer older people support their adult children less than their more affluent counterparts. The authors suggested that a 'warm glow' is not a by-product of these transfers but their intended result.

Killewald et al. (2017) researched changes in the wealth-income correlations in the United States of America by age group over a ten-year period. They found that the association between wealth holdings in one year and levels of income ten years later was positive across the population but stronger among older groups: from a correlation coefficient lower than 0.35 among those aged 25–34 to almost 0.60 among households with a head of household aged between 45 and 54 and slightly lower for the 55–64 age group.

Wealth mobility among the tip of the distributional tail is important due to the possible cultural and political influence of the elites, even in contemporary democratic systems. Consequently, the top 1 per cent and 0.1 per cent wealthiest individuals have started to be studied with keenness. Renewed interest in the topic (classical economists and sociologists such as Marx, Mosca, and Pareto, among others, wrote at length about the role of the elites) was triggered by evidence that wealth has been concentrated in fewer hands across the world since the 1980s. It is generally agreed that, in developed countries, the richest 1 per cent hold about 20 per cent of wealth. The elites have been subject to greater scrutiny, thanks in part to access to administrative data in many countries, and also because wealth inequality has grown the most and the fastest at the very top: the rich have been getting richer during this time.

One academic bone of contention is whether wealth mobility is present across the whole distribution and within the richest echelons. In other words, how high the overall probability of moving up and down the social ladder is and whether there is 'circulation' within the elites (Pareto 1917).

Across the whole distribution, Chetty et al. (2014a) found, for the United States of America, that the probability of moving up in the income distribution is the same for a child born in the 1970s as that of a child born between 1971 and 1986. Alas, given that inequality increased during the period, the rungs have been set wider apart so that the 'birth lottery' is more relevant today for social mobility than in the past.

In turn, looking at the elites, much of the evidence suggests that those at the top have both managed themselves and helped their children to remain at the top: the existence of significant intergenerational transmission of wealth (which is getting increasingly stronger over time), mainly via inter vivos transfers but also by inheritance was found in Norway (Hansen 2014); Sweden (Björklund et al. 2012; Hällsten 2014); Brazil (Medeiros et al. 2015); and, more widely, Latin America (Torche 2014); Germany (Jenderny 2016); and the United States of America (Pfeffer and Killewald 2017) (though it is lower in Denmark (Munk et al. 2016)). There are dissenting opinions: Auten et al. (2013) reported substantial downward mobility (once you are at the top, the only two options are either staying or going downhill of course) in the United States of America among the top 1 per cent richest households: persistence at the top reduced between 2002 and 2006 compared to the 1992–1999 period. Moreover, also using data from the United States of America, Carney and Nason (2018) found substantial circulation among the 1 per cent richest households, a group that is over-represented by small and medium-sized private enterprise owners and managers.

Other studies present a more nuanced version of the circulation of elites story, distinguishing sub-groups within the elites. For example, Korom et al. (2017) presented evidence that it is less likely for descendants of self-made rich entrepreneurs in the United States of America to remain in the top 1 per cent of wealthiest households compared to those belonging to rich dynastic families. In addition, according to Piketty and Saez (2003, 2014), since the end of World War II, there has been a reduction in the relative importance of rentiers at the expense of the 'working rich'. However, Wolff and Zacharias (2009) considered that the sources have not significantly changed. Tuomala et al. (1988) reported that the generation and transmission of wealth in Finland may be different among the 0.1 per cent, 1 per cent, and the 10 per cent richest groups; these authors showed evidence of a reduction in the share of the former group but increasing concentration among the other two groups. The difference resides in the variation in the sources of income and wealth accumulation in each group, with the importance of the main sources among the richest rich, earnings from shares and business assets, waning compared to entrepreneurial profits and housing wealth. In turn, Melldahl (2018) reported that the composition of capital operates as a divisory line within the economic elite in Sweden: between those holding large shares of wealth (rentiers and entrepreneurs) and those earning high-wage incomes. Moreover, reproduction and intergenerational transmission differ by group: wealthy high-earning households favour investing in the human capital of grandchildren (i.e. the 'school-mediated reproduction' strategy (Bourdieu 1989)) more than rentiers, who rest their reproduction strategy primarily on intergenerational wealth transfers.

Notes

1. Lucas (2004)
2. Es dürfte kein Zweifel daran bestehen, daß die funktionelle Einkommensverteilung für die Gesellschafts- und SocialPolitk im engeren Sinne nur von sekundärer Bedeutung ist. Nur wenn man von der funkionellen Verteilung auf die personelle schließen könnte, wäre erstere soziale relevant.
3. See also Piketty (2015), Piketty and Saez (2014), and Piketty and Zucman (2014).
4. 'State of nature' is a technical term used in moral and political philosophy to refer to the situation prior to the existence of institutions, including the State and the legal system. Theories of the state of nature dwell on how individuals in such a situation would come to agreements to form institutions (see, among others, Rawls (2008, Lecture II)). Sometimes, in economics, the expression refers to all the possible though uncertain values that variables could adopt (see, for example, Arrow 1962).
5. It is also known as 'desert responsiveness'.

6. Well-off older persons.
7. Needless to say, in any of these theories there are subtle variations and sub-schools. In contemporary contractualism, from example, apart from the Rawlsian doctrine, another US philosopher, Thomas Michael Scanlon, also developed an influential position.
8. For a theoretical sociological discussion of family and poverty, see Daly (2018).
9. Denmark, Sweden, Switzerland, Austria, Germany, Belgium, the Netherlands, Norway, and Finland—that is, the Rhineland region.
10. See Bourguignon (1979).
11. As Krupp (1967) explains, with i factors in j households, national income can be represented by:

$$Y = \sum_i \sum_j y_{i,j}$$

Personal income distribution focuses on household income:

$$y^i_j = \sum_i y_{i,j}$$

whilst the functional income distribution focuses on factorial income:

$$y^j_i = \sum_j y_{j,i}$$

See also Cowell and Fiorio (2011).
12. See Sydsæter et al. (2012) if you need a refresher reading.
13. Technically, it is sufficient if it is indexed in non-decreasing order.
14. A translation into English can be found in Gini (2005).
15. Also published as Yitzhaki and Schechtman (2013, ch. 2).
16. Source: The World Bank.
17. There are several books and papers introducing the concept of entropy. See Anderson (2007) for a good presentation in the context of scientific knowledge.

References

Albertini, Marco and Jonas Radl (2012). "Intergenerational transfers and social class: Inter-vivos transfers as means of status reproduction?" In: *Acta Sociologica* 55.2, pages 107–123.

Alimi, Omoniyi, David C Maré, and Jacques Poot (2005). "More Pensioners, Less Income Inequality? The Impact of Changing Age Composition on Inequality in Big Cities and Elsewhere". In: 10690.

Almås, Ingvild, Tarjei Havnes, and Magne Mogstad (2011). "Baby booming inequality? Demographic change and earnings inequality in Norway, 1967–2000" In: *The Journal of Economic Inequality* 9.4, pages 629–650.

Almås, Ingvild and Magne Mogstad (2012). "Older or wealthier? The impact of age adjustment on wealth inequality". In: *The Scandinavian Journal of Economics* 114.1, pages 24–54.

Anand, Sudhir (1983). *Inequality and poverty in Malaysia: Measurement and decomposition*. Washington, DC: United States of America: The World Bank.

Anderson, David R (2007). *Model Based Inference in the Life Sciences: A Primer on Evidence*. New York, NY: United States of America: Springer.

Arneson, Richard J (1997). "Egalitarianism and the undeserving poor". In: *Journal of Political Philosophy* 5.4, pages 327–350.

Arrow, Kenneth Joseph (1962). "The Rate and Direction of Inventive Activity: Economic and Social Factors". In: edited by National Bureau of Economic Research. New Jersey, NJ: United States of America: Princeton University Press, pages 609–626.

Atkinson, Anthony B (2015). *Inequality. What can be done?* Cambridge, MA: United States of America: Harvard University Press.

Auten, Gerald, Geoffrey Gee, and Nicholas Turner (2013). "Income Inequality, Mobility, and Turnover at the Top in the US, 1987–2010". In: *American Economic Review* 103.3, pages 168–72.

Bavel, Bas van and Ewout Frankema (2013). *Low Income Inequality, High Wealth Inequality. The Puzzle of the Rhineland Welfare States*. CGEH Working Paper Series 50. Utrecht: The Netherlands.

Beenstock, Michael (2004). "Rank and quantity mobility in the empirical dynamics of inequality". In: *Review of Income and Wealth* 50.4, pages 519–541.

Ben-Porath, Yoram (1982). "Individuals, families, and income distribution". In: *Population and Development Review* 8, pages 1–13.

Benish, William A (1999). "Relative entropy as a measure of diagnostic information". In: *Medical decision making* 19.2, pages 202–206.

Berlin, Isaiah (1969). *Four Essays on Liberty*. Oxford: United Kingdom: Oxford University Press.

Biewen, Martin and Andos Juhasz (2012). "Understanding rising income inequality in Germany, 1999/2000 2005/2006". In: *Review of Income and Wealth* 58.4, pages 622–647.

Björklund, Anders, Jesper Roine, and Daniel Waldenström (2012). "Intergenerational top income mobility in Sweden: Capitalist dynasties in the land of equal opportunity?" In: *Journal of Public Economics* 96.5–6, pages 474–484.

Blaug, Mark (1996). *Economic Theory in Retrospect*. Third. Cambridge: United Kingdom: Cambridge University Press.

Bonvalet, Catherine and Zoé Andreyev (2003). "The local family circle". In: *Population* 58.1, pages 9–42.

Bourdieu, Pierre (1989). *La Noblesse d'État. Grandes écoles et esprit de corps*. Paris: France: Les Éditions de Minuit.

Bourguignon, François (1979). "Decomposable Income Inequality Measures". In: *Econometrica* 47.4, pages 901–920.

Brewer, Mike and Liam Wren-Lewis (2016). "Accounting for changes in income inequality: decomposition analyses for the UK, 1978–2008". In: *Oxford Bulletin of Economics and Statistics* 78.3, pages 289–322.

Butler, Robert N (1989). "Dispelling ageism: The cross-cutting intervention". In: *The Annals of the American Academy of Political and Social Science* 503.1, pages 138–147.

Butrica, Barbara A, Daniel P Murphy, and Sheila R Zedlewski (2009). "How many struggle to get by in retirement?" In: *The Gerontologist* 50.4, pages 482–494.

Cameron, Lisa A (2000). "Poverty and inequality in Java: examining the impact of the changing age, educational and industrial structure". In: *Journal of Development Economics* 62.1, pages 149–180.

Campano, Fred and Dominick Salvatore (2006). *Income Distribution*. New York, NY: United States of America: Oxford University Press.

Cantillon, Sara and Brian Nolan (2001). "Poverty within households: measuring gender differences using nonmonetary indicators". In: *Feminist Economics* 7.1, pages 5–23.

Carney, Michael and Robert S Nason (2018). "Family business and the 1%" In: *Business & Society* 57.6, pages 1191–1215.

Chernozhukov, Victor, Iván Fernández-Val, and Blaise Melly (2013). "Inference on counterfactual distributions". In: *Econometrica* 81.6, pages 2205–2268.

Chetty, Raj et al. (2014a). "Is the United States still a land of opportunity? Recent trends in intergenerational mobility". In: *American Economic Review* 104.5, pages 141–47.

Chu, CY Cyrus and Lily Jiang (1997). "Demographic transition, family structure, and income inequality". In: *Review of Economics and Statistics* 79.4, pages 665–669.

Cohen, Gene D (1994). "Journalistic elder abuse: it's time to get rid of fictions, get down to facts". In: *The Gerontologist* 34.3, pages 399–401.

Corsi, Marcella, Fabrizio Botti, and Carlo D'Ippoliti (2016). "The gendered nature of poverty in the EU: Individualized versus collective poverty measures". In: *Feminist Economics* 22.4, pages 82–100.

Cowell, Frank (2011). *Measuring Inequality*. Oxford: United Kingdom: Oxford University Press.

Cowell, Frank A and Carlo V Fiorio (2011). "Inequality decompositions a reconciliation". In: *The Journal of Economic Inequality* 9.4, pages 509–528.

Dalton, Hugh (1920). "The measurement of the inequality of incomes". In: *The Economic Journal* 30.119, pages 348–361.

Daly, Mary (2018). "Towards a theorization of the relationship between poverty and family". In: *Social Policy & Administration* 52.3, pages 565–577.

Deaton, Angus and Christina Paxson (1994). "Intertemporal choice and inequality". In: *Journal of political economy* 102.3, pages 437–467.

DiNardo, John, Nicole M Fortin, and Thomas Lemieux (1996). "Labor Market Institutions and the Distribution of Wages, 1973–1992: A Semiparametric Approach". In: *Econometrica* 64.5, pages 1001–1044.

Dong, Zhiqiang, Canqing Tang, and Xiahai Wei (2018). "Does population aging intensify income inequality? Evidence from China". In: *Journal of the Asia Pacific Economy* 23.1, pages 66–77.

Essama-Nssah, Boniface and Peter J Lambert (2012). "Influence Functions for Policy Impact Analysis". In: *Inequality, mobility and segregation: Essays in honor of Jacques Silber*. Edited by John A Bishop and Rafael Salas. Bingley: United Kingdom: Emerald Group Publishing Limited, pages 135–159.

Estudillo, Jonna P (1997). "Income inequality in the Philippines, 1961–91" In: *The Developing Economies* 35.1, pages 68–95.

Falkingham, Jane and Christina Victor (1991). "The myth of the woopie?: Incomes, the elderly, and targeting welfare". In: *Ageing & Society* 11.4, pages 471–493.

Fessler, Pirmin, Miriam Rehm, and Lukas Tockner (2016). "The impact of housing non-cash income on the household income distribution in Austria". In: *Urban Studies* 53.13, pages 2849–2866.

Firpo, Sergio, Nicole M Fortin, and Thomas Lemieux (2009). "Unconditional quantile regressions" In: *Econometrica* 77.3, pages 953–973.

Fisher, Irving (1912). *Elementary principles of economics*. New York, NY: United States of America: The Macmillan Company.

Fortin, Nicole, Thomas Lemieux, and Sergio Firpo (2011). "Decomposition methods in economics". In: *Handbook of Labor Economics*. Edited by Orley Ashenfelter and David Card. Volume 4A. San Diego, CA: United States of America: Elsevier, pages 1–102.

Frick, Joachim R et al. (2010). "Distributional effects of imputed rents in five European countries". In: *Journal of Housing Economics* 19.3, pages 167–179.

Fritzell, Johan (1999). "Incorporating gender inequality into income distribution research" In: *International Journal of Social Welfare* 8.1, pages 56–66.

Garner, Thesia I and Kathleen Short (2009). "Accounting for owner-occupied dwelling services: Aggregates and distributions". In: *Journal of Housing Economics* 18.3, pages 233–248.

Gastwirth, Joseph L (1972). "The estimation of the Lorenz curve and Gini index". In: *The Review of Economics and Statistics* 54.3, pages 306–316.

Gini, Corrado (1913). "Sulla misura della concentrazione e della variabilità dei caratteri". In: *Atti del Reale Istituto Veneto di Scienze, Lettere ed Arti* 73, pages 1203–1248.

—— (2005). "On the measurement of concentration and variability of characters". In: *METRON-International Journal of Statistics* 63.1, pages 1–38.

Giorgi, Giovanni Maria and Chiara Gigliarano (2017). "The Gini concentration index: a review of the inference literature". In: *Journal of Economic Surveys* 31.4, pages 1130–1148.
Goldstein, Joshua R and Ronald D Lee (2014). "How large are the effects of population aging on economic inequality?" In: *Vienna Yearbook of Population Research* 12.1, pages 193–209.
Hällsten, Martin (2014). "Inequality across three and four generations in egalitarian Sweden: 1st and 2nd cousin correlations in socio-economic outcomes". In: *Research in Social Stratification and Mobility* 35, pages 19–33.
Hansen, Marianne Nordli (2014). "Self-made wealth or family wealth? Changes in intergenerational wealth mobility". In: *Social Forces* 93.2, pages 457–481.
Heger, Dörte (2018). "Decomposing differences in health and inequality using quasi-objective health indices". In: *Applied Economics* 50.26, pages 2844–2859.
Hinterlong, James and Scott Ryan (2008). "Creating grander families: Older adults adopting younger kin and nonkin" In: *The Gerontologist* 48.4, pages 527–536.
Hong, Baeg Eui and Hye youn Kim (2012). "Trends of income inequality among the elderly in Korea". In: *Asian Social Work and Policy Review* 6.1, pages 40–55.
Jäntti, Markus (1993). *Changing inequality in five countries: the role of markets, transfers and taxes*. LIS Working Paper 91. Luxembourg: Luxembourg: Luxembourg Income Study (LIS).
Jenderny, Katharina (2016). "Mobility of Top Incomes in Germany". In: *Review of Income and Wealth* 62.2, pages 245–265.
Jenkins, Stephen P (1995). "Accounting for inequality trends: decomposition analyses for the UK, 1971–86". In: *Economica* 62.245, pages 29–63.
Johnson, D Gale (1954). "The functional distribution of income in the United States, 1850–1952" In: *The Review of Economics and Statistics* 36.2, pages 175–182.
Juhn, Chinhui, Kevin M Murphy, and Brooks Pierce (1993). "Wage inequality and the rise in returns to skill". In: *Journal of Political Economy* 101.3, pages 410–442.
Kagan, Shelly (1998). *Normative Ethics*. Dimensions of Philosophy Series. Boulder, Co: United States of America: Westview Press.
Kakwani, Nanak C (1977a). "Applications of Lorenz curves in economic analysis" In: *Econometrica* 45.3, pages 719–727.
—— (1977b). "Measurement Of Poverty And Negative-Income Tax". In: *Australian Economic Papers* 16.29, pages 237–248.
Kakwani, Nanak, Adam Wagstaff, and Eddy Van Doorslaer (1997). "Socioeconomic inequalities in health: measurement, computation, and statistical inference". In: *Journal of Econometrics* 77, pages 87–103.
Kang, Sung Jin (2009). "Aging and Inequality of Income and Consumption in Korea". In: *Journal of International Economic Studies* 23, pages 59–72.
Kang, Sung Jin and Robert Rudolf (2016). "Rising or falling inequality in Korea? Population aging and generational trends". In: *The Singapore Economic Review* 61.05. https://doi.org/10.1142/S0217590815500897.

Karunaratne, Hettige Don (2000). "Age as a factor determining income inequality in Sri Lanka". In: *The Developing Economies* 38.2, pages 211–242.
Keister, Lisa A (2014). "The one percent". In: *Annual Review of Sociology* 40, pages 347–367.
Killewald, Alexandra, Fabian T Pfeffer, and Jared N Schachner (2017). "Wealth inequality and accumulation". In: *Annual Review of Sociology* 43, pages 379–404.
Korom, Philipp, Mark Lutter, and Jens Beckert (2017). "The enduring importance of family wealth: Evidence from the Forbes 400, 1982 to 2013". In: *Social Science Research* 65, pages 75–95.
Krupp, Hans-Jürgen (1967). ""Personelle" und "funktionelle" Einkommensverteilung". In: *Jahrbücher für Nationalökonomie und Statistik* 180, pages 1–35.
Kullback, Solomon and Richard A Leibler (1951). "On information and sufficiency". In: *The Annals of Mathematical Statistics* 22.1, pages 79–86.
Lam, David (1984). "The variance of population characteristics in stable populations, with applications to the distribution of income". In: *Population Studies* 38.1, pages 117–127.
Lindert, Peter H and Jeffrey G Williamson (1976). *Three Centuries of American Inequality*. Report IRP-DP-333–76. Washington, DC: United States of America.
—— (2016). *Unequal Gains: American Growth and Inequality since 1700*. The Princeton Economic History of the Western World. Princeton, NJ: United States of America: Princeton University Press.
Listl, Stefan (2011). "Income-related inequalities in dental service utilization by Europeans aged 50+". In: *Journal of Dental Research* 90.6, pages 717–723.
—— (2012). "Income-related inequalities in denture-wearing by Europeans aged 50 and above". In: *Gerodontology* 29.2, e948–e955.
Listl, Stefan and Jr CM Faggion (2012). "Income-related inequalities in chewing ability of Europeans aged 50 and above". In: *Community Dental Health* 29.2, pages 144–148.
Lorenz, Max O (1905). "Methods of measuring the concentration of wealth". In: *Publications of the American statistical association* 9.70, pages 209–219.
Lucas, Robert (2004). "The Industrial Revolution: past and future (2003 Annual Report Essay)", *The Region*, 5:5–20, Federal Reserve Bank of Minneapolis
M, Tribus (1961). *Thermostatics and Thermodynamics. An Introduction to Energy, Information and States of Matter, with Engineering Applications*. Princeton, NJ: United States of America: van Nostrand Company Inc.
Machado, José AF and José Mata (2005). "Counterfactual decomposition of changes in wage distributions using quantile regression". In: *Journal of Applied Econometrics* 20.4, pages 445–465.
Medeiros, Marcelo, Pedro HG Ferreira de Souza, and Fábio Avila de Castro (2015). "O Topo da Distribuição de Renda no Brasil: Primeiras Estimativas com Dados Tributários e Comparação com Pesquisas Domiciliares (2006–2012)". In: *Dados-Revista de Ciências Sociais* 58.1, pages 7–36.

Melldahl, Andreas (2018). "Modes of reproduction in the Swedish economic elite: education strategies of the children of the top one per cent". In: *European Societies* 20.3, pages 424–452.

Mookherjee, Dilip and Anthony F Shorrocks (1982). "A decomposition analysis of the trend in UK income inequality". In: *The Economic Journal* 92.368, pages 886–902.

Morley, Samuel A (1981). "The effect of changes in the population on several measures of income distribution". In: *The American Economic Review* 71.3, pages 285–294.

Motonishi, Taizo (2006). "Why has income inequality in Thailand increased?: An analysis using surveys from 1975 to 1998" In: *Japan and the World Economy* 18.4, pages 464–487.

Munk, Martin D, Jens Bonke, and M Azhar Hussain (2016). "Intergenerational top income persistence: Denmark half the size of Sweden". In: *Economics Letters* 140, pages 31–33.

OECD (2013b). *OECD Framework for Statistics on the Distribution of Household Income, Consumption and Wealth*. Technical report. Paris: France.

Ohtake, Fumio and Makoto Saito (1998). "Population aging and consumption inequality in Japan". In: *Review of Income and Wealth* 44.3, pages 361–381.

Olivera, Javier (2018). "The distribution of pension wealth in Europe" In: *The Journal of the Economics of Ageing*. https://doi.org/10.1016/j.jeoa.2018.06.001.

Oorschot, Wim van (2000). "Who should get what, and why? On deservingness criteria and the conditionality of solidarity among the public" In: *Policy and Politics* 28.1, pages 33–48.

Ourti, Tom Van (2003). "Socio-economic inequality in ill health amongst the elderly: Should one use current or permanent income?" In: *Journal of Health Economics* 22.2, pages 219–241.

Paglin, Morton (1975). "The measurement and trend of inequality: A basic revision". In: *The American Economic Review* 65.4, pages 598–609.

Pamuk, Elsie R (1985). "Social class inequality in mortality from 1921 to 1972 in England and Wales". In: *Population Studies* 39.1, pages 17–31.

Pareto, Vilfredo (1917). *Traité de Sociologie Générale*. Lausanne: Switzerland: Librairie Payot.

Paweenawat, Sasiwimon Warunsiri and Robert McNown (2014). "The determinants of income inequality in Thailand: A synthetic cohort analysis" In: *Journal of Asian Economics* 31, pages 10–21.

Pezzin, Liliana E and Barbara Steinberg Schone (1997). "The allocation of resources in intergenerational households: Adult children and their elderly parents". In: *The American Economic Review* 87.2, pages 460–464.

Pfeffer, Fabian T and Alexandra Killewald (2017). "Generations of advantage. Multi-generational correlations in family wealth". In: *Social Forces* 96.4, pages 1411–1442.

Piketty, Thomas (2014). *Capital in the Twenty-First Century*. Cambridge, MA: Harvard University Press.

—— (2015). "Putting distribution back at the center of economics: Reflections on capital in the twenty-first century". In: *Journal of Economic Perspectives* 29.1, pages 67–88.
Piketty, Thomas and Emmanuel Saez (2003). "Income Inequality in the United States 19131998". In: *Quarterly Journal of Economics* 118.1, pages 1–39.
—— (2014). "Inequality in the long run". In: *Science* 344.6186, pages 838–843.
Piketty, Thomas and Gabriel Zucman (2014). "Capital is back: Wealth-income ratios in rich countries 1700–2010". In: *The Quarterly Journal of Economics* 129.3, pages 1255–1310.
Pyatt, Graham (1976). "On the interpretation and disaggregation of Gini coefficients" In: *The Economic Journal* 86.342, pages 243–255.
Rawls, John (1971). *A Theory of Justice*. Cambridge, MA: United States of America: Harvard University Press.
—— (2008). *Lectures on the history of political philosophy*. Cambridge, MA: United States of America: Harvard University Press.
Regidor, Enrique (2004). "Measures of health inequalities: part 2". In: *Journal of Epidemiology and Community Health* 58.11, pages 900–903.
Rothe, Christoph (2012). "Partial distributional policy effects". In: *Econometrica* 80.5, pages 2269–2301.
Ryan, Paul (1996). "Factor Shares and Inequality in the UK". In: *Oxford Review of Economic Policy* 12.1, pages 106–126.
Sandmo, Agnar (2000). "The Principal Problem in Political Economy: Income Distribution in the History of Economic Thought". In: *Handbook of Income Distribution*. Edited by Anthony B Atkinson and François Bourguignon. Volume 1. Amsterdam: The Netherlands: Elsevier, pages 3–65.
Schwartz, Marlene B et al. (2003). "Weight bias among health professionals specializing in obesity". In: *Obesity* 11.9, pages 1033–1039.
Sen, Amartya (1997). *On Economic Inequality* Clarendon paperbacks. Oxford: United Kingdom: Oxford University Press.
Sen, Amartya K (2000). "Social Justice and the Distribution of Income". In: *Handbook of Income Distribution*. Edited by Anthony B Atkinson and François Bourguignon. Volume 1. Amsterdam: The Netherlands: Elsevier, pages 59–85.
Shorrocks, Anthony F (1980). "The class of additively decomposable inequality measures". In: *Econometrica: Journal of the Econometric Society* 48.3, pages 613–625.
Siegel, Martin, Markus Luengen, and Stephanie Stock (2013). "On age-specific variations in income-related inequalities in diabetes, hypertension and obesity". In: *International journal of public health* 58.1, pages 33–41.
Stiglitz, Joseph, Amartya Sen, and Jean Fitoussi (2009). *Report by the commission on the measurement of economic performance and social progress*. Technical report. Paris: France: Commission on the measurement of economic performance and social progress.

Street, Debra and Jeralynn Sittig Cossman (2006). "Greatest generation or greedy geezers? Social spending preferences and the elderly". In: *Social problems* 53.1, pages 75–96.

Sydsæter, Knut, Peter Hammond, and Anne Strøm (2012). *Essential Mathematics for Economic Analysis*. Harlow: United Kingdom: Pearson.

Theil, Henri (1967). *Economics and information theory*. Studies in mathematical and managerial economics. Amsterdam: The Netherlands: North-Holland Publishing Company.

Tillman, Kathryn Harker and Charles B Nam (2008). "Family structure outcomes of alternative family definitions". In: *Population Research and Policy Review* 27.3, pages 367–384.

Torche, Florencia (2014). "Intergenerational mobility and inequality: The Latin American case". In: *Annual Review of Sociology* 40, pages 619–642.

Tsakloglou, Panos (1993). "Aspects of inequality in Greece: Measurement, decomposition and intertemporal change: 1974, 1982". In: *Journal of Development Economics* 40.1, pages 53–74.

Tung, An-Chi, Chaonan Chen, and Paul Ke-Chih Liu (2006). "The emergence of the neo-extended family in contemporary Taiwan". In: *Journal of Population Studies* 32, pages 123–152.

Tuomala, Matti, Jouko Vilmunen, et al. (1988). "On the Trends over Time in the Degree of Concentration of Wealth in Finland". In: *Finnish Economic Papers* 1.2, pages 184–190.

UNECE (2011). *Canberra Group Handbook on Household Income Statistics, Second Edition*. Technical report. Geneva: Switzerland.

Von Weizsäcker, Robert K (1996). "Distributive implications of an aging society". In: *European Economic Review* 40.3-5, pages 729–746.

Walsh, Kieran, Thomas Scharf, and Norah Keating (2017). "Social exclusion of older persons: a scoping review and conceptual framework". In: *European Journal of Ageing* 14.1, pages 81–98.

Wang, Chen et al. (2018). "Aging and inequality: The link and transmission mechanisms". In: *Review of Development Economics*, pages 1–19. https://doi.org/10.1111/rode.12394.

Wertz, Kenneth L (1979). "The measurement of inequality: comment". In: *The American Economic Review* 69.4, pages 670–672.

Wolff, Edward N and Ajit Zacharias (2009). "Household Wealth and the Measurement of Economic Well-Being in the United States". In: *Journal of Economic Inequality* 7.2, pages 83–115.

Woolley, Frances (2003). "Control over money in marriage". In: *Marriage and the economy: theory and evidence from advanced industrial societies*. Edited by Shoshana Grossbard-Shechtman. Cambridge: United Kingdom: Cambridge University Press, pages 105–128.

Woolley, Frances R and Judith Marshall (1994). "Measuring inequality within the household". In: *Review of Income and Wealth* 40.4, pages 415–431.

Yellen, Janet L (2016). "Perspectives on inequality and opportunity from the Survey of Consumer Finances". In: *RSF: The Russell Sage Foundation Journal of the Social Sciences* 2.2, pages 44–59.

Yitzhaki, Shlomo (1998). "More than a dozen alternative ways of spelling Gini". In: *Research on Economic Inequality* 8, pages 13–30.

Yitzhaki, Shlomo and Edna Schechtman (2013). *The Gini Methodology. A Primer on a Statistical Methodology*. New York, NY: United States of America: Springer Science+Business Media.

Zhang, Juwei and Jing Xiang (2014). "How aging and intergeneration disparity influence consumption inequality in China". In: *China & World Economy* 22.3, pages 79–100.

Zhong, Hai (2011). "The impact of population aging on income inequality in developing countries: Evidence from rural China". In: *China Economic Review* 22.1, pages 98–107.

5

Poverty, Deprivation, and Social Class

Overview

This chapter discusses alternative conceptualisations of poverty, deprivation, and exclusion. It presents multidimensional constructs and topics close to the study of economics and ageing, including financial distress, financial security, and poverty persistence. It describes alternative theories of poverty and the question of social class and later life.

I picked a carton of pomegranate and raspberry fruit juice at a supermarket once. The photograph on the carton showed a bunch of shiny and appetising pomegranate seeds (or arils) and raspberries, but as I cannot take any grapefruit, I checked the list of ingredients, just in case. It did not have any grapefruit, but only 3 per cent of its contents were pomegranate, around the same of raspberry…and almost 50 per cent apple juice from concentrate (plus other bits and pieces)! Basically, I was about to buy an apple juice with a touch of pomegranate and raspberry, not what I had been enticed to purchase. With inequality and poverty studies (as well as 'happiness', as covered in Part I in this volume), sometimes I get the same feeling. A paper seems to be about one thing but a few paragraphs into it I realise it is about something else. Related, yes, but different. My recommendation, therefore, is that before buying into the story as presented in the abstract of a paper or the executive summary of a report, you should go to the data and definitions section and see exactly what the study is about. I know, those sections are usually as boring as the list of ingredients of industrialised fruit juice, but they are as important. Sometimes, the topic of the study is not exactly what it says on the tin (or the carton!).

In a series of papers, Emily Callander and collaborators from the University of Sydney, Australia, explored longitudinal associations between health risks and poverty in later life. They found strong, positive associations with arthritis (Callander and Schofield 2016) and asthma (Callander and Schofield 2015a), as well as longitudinal interactions between poverty and psychological distress (Callander and Schofield 2015b) and self-efficacy (feeling of control, problem-solving, helplessness, etc.) (Callander and Schofield 2017). However, ageing does not inexorably lead to poverty, neither in Europe and Central Asia (Bussolo et al. 2015) nor in developing countries (Castañeda et al. 2018). However, poverty is a reality for millions of older people in these and other regions. Brody (2003, p. 15) remarked that in the United States of America, until as recently as the 1950s and 1960s, 'the phrase "older people" was almost synonymous with "poor people"', for example, during the Great Depression, 'half of all older people were totally dependent economically on their children and another quarter were dependent on public or private welfare'. Older people are still over-represented among the population living in poverty in most countries in the world (UNFPA—HelpAge International 2012), although across the developed countries, this is changing:

- In the United States of America, 35 per cent of people aged 65 years or older were in poverty in 1959; the rate fell to 8.8 per cent by 2015 (10.3 per cent for women; 7 per cent for men), around 4.2 million individuals (Proctor et al. 2016).
- In the United Kingdom, from 28 per cent of pensioners in poverty in 1998, the rate fell to 16 per cent in 2015/2016, just under 2 million people (UK 2018).
- In Australia, just under 30 per cent of people aged over 75 years were living in poverty (measured as living with incomes less than 50 per cent of median household disposable income) in 2014. For the same year, the poverty rates among the over 75-year-olds were 31.6 per cent in Latvia, 30.3 per cent in Mexico, and 23.8 per cent in Switzerland (OECD 2017).
- In a systematic review of number of deaths attributable to social factors in the United States of America, Galea et al. (2011) estimates that of almost 1.8 million deaths in people aged 65 years or older in 2000, over 68,500 were attributable to income poverty. Moreover, the relative risk of mortality for older people in poverty was statistically significant (RR= 1.40 CI:(1.37, 1.43)) compared to older people not in poverty.

There is a huge gender difference in poverty rates in later life: in Norway, merely 1.9 per cent of adults aged over 65 years were in poverty in 2015, compared to 6.3 per cent among women. And similarly in other countries, for example, in 2015 in Spain, the poverty rate among men aged 65 or over was 3.7 per cent and among women of the same age, 6.7 per cent (OECD 2017).

Regarding developing countries, statistics compiled by the World Bank for a group of 27 developing countries showed that the poverty rate of people aged 60 or over averaged almost 36 per cent, roughly the same as that of working-age adults, except that among older people, there was much wider variation: the rates ranged between 24.1 per cent in the Kyrgyz Republic and 46.8 per cent in Thailand (the minimum and maximum rates for adults of working age were 33.7 per cent, in the Philippines, and 40.3 per cent in Iraq, respectively). In addition, there was no correlation between poverty rates for both groups.[1]

However, what are these statistics showing? What is this thing called 'poverty' that they are set to measure? Does the figure about British pensioners refer to same phenomenon as the statistic that the only type of toilet facility that over 50 per cent of people aged 65 or over in Sierra Leone had in 2015 was a communal pit (Kamara et al. 2017)? Or the same as the reported figures about the incidence of poverty among people aged 60 years or older being 80 per cent in Zambia in 2005 (Kakwani and Subbarao 2005) and 74.8 per cent in Honduras in 2003 (United Nations 2015). And if they are not about exactly the same phenomenon, to what extent are they comparable, and are they the result of the same underlying processes?

5.1 An Embarrassment of Definitional Riches?

Poverty is defined in many ways, responding to theoretical approaches and, operationally, to data limitations. Spicker et al. (2013) identified the following twelve 'clusters of meaning' or families of definitions of poverty, which the author argued do overlap with one another but are logically discrete in the sense that they can be separated. Poverty is:

- a material concept: the lack of resources to meet needs
- a pattern of deprivation: the lack of 'basic' needs over a prolonged period of time
- limited command of resources
- low income
- not being able to afford a given standard of living
- the distance below a critical minimum level of income

- an economic position
- a social class
- dependence on social assistance and benefits
- vulnerability to social risks
- lack of entitlements
- social exclusion

In 1999, researchers from the World Bank asked over 20,000 people living in deprived communities in 23 countries how they understood poverty and invited them to engage in discussions about well-being and ill-being; the problems and priorities of different groups; the role of public, private, and civic society institutions; and gender relations within the household and the community. Ten inter-related dimensions emerged:

- livelihood precariousness
- excluded locations
- physical problems
- gender relations, including violence and inequality
- problems in social relations
- lack of security
- experiencing abuse by people in power
- disempowering institutions
- weak community organisations
- limited capabilities

Both the reviews of academic literature and participatory methods convey poverty as 'multidimensional in its symptoms, multivariate in its causes, dynamic in its trajectory, and quite complex in its relation to' life domains such as employment; health; housing; social, cultural and political participation; and so on (Mowafi and Khawaja 2005, p. 260). Despite the multidimensional nature of the concept, most authors define poverty around income. However, there are studies of consumption, financial assets, pension or housing wealth, self-perceived poverty, food insecurity, among other dimensions. Some authors do use a multi-pronged conceptualisation. As in studies of inequality, where the question about 'inequality of what?' is pertinent because the focus could be on inequality of income, financial wealth, health status, subjective well-being, mortality, and so on, the question of 'poverty of what?' is also important and varies among studies.

As mentioned, income-based definitions are, by many economists, the preferred choice of operationalisation: out of fifty-six papers included in a scoping review of the literature, Kwan and Walsh (2018) listed fifty works that conceptualised poverty in later life on the basis on income (or consumption, but with a specific income level).

Income-based poverty definitions can be absolute or relative. The discussion (or 'intense debates' (Sachs 2005, p. 20)) around whether a relative or an absolute definition of poverty should be adopted has been going on since, at least, the late 1970s, and this is not the place to review it, let alone settle it! The consensus seems to be that, in developed countries, poverty must be measured in relative terms, whereas in developing countries, both absolute and relative definitions have a role. Albeit, some authors argue that even the most absolute bare minimum indicator (such as the minimum calories needed to be ingested per day) is socially constructed (yes, even starvation), so that all definitions of poverty are, in fact, ultimately relative.

5.1.1 Equivalisation

As with inequality measures, the choice of unit of analysis is crucial for poverty estimates. Whichever we use, household, family, or individual data, changes the levels of reported poverty. This is particularly important among older people, who are more likely to live alone than other age groups. People living alone do not reap any of the economies of scale that come with sharing resources with someone else. Household needs vary with size, but not proportionally: some costs are fixed (e.g. rent or mortgage payments) and therefore reduce per head. In order to capture the differences in economies of scale derived from the size of the household, income data is adjusted by applying an equivalisation scale.

Equivalisation (or equivalence) scales attempt to make household income comparable across a population by adjusting the magnitudes according to the size of each household: the household equivalent income. A number of equivalisation scales have been proposed in the literature based on estimates of, or assumptions about, a household's elasticity of need with respect to size—that is, how much the needs of a household vary if it increases in size by one additional person, and also considering the age of this additional person (Buhmann et al. 1988). In general, the following formula captures the essence of equivalisation scales:

$$Y_{eq} = \left(\frac{Y}{S}\right)^e \quad (5.1)$$

where the equivalised income Y_{eq} is equal to household income (Y) divided by the size of the household (S) raised to the power e. This coefficient, e, reflects economies of scale within households. If it is set equal to 1, it means that there are no economies of scale and therefore a two-person household needs twice as much income as a one-person household to reach a comparable level of income. If it is set equal to 0, it means that economies of scale are such that no equivalisation is needed. The higher the value of the elasticity e, the smaller the assumed economies of scale. Equation (5.1) is easily modified to accommodate differences in the composition of households: for example, if children are assumed to have different needs than adults of working age or older, the following extension can be used (Deaton and Zaidi 2002):

$$Y_{eq} = \left[\frac{Y}{(A + \alpha_1 \cdot K_1 + \alpha_2 \cdot K_2)} \right]^e \qquad (5.2)$$

where A stands for the number of adults of working age in the household, $K_{1,2}$ represent the number of people of age groups 1, 2 in the household, and $\alpha_{1,2}$ are the equivalisation coefficients for age groups 1, 2, respectively. The official equivalisation scales of income in the United States of America are:

- One- and two-adult households: scale = ((number of adults))$^{0.5}$
- Single parents: scale = ((number of adults) + 0.8 × first child + 0.5 ×otherchildren)$^{0.5}$
- all other households: scale = ((number of adults) + 0.5 × number of children)$^{0.5}$

The 'Oxford scale', also known as the 'OECD equivalence scale', assigns a value of 1 to the first adult household member irrespective of her age, 0.7 to each additional adult and 0.5 to each child. The 'OECD modified equivalence scale' assigns a value of 1 to the first adult household member irrespective of her age, 0.5 to each additional adult, and 0.3 to each child. Another popular equivalence scale, in use in cross-national data sets, consists of dividing the household income by the square root of the number of household members irrespective of their age:

$$Y_{eq} = \left[\frac{Y}{\sqrt{S}} \right]^e \qquad (5.3)$$

It is good methodological practice to carry out sensitivity analyses to test whether the choice of equivalence scale significantly and substantially changes the results or not. Ideally, these sensitivity tests should be implemented for different sub-groups as well. As can be seen, none of these scales adjust for the presence of older people and other groups with different needs in the household.

However, Burkhauser et al. (1996) estimated the relative poverty rates in Germany and the United States of America in the 1980s using a battery of equivalence scales. They found that, among households whose head was 65 years or older, the choice of equivalisation scale did not make much difference except, in Germany, among single households, for whom depending on the scale the equivalised household income was either 103.4 per cent or 72.1 per cent of the median person's equivalised household income.

For this reason, some authors have proposed to have particular needs accounted for. For example,

- Zaidi and Burchardt (2005) estimated disability-adjustment factors by level of disability, given the additional costs (which are not fully compensated by disability benefits[2]) incurred by households in which at least one of its members is disabled. This adjustment takes into account that the equivalence elasticity factor varies with the degree of disability (Morciano et al. 2015). Adjusting for disability, Zaidi and Burchardt found that relative poverty rates (using the 60 per cent median disposable income threshold) in the United Kingdom in 1996/1997 would have stood at 61.6 among pensioner households, against 37.2 per cent excluding the extra costs of disability. Moreover, Morciano et al. (2015) estimated that an older disabled person in the United Kingdom would need a net household income of between 62 per cent and 89 per cent higher than older non-disabled person to reach the same standard of living, a gap that is not considered by the simpler equivalence methods in use.
- Another approach was proposed by Lelli (2005): to use the achievement of a given level of functionings (see Chap. 1, this volume) as the basis for the derivation of equivalence scales. Using data from Italy for 1995, these authors estimated that among retired households using functionings-based equivalence scales, poverty rates would have been 26.1 per cent instead of 14.8 per cent.
- Stewart (2009) proposed the use of subjective evaluations of the personal financial situation (i.e. a person's self-reported level of satisfaction with her income or standard of living) to construct equivalence scales. Using this approach, the author reported that in Great Britain in 2004/2005, poverty

rates among single pensioner households were 2.7 times higher than for pensioner couples, in contrast to 1.6 times higher using the unadjusted equivalence scale.

Household composition is germane to poverty levels among older people, as they are more likely to live alone or with other older people (Evans and Palacios 2015). This reduction in multigenerational households, which in part is a result of economic growth (Aziz et al. 2018), affects the diversity in income sources within older households and accentuates the dependency on one particular source, one which in many developing countries is becoming less generous: pensions (Bussolo et al. 2015). Therefore, it is higher among older people who live alone, followed by those who live with another older person. Living alone is more marked among women given their longer life expectancy, the reduced probability of remarrying or repartnering after widowhood (Doblas et al. 2014; Wu et al. 2014) and, to a lesser extent, following divorce,[3] and that regardless of gender, wealth is a predictor or remarriage or repartnering in later life (Vespa 2012). These factors increase the risk of poverty among older women.

5.1.2 Absolute Poverty

Absolute poverty is related to not being able to afford a certain basket of basic goods (the so-called minimum food basket) or consume at least a given number of calories a day (i.e. the 'caloric sufficient criterion'). The United Nations, in its Report of the World Summit for Social Development, defined it as 'a condition of severe deprivation of basic human needs, including food, safe drinking water, sanitation facilities, health, shelter, education and information' (United Nations 1996, p. 38). The emphasis is on 'physiological efficiency' (Spicker et al. 2013), basic needs, and subsistence. As Schwartzman (no date, p. 2) put it bluntly, it is 'a matter of acute deprivation, hunger, premature death and suffering'.

Absolute poverty is usually operationalised by a minimum income standard (e.g. living with less than X dollars a day). The minimum income level currently in use to estimate the prevalence and incidence of absolute poverty in developing countries (i.e. the 'international poverty line') is US$1.90 per person a day in purchasing power parity terms (see box) (Ferreira et al. 2015).

> **Purchasing Power Parity**
>
> For international comparisons of poverty, an international poverty line is defined in US dollars and then converted into local currencies by means of the respective purchasing power parity (PPP) exchange rate. The PPP is the hypothetical exchange rate that would render a given amount of one currency in a country the same purchasing power as the respective currency in another country.
>
> To illustrate, let's think of two countries, A and B. A is a developed country with a population of 1000 people and B is a developing country with a population of 10,000. Imagine that in country A, 10 per cent of the population has social care needs but, in B, social needs are demanded by 9 per cent of its population. In both countries, the caregiving services are of the same quality. To compute the contribution to the gross domestic product (GDP) of the social care sector, we use the hourly rate for care services. Given that country A is wealthier, social workers are paid more per hour than in country B. Imagine the hourly rates are 5 in the local currency of country A and 2500 in B's local currency, and that the exchange rate stands at 5000 units of the currency in B for each unit of A's currency. (So, applying the exchange rate, we see that a social worker in A earns ten times more per hour than her counterpart in B.) Finally, let's assume that the recipients of care only need two hours a week of services. The GDP of social care in country A is: $5 \times 100 \times 2 = 1000$. And for country B we get: $2500 \times 900 \times 2 = 4,500,000$. Per person, the GDP of social care in local currencies is 1 in A and 450 in B. Using the exchange rate, we see that it is 1 in A but merely 0.09 in B. But it would be incorrect to conclude that the GDP of the social care sector in B amounts to 9 per cent of that in A. We could, instead, use the local hourly rates in A (or, equally, the local hourly rates in B) to value the services in both countries. Let's start with country A's local hourly rates for social care services. The GDP per person for the social care sector in A, measured in A's local rates, is equal to 1, as we have seen. The GDP for the social care sector in B, measured in A's local hourly social care rates, is obtained thus: $5 \times 900 \times 2 = 9000$. Per person, 0.90. Using the local prices in B, we obtain that in A, the GDP for social care per person becomes 500. Therefore, the GDP per person in B is 90 per cent that of A, not 9 per cent!
>
> The international poverty line was set at US$ 1 per person per day at 1985 PPP (Ravallion et al. 1991), adjusted to US$ 1.08 per person a day at 2003 PPP, the median of the lowest ten poverty lines in Bangladesh, China, India, Indonesia, Nepal, Pakistan, Tanzania, Thailand, Tunisia, and Zambia at that time (Chien and Ravallion 2001). It was adjusted again in 2008 to US$ 1.25 per person per day (Ravallion et al. 2009) and in 2015 to its current level of US$ 1.90 per person per day, based on 2011 PPP.

There are also 'quasi-absolute' definitions, which fix the income threshold of one year in real terms and then apply it to subsequent years.

The technical aspects of poverty measurement do matter, because even using the same approach (say, absolute poverty based on international poverty lines), the results may vary substantially depending on changes in one of the underlying assumptions. For example, using the international poverty line approach, Castañeda et al. (2018) estimated that 44 per cent of the people who live in extreme absolute poverty in developing countries are children younger than 15 years old, against 3.8 per cent of people aged 65 or over. Moreover, around 20 per cent of children aged 14 or under live in extreme poverty in contrast to 13.3 per cent of adults aged 65 or over. However, Batana et al. (2013) contended that this approach assumes that every person, regardless of chronological age and sex, requires the same caloric intake and that there are no intra-household economies of scale (i.e. that the cost of living per head is the same for a one-person household than for, say, a couple with three teenage children). Applying these adjustments to 2000 data, these authors found that the absolute poverty rate in developing countries plummeted from 30 per cent to between 3 per cent and 13 per cent, and that the rates among older people became more similar to the general population. It should be apparent that absolute poverty would not have the same policy immediacy and saliency were the reported rates 3 per cent, or even 13 per cent, instead of 30 per cent!

Despite the general agreement that the notion of absolute poverty is not relevant to developed countries, measures of absolute poverty are in use in the United States of America, including indicators specifically prepared to assess poverty levels among older people. One such indicator is the Elder Economic Security Standard.

The Elder Economic Security Standard

When studying poverty rates among retirees or future retirees, the retirement income (current or projected) can be compared to a national threshold calculated similarly to the poverty line or to income levels earned during the employment stage in the life cycle. The latter approach focuses on whether retirement income is or will be sufficient to preserve a certain standard of living—that is, the 'replacement rate' (see Chap. 4 in Volume III). This is important but does not necessarily indicate poverty or risk of poverty. The other method is to establish a 'basic' cost of living according to life circumstances and area of residence.

In the United States of America, there are two federal poverty measures: poverty thresholds and poverty guidelines. Poverty thresholds were set in 1965 on the basis of the cost of food needed to meet the minimum nutritional needs of adults of different ages multiplied by three (as the average household in 1960 spent one-third of its income on food) (Orshansky 1965). The thresholds are updated annually for inflation. The latest figure (for 2016) of poverty threshold for one-person households aged 65 or over is US$ 11,511 a year and for two householders age 65 or older, US$ 14,507 a year. These values are lower than those for younger adults, under the assumption that older people require less calories. Consequently, it may be that an older person living on her own is not classified as poor unlike a younger person living also on her own and with the same household income. In turn, the poverty guidelines, although also vary by household size, do not distinguish between the age of the members of the households. The thresholds are the same throughout the country; the guidelines vary for Alaska and Hawaii. The poverty guideline for one-person households amounted to US$ 12,140 in 2018 and to US$ 16,460 for two-person households (except, as mentioned, in Alaska and Hawaii).

Russell et al. (2006) developed the Elder Economic Security Standard (i.e. the 'Elderly Standard'), which measures the income required to meet the basic needs of older people living independently (i.e. not in intergenerational households) in the community (i.e. non-institutionalised)—see also Mutchler et al. (2015). The Elderly Standard includes the following cost components: housing and related costs, food (at home), medical care, home- and community-based long-term care, private and public transportation, and a host of miscellaneous expenses such as clothing, cleaning products, and so on. Compared to the Federal Poverty Guidelines for 2006, the Elderly Standard excluding long-term care for home owners or people paying off a mortgage was 151 per cent higher for one-person households and 196 per cent higher for two-person households.

MacDonald et al. (2010) estimated the Elderly Standard for Canada in five cities and adjusting for a variety of life circumstances. The authors concluded that 'individual circumstances, rather than age, are the primary drivers in determining the cost of basic needs' [p. 54].

Wallace et al. (2013) applied the Elderly Standard method to data for California, United States of America, for 2007. These authors reported that whilst under the Federal Poverty Guideline, 7.9 per cent of the older population in California were classified as poor, using the Elderly Standard, the proportion rose to 36.4 per cent (one-person and two-person households combined). Among White people of Latin American descent aged 65 or over, it reached almost 60 per cent (the Federal Guideline set the poverty rate among this group at 14.9 per cent).

The Supplemental Poverty Measure

We mentioned above that in the United States of America, poverty is measured in absolute terms, using thresholds set in relation to minimum nutritional needs. However, one increasingly important consumption item for older people, healthcare, is not accounted for in this measure. In the United States of America, there are health insurance programmes for people aged under 65 (Medicaid) and for individuals aged 65 or older (Medicare). These social security interventions are aimed at individuals on low incomes as well as other criteria. Including these cash transfers introduces distortions in the measurement of poverty: as Ruggles (1992, p. 7) remarked, '…some families who are eligible for Medicare or Medicaid theoretically receive resources above the poverty line in medical benefits alone!'. Two approaches to considering medical out-of-pocket expenses are to reduce net disposable income or to increase the poverty lines (i.e. the thresholds) (Short and Garner 2002). For year 2015, the official child poverty rate stood at 20.1 per cent and the elderly (65+) poverty rate was 8.8 per cent. However, subtracting medical out-of-pocket expenses from income reduced the measure of child poverty to 16.1 per cent and increased that of the elderly to 13.7 per cent (Renwick and Fox no date, Figure 6). Hence, incorporating medical expenses would substantially change recorded poverty rates. Moreover, it would also modify the relative situation between age groups (as well as ethnic groups, family types, regions, etc.).

In 2011, the Supplemental Poverty Measure (SPM) was launched (Short 2011), not to replace the official poverty rate (because some pieces of legislation had defined it to establish eligibility and funding distribution for some programmes), but to provide additional information about economic conditions and needs and to guide policy. The SPM differed from the official poverty measure with regard to the measurement units, the poverty threshold, the updating mechanisms, and because the SPM deduced out-of-pocket medical expenses to define income. The SPM consists of cash income plus any in-kind government benefits (such as food stamps or housing subsidies) minus non-discretionary expenditures (say, taxes). Crucially, medical out-of-pocket expenses are included as non-discretionary expenditures and therefore subtracted from cash income to obtain an estimate of the SPM (Bridges and Gesumaria 2013).

Using the SPM, poverty rates among older people were below the overall poverty rate between 2010 and 2015, but in 2016 the measure of poverty for individuals aged 65 or over overtook that for the general population. Fox and Pacas (2018) reported that this surge in poverty among older people was the result of an increase in the share of older people in deep poverty (i.e. below 50 per cent of the poverty threshold) and in persistent (or chronic) poverty (see below), combined with the fact that increases in medical expenses pushed older people just above the threshold into poverty (see also Fox et al. (2015) and Wimer et al. (2016) for trends of poverty rates for older people using the SPM adjustment to historical figures since the late 1960s.).

Whilst the subtraction of medical out-of-pocket expenses from income has gained acceptance among poverty scholars,[4] there have been discussions about whether to include both liquid and near-liquid asset portfolios (Chavez et al. 2018).

5.1.3 Relative Poverty

Relative poverty is related to not being able to afford a living standard deemed socially acceptable in a country or region where a person lives. Townsend (1979, p. 31) adopted a relative approach when he defined the 'poor' as people who 'lack the resources to obtain the types of diet, participate in the activities and have the living conditions and amenities which are customary, or are at least widely encouraged or approved, in the societies to which they belong', that is, individuals whose 'resources are so seriously below those commanded by the average individual or family that they are, in effect, excluded from ordinary living patterns, customs and activities'. This is also the approach adopted by the European Commission, which defined 'the poor' as 'persons, families and groups of persons whose resources (material, cultural and social) are so limited as to exclude them from the minimum acceptable way of life in the Member States in which they live' (*Council Decision of 19 December 1984 on specific Community action to combat poverty (85/8/EEC)* 1984, Art. 1 Section 2).

Relative poverty is operationalised in relation to an income threshold below which a person is classified as living in poverty. This threshold is known as the poverty line. The poverty line *arbitrarily* is set at between 40 per cent or 70 per cent of the median household income in the country; with either 50 per cent or 60 per cent being the most common levels.[5] The median household income refers to the income level that divides the household income distribution exactly in two equal parts: 50 per cent of households earn more than the median and the other 50 per cent earn less. Therefore, using these thresholds, a household is defined as living in relative poverty if they earn less than, say, 60 per cent of the poorer half of the household population. I defined these poverty lines as 'arbitrary'[6] because there is nothing anything intrinsically related to the experience of poverty in any of these thresholds, nor have they been estimated on the basis of rigorous analysis.

Furthermore, poverty estimates vary noticeably according to where the lines are drawn, and given that in most countries there is a large concentration of households around these lines, even a paltry additional sum of money can lead definitionally 'out of poverty' many a household. Again, this is especially the case among retirees: the income distribution of people in retirement (or households whose head is retired) exhibit a 'spike' around a narrow modal interval. In statistical terms, the income distribution of people in retirement is more leptokurtic than that of the general population, so if pensioners experience even minor increases or decreases in their income above or below the modal interval, poverty measures register almost seismic movements (Osberg 2001).

Two standard measures of relative poverty are the headcount rate and the poverty gap.[7]

The Headcount Rate

The headcount rate is the ratio between the number of people or households ('units') below the poverty line and the total population. An income threshold (z) is set beforehand and an observation unit (an individual, household, or family) earning less than that income is classified as poor. Let's assume that there are N units indexed $i = 1, 2, \ldots n$ in a country, earning y_i each. A unit is counted as poor if $y_i \leq z$, so that a dummy variable can be created that is equal to 1 if the unit is classified as poor and 0 otherwise:

$$I(y_i, z) = 1 \text{ if } y_i \leq z$$
$$I(y_i, z) = 0 \text{ if } y_i > z \tag{5.4}$$

The number of units in poverty is:

$$G = \sum_{i=1}^{i=N} I(y_i, z) \cdot n_i \tag{5.5}$$

and the headcount rate becomes:

$$H = \frac{\sum_{i=1}^{i=N} I(y_i, z) \cdot n_i}{\sum_{i=1}^{i=N} n_i} = \frac{G}{N} \tag{5.6}$$

The Poverty Gap

Being below a given poverty line gives no indication of how far a poor unit is from that threshold: the headcount rate is uninformative about how 'deep' into poverty the units classified as poor actually are. The poverty gap is an attempt to indicate how far each unit classified as in relative poverty is below the poverty line. It is based on the difference between the poverty line and a unit's income: $z - y_i$, for each unit i. The poverty gap is calculated thus:

$$P_G = \sum_{i=1}^{i=N} [(z - y_i) I(y_i, z)] \cdot n_i \quad (5.7)$$

Equation (5.7) gives the amount of cumulative income below the poverty gap, the shortfall in income among units below the poverty line. By dividing the poverty gap by the number of units in poverty (i.e. P_G/G), the average gap or shortfall per person in poverty is obtained. Two other indicators are the ratio between the poverty gap and the poverty line (i.e. P_G/z) and the ratio between the poverty gap and the total population (P_G/N).

Other measures in use in the literature include the FGT poverty index, proposed by Foster et al. (1984):

$$\text{FGT} = \frac{1}{N} \sum_{i=1}^{i=N} \left[\frac{(z - y_i)}{z} \right]^\alpha \cdot I(y_i, z) \cdot n_i \quad (5.8)$$

where α is a weight that measures the sensitivity of the distance to the poverty line, that is, the degree of poverty aversion. The poverty gap is, therefore, a particular case in which all units are given the same weight ($\alpha = 1$). The headcount is also a particular case, with $\alpha = 0$. The most common value for the weight is $\alpha = 2$, the squared poverty gap, which gives more weight to improvements in the incomes of the poorest units (Morduch 2005).

Inequality Among Whom?

An important consideration is that relative poverty measures among older people are based on a threshold level estimated for the income distribution of the whole population, not the older population. Let's accept that poverty is relative and that it is measured in relation to an income threshold set at 60 per cent median income. But a question remains: the median income of which income distribution? That is, the median of whose income? Poverty lines are always based on the income of the whole population. So, the approach classifies, say, women aged 75 years or older living on their own in poverty if their income is below *that* poverty line. Unfortunately, studies of poverty in later life have not used a threshold level based on the income distribution of older people alone. Remember that the rationale of using relative poverty measures is to identify those individuals who cannot afford 'the living conditions and amenities which are customary, or are at least widely encouraged or approved'. But are the customary, socially encouraged, or approved conditions and amenities the same for 75-plus-year-old women as for the general population? And, as Bogliacino and Maestri (2014, p. 23) noted:

> The choice of the reference population for the estimation of inequality is particularly important for the study of its social and political impacts, since people's behaviour may respond to changes in income inequality in their reference group, but that may not coincide with the national population.

Veit-Wilson (2002, p. 315) stated that 'most serious scholars' define relative poverty 'in direct terms of socially participatory and non-excluding lifestyles which can only be identified relative to society, time, place and observer'. You may object that equivalisation irons away any differences between households. However, this applies (if at all) to adjustments for economies of scale within households as a result of the sharing of costs, not to adjustments for lifestyles taken almost for granted for an average household in the particular society under study. For example, Fig. 5.1 presents the distribution of equivalised net household incomes after housing costs for all individuals in the United Kingdom and for pensioner households. The official poverty line in 2015/2016 was £255 per week. This was based on the median income for the whole population. Applied to the distribution of income among pensioners' households, the official pensioner poverty rate was 16 per cent. However, using the pensioner poverty line estimated as the 60 per cent median pensioners' income would render 18.7 per cent.

The Sen-Shorrocks-Thon Index of Poverty Intensity

The Sen-Shorrocks-Thon or poverty intensity index applies weights that decrease with the rank order of individuals: each individual is assigned a weight that reflects the distance between her income and the poverty line. Therefore, this index is a weighted sum of the poverty gap ratios, where the weights are defined according to the individual position in the income distribution

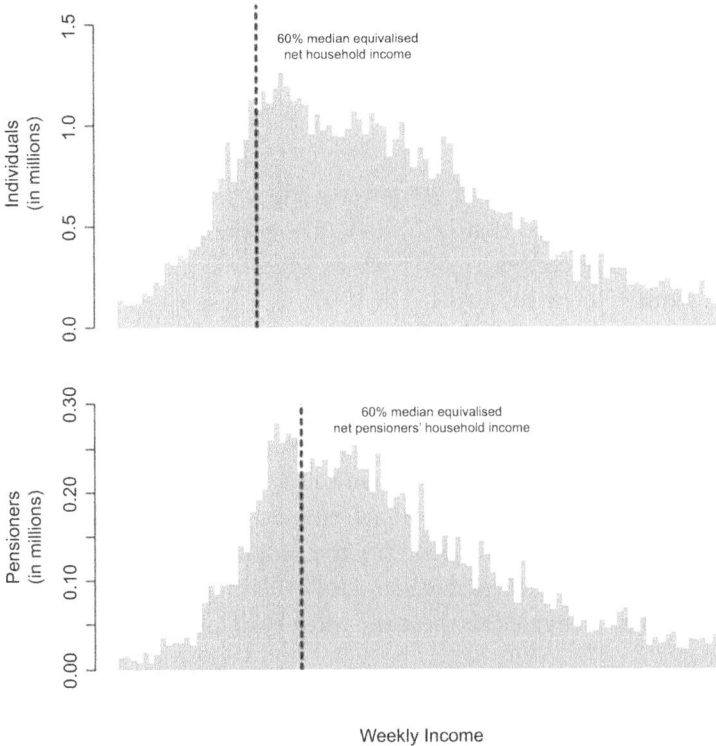

Fig. 5.1 Income distribution (after housing costs) and poverty lines of the whole UK population and pensioner households. *Source: Estimation by the author from the Family Resources Survey, United Kingdom (2015/2016)*

(Aguirregabiria 2006). In symbols,

$$\mathrm{SST} = \sum_{i=1}^{i=P_G} \left(\left[\frac{2(n-i)+1}{n^2} \right] \cdot \left[\frac{(z-y_i)}{z} \right] \right) \quad (5.9)$$

Osberg and Xu (2000) demonstrated that the SST poverty intensity index is equal to the product of the headcount ratio, the average poverty gap between the income of units in poverty and the poverty line, and one plus the Gini index of the poverty gaps of the population in poverty (note that this is not the Gini coefficient of the distribution of income, but of the distribution of poverty gaps)—see also Xu et al. (2001):

$$\text{SST} = \left(\frac{G}{N}\right) \cdot \left(\frac{P_G}{z}\right) \cdot \left[1 + \text{Gini (gap)}\right] \qquad (5.10)$$

where Gini (gap) is the Gini of the poverty gap.

Osberg (2001) showed that in Canada, the poverty intensity index among households whose heads were 65 years or older went down, after taxes and transfers, from 13.6 in 1973 to 1.7 in 1997, as a result of a reduction in the poverty gap from 26.2 to 15.8 and in the poverty rate from 28.4 to 5.4 during the period.

Iceland (2012) reported that, for the United States of America, relative poverty rates among older people were higher than for other adult groups in contrast to absolute poverty rates, because a higher proportion of older people would have incomes above the official poverty lines but below the relative thresholds. This author concluded that 'the proportion of [older] people who did not meet a socially acceptable standard of living in 2000 was higher than that of those who struggled to meet the most basic of physical needs' [p. 43].

5.1.4 Multidimensional Approaches

Many scholars have taken issue with the use of income or wealth as the exclusive poverty measuring elements, propounding that poverty manifests in many other spaces or dimensions. The adoption of a multidimensional definition of poverty must circumvent seven technical problems (Alkire and Foster 2007):

- which dimensions or spaces to include
- which thresholds or cut-off level to set for each dimension
- how to weigh each dimension
- how to allow for interactions between dimensions
- how to identify the multidimensionality (i.e. if, say, fifteen dimensions are included, when a unit is defined as poor? When it falls short in three dimensions? Four? All?)
- how to implement anti-poverty measures, given the multidimensionality of poverty
- how to deal with ordinal data

The United Nations adopted a Multidimensional Poverty Index (MPI) in 2010, revised in 2014, which is based on a family of measures introduced by Alkire and Foster (2007) known as AF indices.[8] It consists of three dimensions: health, education, and standard of living. Health is operationalised by two indicators: nutrition and child mortality. Education is measured with years of schooling and number of children enrolled. The standard of living is captured by means of cooking fuel, type of toilet, drinking water, electricity, floor (poverty is defined by having a home with dirt, sand, or dung floor), and not having at least a radio, TV set, or telephone, or at least a bike, motorbike, car, cart, and so on, or not having a refrigeration, arable land, or livestock (UNDP 2016). The United Nations' version of the MPI does *not* include income among its dimensions. This omission is not due to income being deemed as unimportant or unrelated to poverty, but because the index is considered a complementary measure of income poverty.

The MPI and other AF indices are obtained by means of the 'dual approach' to measuring poverty. Let's denote by $J = 1, 2, \ldots J$ the number of dimensions to be included, and by $i = 1, 2, \ldots n$ the number of units of observation (households, individuals, children only, etc.). With data on these units for all the dimensions, a matrix of $A_{i,j}$ elements can be constructed, denoting the level of achievement of each unit in each of the dimensions (e.g. individual 1 gained a doctorate, owns a car, and individual 2 has not finished primary school and cannot afford a bicycle). For each dimension, a cut-off point z_j is chosen, below which a unit is classified as deprived along that particular dimension. Therefore, a binary variable $D_{i,j}$ can be created which adopts a value equal to 1 if $A_{i,j} < z_j$ and to zero otherwise. Finally, we can think of different weighting factors w_j per dimension to reflect that each dimension may contribute in unequal terms to poverty. The weighted sum of the deprivation variable can now be estimated:

$$c_i = \sum_{i=1}^{i=n} \sum_{j=1}^{j=J} w_j \cdot D_{i,j} \qquad (5.11)$$

For each dimension $D_{i,j}$ classified, each unit as either deprived or not. We now need a poverty cut-off that defines each unit as either poor or not in terms of the number or percentage of dimensions in which it was classified as deprived: if a unit scores more than, say, 50 per cent in the weighted sum of deprivation indicators, c_i, then it is defined as poor. The cut-off points are chosen by means of the inter-relationships among indicators across the dimensions. No matter how careful the selection of indicators, correlations of around 50/60 per cent (e.g. between joblessness and morbidity) cannot be avoided, but the point is that these correlations must not be so high that one of the indicators may be discarded as redundant. A rule of thumb is to choose a cut-off point so that a unit is classified as poor if it is deprived in at least two dimensions (not indicators). Generally, a cut-off point of $k = 0.40$ is adopted, but it may vary across studies.

The dimensions to be included depend on the society and time under study. Stiglitz et al. (2009, pp. 14–15) listed eight 'key dimensions' of well-being, which have informed the choice of dimensions in some studies of multidimensional poverty:

- Material living standards (income, consumption and wealth)
- Health
- Education
- Personal activities including work
- Political voice and governance
- Social connections and relationships
- Environment (present and future conditions)
- Insecurity, of an economic as well as a physical nature

As can be seen, these dimensions are presented in fairly broad terms. The objective is to adjust the list of indicators to the particular circumstances of each country. It goes without saying that the indicators used to measure the dimensions (and sometimes, the list of dimensions itself) change across countries, particularly, depending on the level of economic development. For example, having a home with dirt, sand, or dung floor would be a meaningless and most ineffective measure in Canada where, perhaps, not having a refrigerator and a kitchen sink with a tap/faucet would be more attuned with the socio-economic reality. Someone told me that on studying 'cold homes' in England and Finland, the researchers (based in England) were astonished to find the relatively large proportion of houses in Finland without double glazing, a marker suggestive of possible relative poverty in England, only to realise (later) that not having *triple* glazing is an indicator of financial difficulties in that country. For this reason, the contents of the MPI vary across countries, which makes it a relevant measure but hardly comparable internationally.

Alkire and Foster (2007) estimated multidimensional poverty for Indonesia in 2000 and the United States of America in 2004. For Indonesia, eight dimensions were selected: expenditure, body mass index, years of schooling, cooking fuel, drinking water, sanitation (i.e. where householders go to the toilet), sewage disposal, and solid waste disposal. For the United States of America, four dimensions were chosen: income against poverty line, self-reported health, health insurance, and years of schooling. Not only do the lists of dimensions differ, but the poverty cut-off point for schooling in Indonesia was set at six years, roughly the duration of primary school studies, but in the United States of America, having received a high school diploma or not was the dividing line. In developed countries, multidimensional poverty has translated into the notions of multiple deprivation, social exclusion, and consistent poverty.

Multiple Deprivation

Multiple deprivation is a count of items units go without due to financial reasons, health, disability, or social exclusion. It is important to note that reasons why people cannot buy or experience certain items go beyond income and wealth for the lack to be indicative of deprivation. Only if a respondent to a survey says that she does not have or experience an item because those items are irrelevant to her or because she does not want them or like them, the lack is not considered a marker of deprivation.

In 2014, the European Union prepared a module on material deprivation (MD) to measure social exclusion and poverty. The following list of adult and children's items at household and personal levels was drawn up, whose unaffordability would be indicative of material deprivation (*Commission Regulation (EU) No 112/2013 of 7 February 2013 implementing Regulation (EC) No 1177/2003 of the European Parliament and of the Council concerning Community statistics on income and living conditions (EU-SILC) as regards the 2014 list of target secondary variables on material deprivation* 2013):

(a) Adult items at household level:

- Financial stress
    - Replace worn-out furniture

(b) Adult items at personal level:

- Basic needs
    - Replace worn-out clothes by some new (not second-hand) ones
    - Two pairs of properly fitting shoes (including a pair of all-weather shoes)

- Leisure and social activities
  - Get-together with friends/family (relatives) for a drink/meal at least once a month
  - Regularly participate in a leisure activity
  - Spend a small amount of money each week on yourself
- Durables
  - Internet connection for personal use at home
- Mobility
  - Regular use of public transport

(c) Children's items collected at household level:

- Basic needs
  - Some new (not second-hand) clothes
  - Two pairs of properly fitting shoes (including a pair of all-weather shoes)
  - Fruits and vegetables once a day
  - One meal with meat, chicken, or fish (or vegetarian equivalent) at least once a day
- Educational or leisure needs
  - Books at home suitable for their age
  - Outdoor leisure equipment
  - Indoor games
  - Regular leisure activity
  - Celebrations on special occasions
  - Invite friends round to play or eat from time to time
  - Participate in school trips and school events that cost money
  - Suitable place to study or do homework
  - Go on holiday away from home at least one week per year

A revision was proposed in 2017 (Guio et al. 2017), which recommended the following condensed list of items:

- Some new clothes
- Two pairs of shoes
- Getting together with friends
- Leisure activities

- Pocket money
- Unexpected expenses
- Holiday
- Replace worn-out furniture
- Avoid arrears

A study on multidimensional poverty using data from the United Kingdom for 2009/2010 used the following indicators to measure poverty among pensioner households (Wood et al. 2012):

- Financial
    - Low income (70 per cent of median income)
    - Bill payments arrears
    - Subjective financial situation
- Education
- Material deprivation
    - Lack of consumer durables
    - Missing out on social and leisure activities
    - Car ownership
- Housing
    - Tenure
    - Overcrowding
    - Fuel poverty[9]
- Health and well-being
    - Physical health
    - Mental health
- Local area and social networks
    - Neighbourhood deprivation
    - Level of support from neighbours
    - Level of support from family
    - Participation
    - Interest in politics
- Whether household contains a single adult or multiple adults

Wood et al. used statistical technique known as 'latent class analysis' and identified five different groups of pensioner households on low income, which varied according to educational attainment, ownership of durable goods and cars, housing tenure, degree of interest in politics, level of deprivation of the area of residence, and so on, but they all earned an income below 70 per cent of median household income. This study, therefore, combines a measure of income poverty with multiple deprivation indicators. The five latent classes are:

- *stoics*, about 33 per cent of all pensioner households in poverty. Renters with low qualifications living in more deprived neighbourhoods, though reporting not struggling financially
- *coping couples*, about 23 per cent of low-income pensioner households. Extremely low income but little disadvantage; home owners, high qualifications and living in least deprived neighbourhoods
- *cheerful grans*, about 20 per cent of low-income pensioner households. Mostly female living alone, more likely home owners in less deprived neighbourhoods, with good physical and mental health
- *trouble shared*, about 18 per cent of pensioner households in poverty. Living in more deprived areas, home owners, but reporting struggling financially, and have some mental and physical health problems
- *left alone*, about 8 per cent of low-income pensioner households. Extreme deprivation, struggling to pay bills, living alone, half renters, experiencing fuel poverty, and high prevalence of mental and physical health problems

I mentioned in the previous section that some researchers have obtained different estimates of relative and absolute poverty rates among older people in the United States of America. This discrepancy was also reported for the United Kingdom using a relative poverty indicator and a multiple deprivation measure: with data for pensioners aged 65 or over from 2011/2012, Kotecha et al. (2013) found that 12 per cent were living below the poverty line, 7 per cent were materially deprived, and merely 2 per cent were in both categories.

Social Exclusion

Social exclusion refers to not being able to participate in society. It is generally considered a wider concept than poverty, denoting other disadvantages related to social participation and marginalisation. Usually, poverty is understood as only one of the constitutive elements of social exclusion.

Burchardt et al. (2002) identified four domains from which individuals may be socially excluded: consumption, production, political engagement, and social interaction. In turn, Ogg (2005) identified the following constituent elements: social interaction with friends and relatives, participation in social activities, self-rated physical health and mental health, self-rated income, and quality of the local neighbourhood. The risk of social exclusion is particularly acute in later life because key transitions which are more prevalent in old age, such as retirement or widowhood, are associated with experiencing social exclusion. Furthermore, ageism drives older people into withdrawal and marginalisation. Other predictors of social exclusion in later life include having been socially excluded earlier in the life course, living alone, renting, low income, dependency on benefits, gender, ethnicity, and neighbourhood conditions such as poor housing and transport services.

The definitions of social exclusion usually combine political, social, and economic factors. This is also present in the definition proposed by Estivill (2003, p. 19):

> An accumulation of confluent processes with successive ruptures arising from the heart of the economy, politics and society, which gradually distances and places persons, groups, communities and territories in a position of inferiority in relation to centres of power, resources and prevailing values.

Another element that is frequently included, which is not always the case in operational definitions of multidimensional poverty, is the characteristics of the local area or neighbourhood of residence (Ogg 2005; Scharf et al. 2005; Scharlach and Lehning 2013; Walsh et al. 2017).[10] According to Buffel et al. (2013), the amenities and characteristics of the local area gain in importance in later life because:

- older people tend to spend more time in their homes than younger people
- when not at home, older people tend to spend more time in their locality than younger people
- older people tend to depend more on local relationships for support than younger people
- older people tend to exhibit less geographical mobility so they would have resided, on average, longer in the same locality than younger people
- older people tend to be more emotionally attached to their community than younger people

Consistent Poverty

Another approach that combines a measure of headcount poverty with two or more deprivation items is the 'consistent poverty' metrics. For developed countries, a unit is classified as 'consistently poor' if its income (usually the variable of choice in this case) lies below the poverty line *and* it cannot afford at least two of the following items[11]:

- Two pairs of strong shoes
- A warm waterproof overcoat
- Buy new not second-hand clothes
- Eat meat with meat, chicken, fish (or vegetarian equivalent) every second day
- Have a roast joint or its equivalent once a week
- Had to go without heating during the last year through lack of money
- Keep the home adequately warm
- Buy presents for family or friends at least once a year
- Replace any worn-out furniture
- Have family or friends for a drink or meal once a month
- Have a morning, afternoon, or evening out in the last fortnight for entertainment

Whatever the items included in the lists, there are three approaches about how to assess their relative importance to deprivation or consistent poverty (Hulme and McKay 2013):

- A simple count based on equal weights, so that, say, going without a roast joint or its equivalent once a week is as important as going without heating the home
- Prevalence weights, based on how extended the goods or experiences are in the society. The assumption is that the more prevalent an item is, the more necessary it becomes and therefore the more deprived become those people who have to do without them. Here we can see how the definition of a 'necessity' is a social construct. In 1972, only 42 per cent of households in Great Britain had a telephone. In turn, in the United States of America, by 1970 only 13 per cent of households did *not* have a phone. Whether having a telephone in the home was a 'necessity' or not in Great Britain in the early 1970s, it was much less prevalent than in the United States of America; therefore its absence, even due to financial reasons, would not have been considered a marker of deprivation according to this approach, although it

could be argued that it would have been so in the United States of America. By 1995, in turn, the proportion of households with a telephone in Great Britain had gone up to 93 per cent; by then, going without it would have been considered an element of deprivation.[12]

- Differential weighting, where the focus is not on the list of items but on the list of reasons as to why the respondent says she does not have or experience each item. Going without an item because she does not have the money to buy it gets a higher weight than health or disability reasons.

Poverty multidimensionality entails that units may be classified as poor on one dimension but not on another. For example, in their study on multidimensional poverty in Italy in 2004, Coromaldi and Zoli (2012) reported that older people exhibited better scores in finance-related indicators (e.g. arrears on utility bills, capacity to afford a meal, transport or medical treatment; possession of a dishwasher, a computer or a car; and housing-related problems such as noise, pollution, or crime) than younger individuals, but worse values in housing facilities (e.g. having a shower or bath or hot water) and health. Overall, the authors concluded that poverty was less prevalent among older people than younger individuals in 2004 in Italy. Furthermore, Suppa (2015) analysed multidimensional poverty in Germany in 2001–2002, 2006–2007, and 2011–2012 and reported that all the indicators worsened with age. Moreover, the relative contribution of each dimension also changed with chronological age: in later life, deprivation scores in health and social participation become more relevant for multidimensional poverty.

Studies of poverty in later life show a snapshot of the older population in a country in one period of time. It may be argued that an element of the poverty levels in certain dimensions is the result of cohort effects and that therefore poverty in later life will improve. One such dimension is education. As already mentioned, in developed countries, multidimensional poverty studies include having a high school diploma as a watershed to count individuals as poor with regard to education. With increasing numbers of younger people finishing not only high school but university studies, an immediate conclusion would be that there will be fewer poor older people by the time currently younger cohorts reach later life. However, poverty is not only multidimensional but dynamic: socially defined necessities change over time, and so do the definition of poverty.

5.1.5 Mortality-Adjusted Poverty Rates

When studying poverty among older people, it must be taken into account that measures of poverty in later life may be biased as the life expectancy of people in poverty is shorter than that of people not in poverty, and for many, it may not be long enough for them to reach the category of 'older people': poverty is a key determinant of gaps in life expectancy in a population. This bias is the cause of the mortality paradox.

The mortality paradox refers to the statistical artefact that renders lower poverty rates the worse the survival conditions of the poor (Kanbur and Mukherjee 2007; Lefebvre et al. 2013).

Ponthière (2017, Box 9) presented a simple algebraic framework to explain the bias in poverty measures due to excess early mortality among poor older people. It consists of a bias that drives *down* the poverty rates among older people against other age groups. Imagine a number of n_p individuals are classified as living in poverty in a population of size n, so that $n_a = n - n_p$ do not live in poverty. The population is divided between young (y) and older individuals (o). We assume that the life expectancy of younger poor and non-poor people, which we denote by l_p and l_a, respectively, varies according to whether they live in poverty or not such that $l_p < l_a$.

The poverty rate among young people is:

$$\frac{y_p}{y_p + y_a} \tag{5.12}$$

In turn, the poverty rate among older people is:

$$\frac{y_p \cdot l_p}{y_p \cdot l_p + y_a \cdot l_a} \tag{5.13}$$

Equation (5.13) is equal to:

$$\frac{y_p}{y_p + y_a \cdot \frac{l_a}{l_p}} \tag{5.14}$$

Comparing Eqs. (5.14) and (5.12), we can see that

$$\frac{y_p}{y_p + y_a \cdot \left(\frac{l_a}{l_p}\right)} < \frac{y_p}{y_p + y_a} \tag{5.15}$$

Therefore, the poverty rate in later life is lower than the poverty rate at younger ages, simply because of the mortality differential against the poor.

Solutions to this paradox rest on the use of lifetime income profiles extended to account for early mortality among the poor, that is, lifetime incomes *as if* the prematurely deceased individuals were alive and were earning an income. The exact adjustments can be seen in Lefebvre et al. (2013, 2014), where the 'missing' older people were factored in the poverty measures. Lefebvre et al. and Lefebvre et al. used estimates of the (non-linear) relationship between income and life expectancy at older ages separately for men and women in two regions in Belgium and adjusted poverty rates by incorporating the 'missing' individuals. Using a relative income definition of poverty, poverty rates after the adjustment fell compared to official non-adjusted poverty rates. However, using an absolute poverty definition, the adjusted poverty rates were higher than the non-adjusted measure by about 6–7 points.

In turn, Lefebvre et al. (2018) looked at poverty rates among older people in eleven European countries from 2007 using the FGT measures (see above) and two alternative procedures to estimate the lifetime income of early deceased individuals. In all cases, adjusted poverty rates were higher than non-adjusted rates, although the gaps varied across country, gender, definition of poverty and of lifetime income. These authors concluded that the missing persons' bias was relevant even for higher income countries, as 'taking the "missing poor" and "hidden poverty" into account creates a much greater contrast with the standards of the surviving populations' [p. 457].

The conclusion in Lefebvre et al. (2013) is worth repeating:

> Suppose that a government can choose between two policies: on the one hand, a transfer program towards the elderly poor; on the other hand, a free access to health care services below some income level. Undoubtedly, the first policy would reduce the standard old-age poverty rate, but the second policy, by increasing the number of poor persons surviving to the old age, would raise the old-age poverty measure. Relying on standard poverty measures would thus favor the first policy. On the contrary, if one uses adjusted old-age poverty measures, the first policy, by focusing on the surviving old, would only reduce poverty to a small extent, whereas the second policy, by preventing lots of premature deaths, would strongly reduce the (so-measured) poverty. Thus the way in which poverty is measured is far from neutral regarding the assessment of social policies.
>
> (Lefebvre et al. 2013, pp. 310–311)

5.1.6 Subjective Poverty

Another approach to defining and measuring poverty is to ask people the so-called minimum income question: how much money/income they think is needed to 'make ends meet'. As we see, it is not a self-reported poverty status; the question is not about whether the respondent feels or thinks she is poor. Neither the respondent is invited to rank herself along a 'poverty ladder' as in happiness studies (see Chap. 1 in this volume). Here the focus is on a minimum sum below which the respondent understands could suffer from financial distress (see below). Therefore, in order for this question to be the basis of a poverty measure, the reported minimum level is compared with the equivalised household income of the respondent and a subjective poverty line is derived.

One problem with this approach is that it may lead to inconsistencies because individuals on the same level of income may report different levels as the absolute minima. Some studies found that self-reported minimal levels increase with household income: in fact, the association between household income and self-reported minimum income follows a non-linear, positive but diminishing pattern.

In Fig. 5.2, z^* is the social subjective minimum, which corresponds to the 'subjective poverty line' (Pradhan and Ravallion 2000). Econometric analyses allow to estimate a social subjective minimum from survey data (Goedhart et al. 1977).

Several studies compared poverty rates derived from relative and subjective measures; some of these studies focused on older people. Three examples:

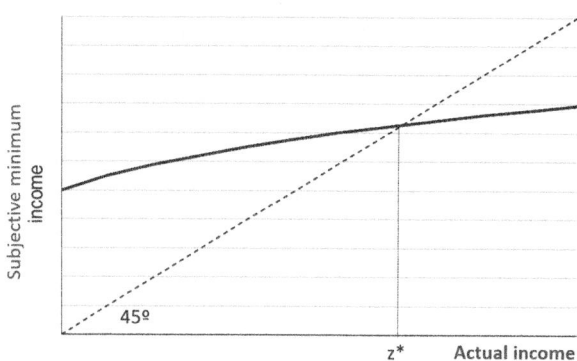

Fig. 5.2 Subjective poverty line. *Source: Figure is illustrative, prepared with mock data*

- Using 1995 data from the United States of America, Garner and Short (2003) reported that whilst the official poverty rate (i.e. Eq. (5.1.1)) among people aged 65 or over living alone was 16.1 per cent and using a poverty line of 50 per cent median income 23.5 per cent, the subjective poverty rate reached 38.3 per cent. For two-person households with the head aged 65 years or over, the official rate was 4.9, using the 50 per cent median income line, it amounted to 14.4 per cent, but with the subjective measure, the poverty rate was 18.1 per cent.
- In a comparative study using 2001 data from Madagascar, Lokshin (2004) reported that older people living on their own had the lowest objective poverty rate (based on caloric intake) of other household types, but that 53 per cent of these people considered their levels of spending on food or clothing inadequate. Though their subjective poverty indicator was lower than for other household types, older people living alone exhibited the widest discrepancy between objective and subjective measures of poverty.

Some studies focusing on the effects of poverty on other outcomes also adopted a comparative approach. For example, Adena and Myck (2014) looked into poverty as a determinant of changes in health among people aged 50 or over in twelve European countries between 2010 and 2014. The findings suggest that poverty defined as relative income had no statistically significant effects on health, but that subjective poverty significantly increased the probability of transitioning into poor health and reduced the probability of recovering. Furthermore, subjective poverty was highly associated with mortality.

5.1.7 Financial Distress

A measure that combines income and wealth and is based on the life-cycle framework is the financial distress indicator, also known as financial fragility. The life-cycle hypothesis states that economic agents seek to smooth consumption to maintain a similar standard of living over their lives. This goal would lead them to save during income-generating periods to fund consumption during periods not in paid employment, particularly in retirement. Therefore, at any stage in the life cycle, an economic agent may need to draw on their savings (i.e. accumulated wealth) to face unexpected expenditure obligations or an interrupted generation of income. Financial distress measures the relationship between wealth and income.

To serve as an element in the measurement of financial distress, wealth has to be computed net of debt, or if a gross wealth estimate is used, then gross debt is also to be included in the analysis. The importance of debt as a financial stressor is not only a question of the amount owed, but of its time to maturity, collateral, and the ratio between instalments and income.

Financial distress is defined as the ratio between net wealth and income. According to Christelis et al. (2009, p. 369), it 'indicates households' capacity to handle future income declines, either expected or unexpected…' and is a measure of risk of poverty: '…households with low net worth income ratios are at risk of being poor in the future and are more vulnerable to shocks' [p. 369]. The operationalisation demands decisions around timing and ratio levels. In practice, then, financial distress is measured as financial wealth net of non-mortgage debt that does not exceed three months of household income. Using data for eleven European countries from 2004, Christelis et al. (2009) reported that around 60 per cent of households headed by a person aged 65 or over in Greece, Italy, and Spain were in financial distress. Furthermore, the authors found significant differences by housing tenure: renters were more exposed to poverty risk than home owners; in Denmark, this gap was most noticeable with almost 60 per cent of tenants under financial distress against 30 per cent of outright owners.

Cavasso and Weber (2013) modified the measure of financial distress to the one proposed by Christelis et al. by adding to the relationship between net financial wealth and income to be lower than three months the condition that household per capita income is not in the top third of the income distribution. The authors applied this alternative measure to data of households headed by a person aged 65 or over from 16 European countries for 2011 and reported significant differences across countries and between renters and home owners. Around 10 per cent of home owners were in financial distress in Sweden, Denmark, and Switzerland but over 50 per cent of home owners were at risk in Estonia, Poland, Hungary, and Slovenia. Portugal and Spain also showed rates of financial distress exceeded 50 per cent for renters.

Bonfatti et al. (2013) applied the definition by Cavasso and Weber to analyse changes in financial fragility of older European households between 2011 and 2013. They reported that the proportions of households in distress remained fairly stable across countries, although within each country there was a sizeable transitioning in and out of financial fragility. The following characteristics of the household head were associated with the probability of falling into financial distress: being a woman, being a foreigner, having poor health, low educational attainment, small social network or low income, renting, not being in paid employment, and age: the older households, the less likely the household would have transitioned into financial distress. Related to this final finding, the authors reported that 'the presence of a partner and the size of the social network are significant protective factors only for the older group' [p. 232].

Moreover, Brunetti et al. (2016) carried out a study of financial fragility in Italy using a different definition of distress: 'to be able to afford expected expenses, but not to have a sufficient liquidity buffer to face unexpected ones'. Not exclusively focused on older households, Brunetti et al. reported that the probability of financial fragility increased when the age of household head exceeded 40 years old.

5.1.8 Financial Security

Financial insecurity (sometimes referred to as income or economic insecurity) was included by Walsh et al. (2017) as one of the dimensions of inequality in later life. According to Stiglitz et al. (2009), older people are exposed to two different but associated risks regarding the uncertainty about the future needs and the resources available to meet them: the risk of falling into poverty because of low retirement income and the risk of increased volatility in pension income as a consequence of pension reforms in most developed and some developing countries that have shifted provision from government to the private sector and the risks from governments and private firms to individuals. Income security is about the level of income and its adequacy. Millions of older people around the world would not have enough resources to survive without family arrangements, charitable work, and other informal arrangements in existence. Income security or income maintenance is the policy objective of providing or guaranteeing that every older person has enough income to cover their needs. The most widespread mechanism is the pension system.

5.1.9 Chronic or Persistent Poverty

I mentioned already that most measures of poverty offer but a snapshot of the condition in one point in time. However, a lifetime in poverty is not the same as a short spell due to particular circumstances. This may sound rather obvious, but the distinction between persistent (or chronic) and transient poverty is not captured by all the measures of poverty seen above. Chronic or persistent poverty is poverty (generally, measured in absolute terms) experienced throughout a prolonged and uninterrupted period of time.

Three types of chronic poverty have been identified (Shepherd 2013):

- Long-term poverty: poverty experienced for so long that unless external conditions change, individuals will remain in poverty.
- Life-course poverty: poverty experienced throughout the whole life course.
- Intergenerational poverty: poverty transmitted from parents or other carers to their children.

Concerning the spells in poverty, two important aspects are their timing and contiguousness. Timing is important because for the same duration of time in poverty, spells that took place mostly in childhood or early youth may have different effects compared to experiencing most of poverty closer to or in old age. In this respect there are two main approaches in the literature. One strand understands that poverty in childhood and early youth have more deleterious effects given their greater relevance for the rest of the life course. The opposing view considers that spells closer in time are worse than earlier, on the basis of the 'loss aversion' effect propounded by behavioural prospect theory (see Chap. 8 in this volume). Contiguousness (or duration sensitivity) refers to the duration of the poverty spells. For example, imagine that over seven years, an individual has spent four consecutive years in poverty whilst another individual has spent the first two and the last two years in poverty. The cumulative hardship hypothesis conjectures that the former individual would be more deeply affected than the other individual, all else equal, given that the gaps out of poverty could offset some of the damaging effects of experiencing an interrupted long spell in poverty. With timing and contiguousness, several trajectories of poverty can be distinguished (over seven years, there are 128 possible trajectory combinations!). Table 5.1, in which 0's stand for a period not in poverty and 1's for periods in poverty, shows five of these trajectories:

In Table 5.1, five individuals $a-e$ spent four periods out of seven ($t1-t7$) in poverty. Early scarring effects would give a higher weight to individual a than b, or to individual d than c or e. Loss aversion would give the opposite.

Table 5.1 Different poverty trajectories

	t1	t2	t3	t4	t5	t6	t7
a	1	1	1	1	0	0	0
b	0	0	0	1	1	1	1
c	1	0	0	0	1	1	1
d	1	1	1	0	0	0	1
e	1	1	0	0	0	1	1

Regarding contiguousness, individuals a and b, having spent four consecutive periods in poverty, would be worse off than the other three individuals, and individual e would be given the lowest weight as the only one who has never been more than two consecutive periods in poverty.

Different indicators of lifetime poverty have been developed: the Rodgers-Rodgers (RR); Foster (F); Calvo-Dercon (CD); Bossert-Chakravarty-D'Ambrosio (BCD); Hoy-Zheng (HZ); Gradin-Del Rio-Canto (GDC); and Mendola-Busetta indicators. They differ primordially in assumptions and axioms related to duration, timing, contiguousness, and so on. For example, CD incorporates duration sensitivity as well as a time discount factor that reduces the impact of poverty spells experienced in the past; HZ, in contrast, gives more weight to poverty in early stages of the life course.

The Rodgers and Rodgers Indicators of Transient and Chronic Poverty

Rodgers and Rodgers (1993) focused on the notion of permanent or lifetime income. A lifetime in poverty would consist in a permanent income below the poverty line. The starting point is the construction of the average annual poverty index: the weighted sum of all the poverty indices recorded for each individual over her lifetime. The authors recommended the FGT index (see Eq. (5.8)) as a measure of poverty. The following indicator captures the average annual poverty index:

$$A(T) = \frac{1}{(N \cdot T)} \sum_{t=1}^{t=T} \sum_{i=1}^{i=N} \left[\frac{(z - y_i)}{z} \right]^\alpha \cdot I(y_i, z) \cdot n_i \qquad (5.16)$$

For each individual, the lifetime income y^* is estimated following the life-cycle hypothesis. Chronic poverty can be measured as a lifetime income below a lifetime poverty line:

$$C(T) = \frac{1}{(N)} \sum_{i=1}^{i=N} \left[\frac{(z - y_i^*)}{z} \right]^\alpha \cdot I(y_i^*, z) \cdot n_i \qquad (5.17)$$

However, over a lifetime of $t = 1, 2, \ldots T$ periods, each individual may have spent a given number of periods in poverty: some people never experience poverty, some spend all their lifetime in poverty, and many others still live through spells in poverty over part of their life course (especially in their youth:

Mendola et al. (2009), in their study of youth poverty in eleven European countries, reported that an extraordinary 57 per cent of Europeans experience relative poverty at some point in their lives). Therefore, in addition to chronic poverty, transient or temporary poverty must also be measured. Rodgers and Rodgers suggested that the difference between the average annual poverty index and the chronic poverty measure would capture transient poverty over the T periods:

$$\text{Tr}_P(T) = A_P(T) - C_P(T) \qquad (5.18)$$

In Rodgers and Rodgers (2010), the authors applied their method to data from Australia between 2001 and 2007. However, instead of the FGT index, they opted for measuring poverty using a poverty line set at a given proportion of median lifetime income. Older people exhibited the highest chronic poverty rates of all age groups, though not the highest average annual poverty rates (younger people transitioning into adulthood had higher average annual poverty rates than older people, but this was mostly transitory poverty). The second age group in terms of chronic poverty were the adults transitioning into old age. Interestingly, excluding imputed rental values, chronic poverty rates among older people were three times higher than after including imputed rental values. These results were consistent irrespective of the thresholds used to set the poverty lines, and even though they were sensitive to the equivalence scale, older people still exhibited the highest rates of permanent poverty.

The Foster Indicator of Chronic Poverty

Foster (2009) proposed an indicator of chronic poverty based on a dual cut-off approach: an income and a duration cut-off. The income cut-off uses the FGT index to identify units living in poverty. The duration cut-off uses a fixed duration of poverty over time (or a proportion of the whole period under study) to identify those units who have spent a longer time in poverty over their lifetime than the cut-off; these units are classified as chronically poor.

The Calvo-Dercon Indicator

Calvo and Dercon (2009) developed a series of chronic poverty indicators according to different axioms. The authors incorporated considerations of uncertainty in forward-looking measures of poverty over time. That is, not only to look back to past trajectories but to forecast or project future individual trajectories. This indicator is based on a number of assumptions:

- Periods not in poverty may offset spells of poverty.
- Periods in which a unit is 'just so' in poverty according to the chosen definition may offset periods in which the shortfall is much deeper.
- Timing of the loss aversion type rather than the cumulative disadvantage, so that the measure includes a discount factor that reduces the impact of past poverty spells.
- Poverty vulnerability increases with uncertainty even if expected poverty rates remain unaltered: risk is bad *per se*.

The CD forward-looking measure responds to this specification:

$$\text{CD}_T = \sum_{t=1}^{t=T} \beta_{T-t} \cdot \left[1 - E\left(\frac{\tilde{y}_t}{z}\right)^\alpha \right] \quad (5.19)$$

where E is the mathematical expectation of future poverty rates $\tilde{y}_t = \text{Min}[y_t, z]$; $\beta > 0$ is the time discount factor and $0 < \alpha < 1$ is a risk coefficient.

The Bossert-Chakravarty-D'Ambrosio Indicator

Bossert et al. (2012) proposed an indicator of poverty persistence that consists of the weighted average of the individual normalised poverty scores. The weighting factors correspond to the length of the poverty spells so that consecutive spells get a higher weight than interrupted periods in poverty for any equal length of time in poverty. In symbols,

$$\text{BCD}_i^\alpha = \left(\frac{1}{T}\right) \cdot \sum_{t=1}^{t=T} \left(l^t \cdot p_{i,t}^\alpha \right) \quad (5.20)$$

where l is the length of each spell in poverty and α is a parameter that reflects whether the distance to the poverty line is considered or not. If $\alpha = 0$, only the number of periods spent in poverty are taken into consideration irrespective of the poverty intensity (i.e. distance to the poverty line); with $\alpha = 1$, the intensity is included; and with $\alpha = 2$, the poorer the individual, the higher is the weight given to her.

The Hoy-Zheng Indicator

Hoy and Zheng (2011) proposed another indicator of lifetime poverty. The HZ indicator consists of the weighted sum of all spells of poverty (measured against the poverty line but based on consumption, rather than income, levels) over the life course of an individual plus a measure of retrospective permanent or lifetime consumption. The procedure is to measure poverty levels at each period for each unit (usually, individuals) and aggregate them into a 'poverty experience' indicator and to combine this indicator with a measure of lifetime consumption. The latter accounts for periods out of poverty that may offset the spells in poverty. Hoy and Zheng recommended the application of greater weights to spells of child poverty. They added that greater weights could also be applied to poverty spells in later life. The suggested structure of the weighting factors over the life course not only recognises the stronger impact of child (and potentially old-age) poverty, but also that longer spells in poverty may be more detrimental than shorter spells adding up to the same period in poverty as a whole. For example, living for five consecutive years in poverty against living in poverty during three months every two years over a period of thirty years.

Mendola-Busetta Poverty Persistence Index

The MB poverty persistence index, introduced by Mendola et al. (2011), is based on the cumulative hardship hypothesis. The procedure consists of producing sequences of periods not in poverty (given a value equal to 0) and in poverty (equal to 1) for each unit observed over time. Let's assume we get the following sequences:

- A: 11110
- B: 01111
- C: 10111
- D: 11101
- E: 11011

The MB index captures the distance between any two periods spent in poverty and the 'poverty permanence probability', which is the ratio of the number of units classified as poor in two consecutive periods to the size of the whole population. For example, the distance between periods in poverty for individuals A and B is different than for the other three individuals, because the latter spent one year out of poverty in-between spells of poverty. Therefore, there are gaps in the poverty sequences of individuals C, D, E. In the table, the poverty permanence probability is the same for all periods because, in any two consecutive periods, there are always three out of five individuals in poverty, but of course this is very unlikely in the real world. The MB index is calculated according to this formula:

$$\text{PPI}_i = \frac{\sum_{i=j, k \in s^{(i*)}}^{i=T} (d_{j,k} + 1)^{-p_{j,k} \cdot (o_{j,k}+1)} \cdot \left\{ \frac{\left[\frac{(z_j - y_j)}{z_j} + \frac{(z_k - y_k)}{z_k} \right]}{2} \right\}}{\binom{T}{2}} \quad (5.21)$$

where d is the distance, p is the poverty permanence probability, o corresponds to the number of waves out of poverty, and s is the vector of spells in poverty.

Busetta and Mendola (2012) used the MB index in a study of poverty persistence in eleven European countries from 1994 to 2001. The authors reported that poverty tends to be more prevalent during childhood and youth, and later life, but that the patterns vary across countries: Denmark, France, Belgium, Greece, and Germany presented higher poverty persistence among older people whilst children and younger people were more likely to be in persistent poverty than older people in Italy and Spain (no clear pattern by age was found for Ireland and Portugal). Moreover, Belgium and Denmark exhibited increasing poverty levels with age.

The Gradin-Del Rio-Canto Indicator

Gradín et al. (2012) developed a chronic poverty indicator whose estimation follows a two-step procedure. First, poverty levels of units are aggregated by period and, second, these aggregated unit indices are further aggregated for the whole population. The GDC indicator captures poverty intensity (i.e. poverty gap or distance to poverty line), duration, and contiguousness. It also includes two types of aversion to inequality: that it is preferable that poverty is equally distributed among the poor instead of having inequality within the population in poverty, and that it is preferable that each unit in poverty does not experience variability in poverty levels. This indicator is expressed thus:

$$\text{GDC} = \begin{cases} \frac{1}{N} \cdot \sum_{i=1}^{i=N} \left[\frac{1}{T} \cdot \sum_{t=1}^{t=T} g_{i,t}^{\gamma} \cdot \left(\frac{s_{i,t}}{T} \right)^{\beta} \right]^{\alpha} & \text{if } \alpha > 0 \\ \frac{q}{N} & \text{if } \alpha = 0 \end{cases} \quad (5.22)$$

where g is the poverty gap; γ is a parameter that reflects the degree of aversion to individual (i.e. each unit's) inequality over time; s captures the duration of poverty spells; $\beta > 0$ is the contiguousness or duration factor giving higher values to larger uninterrupted spells; α reflects the degree of aversion to inequality among poor units in each period; and q corresponds to the number of inter-temporally poor units. Regarding the poverty gap, we have

$$g_{i,t}^\gamma = \begin{cases} \left(\frac{z_{i,t} - y_{i,t}}{z_t}\right)^\gamma & \text{if} \quad y_{i,t} < z_{i,t} \\ 0 & \text{otherwise} \end{cases}$$

Bayaz Ozturk and Macdonald (2017) applied the GDC in their study of poverty among older households in the United States of America between 2001 and 2009, broken down by ethnic composition (White, Black, and Hispanic households). The authors presented different results according to varying the values of the three sensitivity parameters (i.e. γ, α, and β). An increase in the coefficient of aversion to inequality within the poor (α), for example, the GDC index among Black and Hispanic households increases but among White households goes down. Similarly if the duration of poverty spells (β) is given a greater value.

5.2 Theories of Poverty

> One of the more vexing questions in the theory of income distribution concerns itself with the intergenerational transmission of inequality via human capital.
> (Blinder 1974, p. 22)

We saw in the previous chapter that poverty is sometimes measured with reference to income, but that other approaches are also in use. It should come as no surprise that this is also the case with theoretical developments of the causes of poverty. Regarding poverty in later life, some of the most promising approaches point to institutional, structural, and socio-demographic determinants of poverty during the life course.

One such approach was proposed by Jäntti and Danziger (2000), who developed an analytical framework to understand poverty based on stock and flows between households and three key sectors[*]: labour market, the public sector, and the capital markets. These authors distinguished between one-person and two-person households with and without children, and finally, within one-person households, between single men and women: the intersection between income, wealth, and household composition is crucial to understand poverty in later life.

Similarly, in their literature review of gerontological and social work studies on poverty, Rissanen and Ylinen (2014) identified several variables and processes that have been associated with poverty in later life. These authors grouped the causes or sources of poverty into three levels: individual, community, and political [Table 1]:

(a) Individual level

- Women
  - Longevity
  - Singlehood or widowhood
  - Earlier domestic roles (e.g. childcare)
- Weak labour market position (e.g. part time or temporary)
- Low educational level
- Lack of motivation or skills to engage in money-generating activities
- Control over household resources
- Demanding caregiving responsibilities
- Health conditions
- Health and social care expenses

(b) Community level

- Rurality
- Belonging to a minority group
- Ageism in the workplace

(c) Political level

- Later life not a priority for policies and programmes
- Budget cuts in related policy areas (e.g. housing, healthcare, long-term care)
- Unclaimed benefits

The classification suffers from juxtaposing grey areas, as almost any attempt at organising in clear-cut categories an inter-related, complex and multidimensional family of concepts from several disciplines and sub-disciplines. However, the list is comprehensive enough to guide us through the discussion.

- First, we see once again that some of the processes and causes of poverty in later life identified by Rissanen and Ylinen would find their origin earlier in the life course. In reality, most of the factors leading to poverty in later life originate much earlier in the life course, and it should not be overlooked that, by the time they retire, older people 'lack opportunity to change their finances' (Ghilarducci 2004, p. 7). More generally, as Marchand and Smeeding (2016) pointed out, there is a huge difference between the economics of the *aged* and the economics of *ageing*.
- Second, the several causes of poverty in later life are beyond the control and responsibility of the individuals. I mentioned that the categorisation presented by Rissanen and Ylinen is not immune to criticism: for example, to what extent can the feminisation of childcare responsibilities be classified as an 'individual' process? The interaction (or not) between agency and social structure, between the individual and social forces, is one of the central topics in sociological thinking. Agency—and its associated notions of control and mastery—are, as Moen (2013, p. 193) explained, 'institutionalized within the social organization of and power distribution in roles and relationships' and, as Settersten and Gannon (no date, p. 37) pointed out, these notions and the embedding of 'agency within structure bring significant challenges because they demand that boundaries between disciplines be crossed …[and] a critical evaluation of the unique nature and effects of structure agency dynamics within and across distinct life periods'. However, artificially separating 'individual' and 'social (community, political)' variables as if demarcating two parallel realms obliterates the 'creative tension' (Walker 2006) between structure and agency, between the social and the individual, and how this tension plays out in later life. Moreover, even characteristics such as gender and age, seemingly incontestably individual in nature, are in fact social constructions (see Chap. 1 in Volume I for a discussion of the different notions of age) that come with a gamut of attached expected social roles. Hence, as already mentioned, Rissanen and Ylinen's attempt is useful not as a faultless categorisation but as a handy list of factors associated with poverty in later life. However, to understand, say, why women who assumed childcare and other household responsibilities earlier in their life course are over-represented among poor older people worldwide, we need to go beyond individual characteristics. For example, Smeeding and Sandstrom showed how highly inter-related poverty, gender, and living arrangements are among older people: older women living alone

are more likely to live in relative poverty than older men and older couples. As Piketty asserts, there is nothing automatic, unavoidable, or inexorable about this outcome. Moreover, these findings suggested that marital status *per se* is not relevant: whether older people live alone as a consequence of widowhood, divorce, or singlehood, it is the combination of living arrangements and gender with its knock-on effects on the labour market, pension contributions, and socially defined caring responsibilities which conspire against the financial security of a disproportionate number of older single women—see also Saunders and Smeeding (1998), Smeeding (1999), Smeeding et al. (2008), and Veall (2007) and the special issue on gender and ageing of the journal *Feminist Economics* (Volume 11, Issue 2, 2005).
- Third, some of the factors listed above have a strong cohort dimension: low educational level may have contributed to low employment income and savings in the past, but with increasing average educational attainment in both developing and developed countries since the 1950s, this is becoming less relevant.[13]

Despite the efforts of many social scientists, theories of poverty focusing exclusively on the individuals are influential in academic and policy circles. They also reflect widespread public opinion: in 2012, an opinion poll in the United Kingdom found that 26 per cent of individuals surveyed believed that poor people lived in poverty as a consequence of laziness or lack of willpower (BritainThinks 2012). A similar opinion was held by over 30 per cent of individuals in 2007 in Malta, Latvia, Lithuania, and the Czech Republic (Kallio and Niemelä 2014). In a nutshell, a non-negligible percentage of public opinion think that the poor 'choose' poverty. And that poverty is a choice or the outcome of choice is one of the explanations that social sciences have produced to account for this phenomenon. This view attributes causality and places the normative onus upon the individuals living in poverty (Dixon 2012; Dixon et al. 2005). Almost without saying, there is an alternative view that focuses on structural factors outside the individuals, and attempts at combining both approaches.

Interestingly, theoretical developments in this area mirror a division in public opinion and lay understanding of the causes of poverty. Sociologists have developed the sub-field of 'stratification beliefs' (Feagin 1975; Hunt and Bullock 2016; Kluegel and Smith 1981), which studies the causal attributions for poverty by the 'general public'. Three types of causal attributions are usually identified:

- Individualist: the causes of poverty lie in the individuals who live in poverty, their traits and culture, and so on
- Structuralist: the causes of poverty lie outside the individuals who live in poverty, in social structures and institutions
- Fatalist: a less important view from a theoretical perspective, it assigns to luck or destiny an individual's 'lot' in life

This chapter briefly overviews some of these theoretical developments. I encourage the reader to keep in mind the following warning by Piketty (2014, p. 20):

> ...one should be wary of any economic determinism in regard to inequalities of wealth and income. The history of the distribution of wealth has always been deeply political, and it cannot be reduced to purely economic mechanisms.

5.2.1 Individualist Approaches

Piachaud (1981, p. 421) pointed out:

> To **choose** not to go on holiday or eat meat is one thing: it may interest sociologists, but it is of no interest to those concerned with poverty. To have little or no **opportunity** to take a holiday or buy meat is entirely different.

If relative poverty is a choice, should it be called poverty given this term's connotation of want and deprivation? This is a very pertinent question for studies of poverty in later life, as many older people on low incomes seemingly decide to go without certain items out of unfettered choice. But are they truly free to choose?

One influential theory that focuses on the individual proposes that poverty is a sub-culture. It contends that not every impoverished person shares the characteristics that make up this sub-culture, but for those who do, poverty is a way of life and, as such, it becomes an intergenerational trap. This is the 'culture of poverty' theory.

Sub-culture of Poverty

The culture (or, more appropriately, a sub-culture) of poverty theory posits that people living in poverty develop social and psychological traits that make poverty persistent across generations (Lewis 1961, 1966; Lewis et al. 1966). As a result of his anthropological studies,[14] Lewis et al. listed several personal characteristics of people embedded in a culture of poverty, including the following:

- low educational attainment
- social, cultural, and civic participation
- financial insecurity
- labour precariousness
- low income
- unemployment spells
- lack of savings
- high indebtedness
- overcrowded housing conditions
- proclivity to physical violence and authoritarian attitudes
- early sexual initiation
- high rates of family dissolution and abandonment
- strong orientation to the present
- weak self-control and capacity to defer or delay gratification
- fatalism, resignation, and defeatism
- lack of class consciousness and sense of history

This approach distinguishes between people living in poverty and those living in a sub-culture of poverty: not all people in a situation of poverty develop or are immersed in a sub-culture of poverty because the latter is a way of life. One relevant corollary of this theory is that there would be little social mobility within those in the sub-culture of poverty; in other words, that there would be a strong intergenerational transmission of poverty. Empirical studies of poverty at household level have produced mixed results (see next chapter): whilst some authors found no evidence of poverty traps over time (e.g. Naschold (2013), in Pakistan and Ethiopia), other authors reported persistence in the dynamics of low income and asset accumulation among households in places like rural Kenya and Madagascar (Barrett et al. 2006), South Africa (Adato et al. 2006), or Ecuador (Araujo et al. 2017). However, mirroring the proviso that not every person in poverty lives in a sub-culture of poverty, attention must be paid (which has not been the case in econometric studies) that the culture of poverty theory does not predict low mobility within the poor in general.

This is a theory about intergenerational poverty persistence. One of its main criticisms is that it basically reflects the views of Lewis et al. and others about what keeps families locked in poverty over generations. Whilst the culture of poverty theory focuses on characteristics that individuals assimilate and reproduce, an alternative theory proposes that social and parental investment on children is a key determinant of children's educational attainment and internalisation of life skills, and that it is these skills and the level of education achieved which in turn determine whether an individual will live in poverty or not. Moreover, the theory conjectures that low social and parental investment on children is more likely in poor families and neighbourhoods, which socially reproduces patterns of poverty and contributes to its intergenerational transmission (Haveman and Wolfe 1995).

Behavioural Approaches

The culture of poverty is firmly inscribed within the theoretical frameworks that attribute causality to individuals living in poverty. Behavioural economics also places its emphasis on the individuals. However, instead of focusing on cultural traits, this approach is concerned about an individual's behavioural aspects and condition (Watts 1968). According to this literature, lack of hope, myopia, lack of willpower, and lack of aspirations would be among the behavioural characteristics of individuals living in poverty (Dalton et al. 2016).

Dalton et al. (2016) proposed a theory of poverty persistence based on the notion of 'aspirations failure' derived from behavioural economics (see Part III in this volume). The theory is founded on three premises:

- How high or how much an individual aspires to achieve, earn, and so on, acts as a reference point. According to prospect theory, losses relative to a reference point would have a greater impact on an agent's utility than gains of equal amount. Therefore, higher aspirations may thrust individuals towards exerting more effort to achieve goals, but may also have negative outcomes on utility, which—if realised—will feel greater.
- Aspirations and effort mutually reinforce one another.
- Despite the positive feedback effect between aspirations and effort, decisions about how much effort to exert are generally taken with aspirations as given: the feedback mechanism is not internalised.

The third premise is the key to the aspirations failure that lies at the heart of this theory. The theory posits that every economic agent is afflicted by this failure, regardless its level of income or wealth. However, the poor are more likely to exhibit this decision-making bias given the external constraints they face; in particular, their lower wealth would reduce any marginal benefits derived from a given level of effort, and due to the feedback effect, the reduced effort would lead to reduced aspirations, which would lead to diminished effort, and so on, entrapping the individuals in a condition of poverty.

Dalton et al. pointed out that according to their theory, poverty reduces aspirations, not that the poor live in poverty because of low aspirations. There is a feedback mechanism behind the explanation, but in the beginning, poverty would be the chicken and the aspirations, the egg.

Poverty, Command, and Choice

Watts (1968) proposed an economic approach to poverty as a constraint of the choice set, that is, as a reduced level of command over goods and services. This approach is less interested in the realisation of choices (i.e. whether a person buys a second pair of shoes or not) and more concerned about the command over resources to exert choice. Current income is one decisive factor behind this command, but not the only source: according to Watts, permanent or lifetime income is a much better indicator of degree of command over goods and services. Therefore, this approach focuses on poverty as a long-term condition over the life course. However, it does not provide an answer as to why the command over resources is constrained.

Psychological theories of poverty (Turner and Lehning 2007) do offer explanations based on individual characteristics: from differences in intelligence or developmental views on differential cognitive skills according to social class background to social selection and social drift interpretations, there is an emphasis on placing somehow the causes 'inside' the individuals.

Another strand of the literature emphasises the existence of exposure factors acting as threats, older people would act on coping mechanisms to avoid falling into poverty. Though many older people would be at risk of poverty—that is, exposed to the threats—only those without enough coping mechanisms to withstand the threats would fall into poverty. Among the changes that may lead to poverty in later life, a substantial fall in income as a result of retirement is the most important. Coping mechanisms that may compensate for the loss in income in later life are classified into individual capacities, social networks, and social protection (Shröder-Butterfill and Marianti 2006) and include education, skills, health, and adaptation capability, financial support from family members, downsizing and equity release, reduced consumption, and government transfers and benefits.

5.2.2 Structuralist Approaches

Looking beyond individual causes, many authors have sought explanations for poverty and exclusion in the social environment. Regarding income poverty and poverty persistence, a number of institutional factors have been identified, including market failures that lead to lack of access to credit and insurance and barriers to employment; government failure and coordination problems; pension reforms and the erosion of pension savings; and low human capital, conducive to low productivity and labour income. To this list, some authors have added the role of globalisation in the generation and reproduction of poverty.

Concerning social exclusion, Jehoel-Gijsbers and Vrooman (2008) proposed that it has two aspects: economic structural exclusion and socio-cultural exclusion. Each aspect consists of two dimensions. Economic-structural exclusion consists of material deprivation and social rights. Socio-cultural exclusion is composed of social participation and normative integration. In turn,

- Material deprivation refers to being excluded from goods and services.
- Social rights exclusion refers to inadequate access to public services and includes waiting lists, financial barriers to access, legal impediments, or lack of safety in public areas.
- Exclusion from social integration comprises not being able to take part in social networks, leisure and cultural activities, social isolation, and so on.
- Lack of normative integration includes low compliance with norms and values associated with active social citizenship, abuse of social security system and other deviating and criminal behaviour, and lack of involvement in society at large.

Furthermore, Walsh et al. (2017) synthesised the literature of social exclusion in later life into six broad themes or domains:

- material and financial resources
- social relations
- services, amenities, and mobility

- civic participation
- neighbourhood and community
- socio-cultural aspects of society

Despite these efforts and the relative common ground in terms of the identification of domains of social exclusion, according to Walsh et al. (2017, p. 87)

> there is in fact a tendency to neglect a detailed theoretical explanation of why exclusion occurs in old age. This is in terms of: how macro, meso and micro factors combine and interact to construct or protect against multidimensional old-age exclusion; how ageing as a life-course process can increase susceptibility to multi-dimensional exclusion; and how outcomes in particular domains function as components in other forms of exclusionary processes to construct multidimensional old-age exclusion.

One structuralist theory of poverty in later life is the 'structured dependency' theory (Townsend 1981). Its main contention is that inequality and poverty is mostly the result of political, bureaucratic, and economic forces (institutions and rules) that push older people into a condition of social exclusion by negating sufficient resources and social opportunities. Among the main structural mechanisms, Townsend (1981, p. 5) listed the low levels of replacement income propitiated by the pension system, the redundancy created by the institutionalisation of retirement, and the influence in resources and attitudes conducive to social dependency by the social care system. In Townsend's words:

> …the imposition, and acceptance, of earlier retirement; the legitimation of low income; the denial of rights to self-determination in institutions; and the construction of community services for recipients assumed to be predominantly passive

5.2.3 Intergenerational Income Elasticity

Many of the drivers behind poverty in later life have their origin much earlier in the life course. That is one of the messages from structuralist approaches such as the structured dependency theory. Therefore, it is important to consider the intergenerational transmission of income and wealth, because if the chances of upward social mobility are curtailed, then it is more likely that a child born in poverty becomes, if survives into later life, an older person in poverty.

One of the most important parameters in studies of intergenerational transmission of wealth and income is the intergenerational elasticity. It is usually estimated for income. The intergenerational income elasticity is the percentage difference in income in one generation divided by the percentage difference in income in the previous generation. Solon (2004) presented a simple theoretical model according to which the intergenerational income elasticity is greater:

- the greater the heritability coefficient
- the more productive the human capital investment
- the greater the earnings return to human capital investment
- the less progressive the public investment in children's human capital

Imagine a family i with an older person born in $t-1$ and her child, born in t (the time periods in this model correspond to generations). The parent has to allocate her income in $t-1$ ($y_{i,t-1}$), net of taxes τ between consumption C in $t-1$, and investing in her child's human capital (i.e. health or education), I. The budget constraint can be represented then by:

$$(1-\tau) \cdot y_{(i,t-1)} = C_{(i,t-1)} + I_{(i,t-1)} \qquad (5.23)$$

The government levies the taxes to fund public investment in the human capital of children, G. The child's human capital, h, is a function of parental and government investments and can be denoted thus:

$$h_{(i,t-1)} = \theta \cdot \log\left(I_{(i,t-1)} + G_{(i,t-1)}\right) + e_{(i,t)} \qquad (5.24)$$

where $\theta > 0$ corresponds to the marginal product of human capital investment and is assumed to be positive (additional investments increase a child's human capital) and e denotes the human capital endowment of the child. Endowment includes cognitive ability, physical appearance, attitudes, family connections, height, attractiveness, IQ, skin colour, birth order, personality traits, and so on (d'Addio 2007). A child's endowment is assumed to be positively correlated to her parent's endowment:

$$e_{(i,t)} = \delta + \gamma \cdot e_{(i,t-1)} + v_{(i,t)} \qquad (5.25)$$

where δ represents the level of endowment not transmitted from the parents and $0 < \gamma < 1$ is the heritability coefficient; v is a stochastic error term.

A child's income depends on her human capital as shown in Eq. (5.26), with $p > 0$ denoting the income returns to human capital investment and μ is the level of income independent from a child's human capital.

$$\log y_{(i,t)} = \mu + p \cdot h_{(i,t)} \tag{5.26}$$

The model focuses on the parental decision: the optimal level of consumption and investment in her child's human capital. In other words, the model assumes a member of the $t - 1$ generation has a utility function represented by:

$$U_i = (1 - \alpha) \cdot \log C_{(i,t-1)} + \alpha \cdot \log y_{i,t} \tag{5.27}$$

where $0 < \alpha < 1$ is a measure of altruism. Replace first Eq. (5.23) in Eq. (5.27):

$$U_i = (1 - \alpha) \cdot \left[(1 - \tau) \cdot y_{(i,t-1)} - I_{(i,t-1)}\right] + \alpha \cdot \log y_{i,t} \tag{5.28}$$

Now, replace $\log y_{i,t}$ with Eq. (5.26):

$$U_i = (1 - \alpha) \cdot \left[(1 - \tau) \cdot y_{(i,t-1)} - I_{(i,t-1)}\right] + \alpha \cdot \left[\mu + p \cdot h_{(i,t)}\right] \tag{5.29}$$

Finally, replace $h_{(i,t)}$ with the expression in Eq. (5.24). We get:

$$U_i = (1 - \alpha) \cdot \left[(1 - \tau) \cdot y_{(i,t-1)} - I_{(i,t-1)}\right] \\ + \alpha \cdot \left\{\mu + p \cdot \left[\theta \cdot \log\left(I_{(i,t-1)} + G_{(i,t-1)}\right) + e_{(i,t)}\right]\right\} \tag{5.30}$$

This is the key equation in the model. A parent maximises this utility function with respect to investment in her child's human capital:

$$\frac{\partial U_i}{\partial U_{(i,t-1)}} = -\frac{(1 - \alpha)}{\left[(1 - \tau) \cdot y_{(i,t-1)}\right] - I_{(i,t-1)}} + \frac{\alpha \cdot \theta \cdot p}{(I_{(i,t-1)} + G_{(i,t-1)})} = 0 \tag{5.31}$$

Solving Eq. (5.31), the following expression can be obtained:

$$I_{(i,t-1)} = \left[\frac{\alpha \cdot \theta \cdot p}{1 - \alpha \cdot (1 - \theta \cdot p)}\right] \cdot (1 - \tau) \cdot y_{(i,t-1)} - \left[\frac{1 - \alpha}{1 - \alpha \cdot (1 - \theta \cdot p)}\right] \cdot G_{(i,t-1)} \tag{5.32}$$

Equation (5.32) shows that if government spending on human capital of the younger generation is held constant, parents with higher income levels will invest more in their children's human capital. Also, more altruistic parents will invest more, and so will any parent the higher the income returns to human capital investments (for their children). In turn, if taxes are held constant (i.e. $\tau = 0$), public investment in human capital partly crowds out parental investments.

Equation (5.32) does indicate that higher income households invest more in their offspring education and health, *ceteris paribus*, but how does this affect the intergenerational transmission of income? Here's where the intergenerational income elasticity needs to be obtained. Using Eqs. (5.24), (5.26), and (5.32), we obtain the following approximation[15]:

$$\log y_{i,t} \cong \mu + \theta \cdot p \cdot \log \left[\frac{\alpha \cdot \theta \cdot p \cdot (1-\tau)}{1 - \alpha \cdot (1 - \theta \cdot p)} \right] + \theta \cdot p \cdot \log y_{(i,t-1)}$$
$$+ \theta \cdot p \cdot \left\{ \frac{G_{(i,t-1)}}{[(1-\tau) \cdot y_{(i,t-1)}]} \right\} + p \cdot e_{i,t} \quad (5.33)$$

Solon assumes that government policy is such that the ratio of public investment on children's human capital to parental net income decreases with parental net income:

$$\frac{G_{(i,t-1)}}{[(1-\tau) \cdot y_{(i,t-1)}]} \cong \varphi - \gamma \cdot \log y_{(i,t-1)} \quad (5.34)$$

where $\gamma > 0$ is the measure of policy progressiveness. Replacing Eq. (5.34) into Eq. (5.33), we obtain:

$$\log y_{i,t} \cong \mu + \theta \cdot p \cdot \log \left[\frac{\alpha \cdot \theta \cdot p \cdot (1-\tau)}{1 - \alpha \cdot (1 - \theta \cdot p)} \right] + \theta \cdot p \cdot \log y_{(i,t-1)}$$
$$+ \theta \cdot p \cdot \varphi - \theta \cdot p \cdot \gamma \cdot \log y_{(i,t-1)} + p \cdot e_{i,t}$$
$$\log y_{i,t} \cong \mu + \theta \cdot p \cdot \log \left[\frac{\alpha \cdot \theta \cdot p \cdot (1-\tau)}{1 - \alpha \cdot (1 - \theta \cdot)} \right] + \theta \cdot p \cdot \varphi$$
$$+ \theta \cdot p \cdot (1 - \gamma) \cdot \log y_{(i,t-1)} + p \cdot e_{i,t} \quad (5.35)$$

A long one, I admit, but do not despair! In Eq. (5.35), everything other than the last two terms are simply (well, 'simply') fixed coefficients, so they can all be replaced by one constant, say μ^*:

$$\log y_{i,t} \cong \mu^* + \theta \cdot p \cdot (1-\gamma) \cdot \log y_{(i,t-1)} + p \cdot e_{i,t} \quad (5.36)$$

This equation not only looks much better than the previous one; it is a well-known econometric model (remember that $e_{i,t}$ is defined in Eq. (5.25)):

$$y_t = \beta \cdot y_{t-1} + \varepsilon_t,$$
$$\varepsilon_t = \rho \cdot \varepsilon_{t-1} + u_t \quad (5.37)$$

In Eq. (5.36), the coefficient of interest is the expression $\theta \cdot p \cdot (1-\gamma)$, which Greene (2003) explains, is obtained by:

$$\frac{(1-\gamma) \cdot \theta \cdot p + \lambda}{1 + (1-\gamma) \cdot \theta \cdot p \cdot \lambda} \quad (5.38)$$

This expression—let's call it β—is the intergenerational income elasticity. β increases with the heritability coefficient (λ), the marginal product of human capital investment (θ), the investment returns to human capital (p), and diminishes with the progressiveness of public policy (γ). Solon obtained similar relationships for the variance of parental income in each period, which, as we saw in Chap. 4, is a measure of inequality in the income distribution. Therefore, inequality also increases with the heritability coefficient, the productivity of and returns to human capital, and diminishes as public policy becomes more progressive.

5.2.4 Equal Burden-Sharing

On 4 October 2006, Ben Bernanke, the then Chairman of the Federal Reserve, gave a speech before *The Washington Economic Club*, in Washington, DC, United States of America (Bernanke 2006). His topic was whether future generations will be treated fairly in the face of what he saw were 'severe fiscal challenges' resulting from the 'coming' demographic transition. The big question, he said, was how 'the burden of an aging population is to be shared between our generation and the generations that will follow'. Part I in Volume II discusses the fiscal implications of population ageing. Here I want to focus on one principle of intergenerational justice Bernanke introduced in that speech:

> …that the current generation and all future generations experience the same percentage reduction in per capita consumption relative to a baseline scenario without population ageing

He termed it 'equal burden-sharing'; the idea is as follows. It is based on a hypothetical counterfactual (apart from a number of assumptions[16]). The starting premise is that population ageing from below (see Chap. 5 in Volume I) will reduce the labour force. Production and consumption per capita, and therefore living standards, will likely fall as a consequence, but a key behavioural parameter is how much the current generation approaching retirement (i.e. the Baby Boomers) is saving.

The life-cycle hypothesis suggests that they should be saving the amount required to smooth out consumption levels over their life courses. The point Bernanke brought home is that saving rates are, in comparison, painfully low in the United States of America. This means that Baby Boomers are maintaining, on average, higher consumption levels than optimal. Will they suffer from a substantial drop in their living standards and perhaps destitution in their later lives? Well, no if they can get away with it by making the currently younger and future generations transfer the necessary funds to make up the difference: if cometh the hour, cometh the youth. By transfers, Bernanke means taxes. However, this would require that the younger generations should save not only more than the Boomers currently do, but more than what would be optimal to them. The increase in savings required from the younger generations would not add a single coin to their retirement income: it would be transferred to their parents and grandparents and the grandparents of their grandparents, and so on. That's not fair, says Bernanke.

What if the different generations shared the cost of funding, the increase in the proportion of retirees in the population, and the reduction in the labour force and, possibly, productivity? How much more would the Boomers have to save towards *their* retirement? How much would their consumption fall? And, by extension, how much less would the younger generations need to transfer and how much more would they be able to consume? 'Rough' estimations indicated that if nothing changed between 2005 and 2025, the consumption per capita of the future generations would be 14 per cent less than if there were no demographic change (and remember that the 'nothing changes' scenario implies that the Boomers would not reduce their consumption per capita). Here is the unfairness in the intergenerational sharing of the macroeconomic 'burden' of population ageing. However, an increase in national saving (mostly by Boomers) of 3 per cent would be fair: then the Boomers and the younger generations would experience a reduction in consumption per capita of 4 per cent. In turn, Sheiner et al. (2007) estimated that an immediate increase in the retirement age from 65 years to 67 years would bring down the fall in consumption per capita by 2025 (assuming that nothing else changed) from 14 per cent to 1.5 per cent; if, instead, the retirement age were set at age 70, consumption per person would increase by 7.7 per cent by 2025 compared to the no demographic change scenario.

5.2.5 The Great Gatsby Curve

The Great Gatsby curve is a graph that depicts the intergenerational income elasticity on the vertical axis and income inequality (generally, using the Gini coefficient) on the horizontal axis (Krueger 2012). The graph shows, for developed (and also developing) countries, a strong positive association: places with greater inequality tend to have lower intergenerational mobility (Corak 2013).

In the model by Solon, we concluded (see Eq. (5.38)) that the higher the investment returns to human capital (p), the greater is the intergenerational income elasticity, β. According to Eq. (5.36), this will increase the income of children from better-off parents more than the income from poorer households. This is one theoretical explanation for the curve: with greater inequality, well-off families have more incentives to invest more on their children's human capital than poorer families, thus reducing intergenerational mobility. Of course, social capital may also play a part: if who you know is important for getting ahead economically and socially, it is more likely that well-off families are better connected and have an edge in this regard. The association has also been reported within the United States of America, where a state-level study found that intergenerational mobility is lower, the greater the income inequality (Chetty et al. 2014b)—for an opposed view, see Bloome (2014). Another mechanism linking inequality to mobility is spatial segregation, that is, the degree of inequality across neighbourhoods (Durlauf 2018).

With regard to developing countries, the role of these (and other) drivers are orders of magnitude greater than in developed countries.

5.2.6 Anti-poverty Role of Pension Income in Low-Income Developing Countries

Part II in Volume III covers the topic of economics of pensions. Here I only touch upon social pensions as anti-poverty measures, that is, non-contributory pension systems designed primordially to keep out of absolute poverty older people who have not made contributions to pension systems.

Non-contributory pensions are an essential part of 'social safety nets' (also known as 'safety nets', 'social assistance', or 'social transfers') in developed and developing countries alike. Social security benefits have been credited, for example, with the dramatic reduction in poverty in later life in the United States of America since the 1950s (Engelhardt and Gruber 2006; Van de Water et al. 2013). In both groups of countries, the main beneficiaries are individuals in the lowest income deciles (The World Bank 2015). Most middle-income developing countries have a system of social pensions in place, and sub-Saharan low-income countries host some of the largest schemes.

Data from the World Bank show that old-age social pensions represent the highest proportion of spending on cash transfers worldwide, more than poverty targeted transfers, cash benefits for families and children, education benefits, and so on: over 24 per cent of all cash transfers correspond to non-contributory pensions (The World Bank 2015, Figure 2.6). However, in low-income countries, the coverage of non-contributory pensions varies from just over 20 per cent of the population aged 60 or over in Mozambique to over 80 per cent in Botswana or Seychelles, and even over 100 per cent in Mauritius and Namibia. All in all, around 50 per cent of older persons above statutory pensionable age benefit from a social pension worldwide (Department 2014).

Even in countries with universal—or almost—coverage, the eligibility criteria (including minimum chronological age) differ, the level is not adequate, and indexation schemes in place suffer from uncertainty.

Minimum qualifying age is important given the lower life expectancy in many low-income countries (and especially of individuals from poor backgrounds in low-income countries): in the Philippines, where the life expectancy at age 60 is 78,50 for women and only 75,33 for men, social pension starts at age 77, which means that on average women would receive these transfers for less than two years and men would not receive any. Worse still is the situation in the Dominican Republic,[17] where the minimum age was set at 90 (yes, NINETY) years (Böger and Leisering 2017, Table 1). Social pensions cover roughly a mere 5 per cent of consumption in low-income countries and about 20 per cent of consumption in lower middle-income countries.

Despite these grave pitfalls, in poorer countries non-contributory old age pensions have positive intergenerational consequences and indirect economic benefits. For example, in many rural areas, social pensions have allowed younger, economically active adult household members to migrate in search of jobs leaving children under the care of grandparents. The cash transfers from social pensions plus the transfers from younger adults represent a relatively substantial inflow of resources to poor rural areas in low-income countries contributing to food security and to averting deep poverty (Bird 2013).

Health problems are a major barrier to social mobility among the poor; in fact, according to Harper (2004), it is the biggest risk to which poor households are exposed in the developing world. Social pensions also indirectly contribute towards increasing employment chances of younger generations through improvements in health. Inter-vivos transfers have been found to be associated with positive health impacts and, via health, with increasing social mobility among the poorest members of a society.

5.3 Social Class and Later Life

Social class is a concept with a 'peculiar explosiveness' (Dahrendorf 1959). The American Psychological Association created a task force in 2005 to study socio-economic status (SES). It published a report in 2007, where it recognised socio-economic factors and social class as 'fundamental determinants of human functioning across the life span' (American Psychological Association, Task Force on Socioeconomic Status 2007, p. 1). However, in contemporary mainstream economics, the notions of social class or SES hardly register at all: inequalities in income or wealth, income levels (or 'brackets') or poverty, yes, but the 'lumpenproletariat', the 'bourgeoisie', the 'working class', or the 'landed gentry' are confined to the conceptual armoury of scholars in economic history and history of economics—and SES is, at best, included an independent variable in econometric models, but rarely becomes the central focus of study. Sure, other disciplines, such as social epistemology, for example, also consider SES solely as a covariate in their empirical work, but it is more difficult to understand that socio-*economic* status is not a central subject of analysis in contemporary mainstream economics. In contemporary sociology, in contrast, the concept is hotly debated. But what is social class? And SES?

Classical sociology developed three distinct views of social class:

- The Marxist tradition understands social class as the result of structural relations that determine the access to means of production and the role and position of an individual in an economic system and mode of production.
- The Weberian tradition also understands social class as the result of the position of an individual in the production process, but rather than emerging from structural relations, social class is defined by life chances and opportunities created and made available by an economic system and mode of production.
- The Functionalist tradition understands social class as the result of the necessity within an economic system and mode of production to allow for the division of labour and generate a mechanism of incentives and rewards to motivate individuals to strive towards power and status conducive to order, cooperation, and stability.

One operational approach considers social classes as positions in a social structure, generally linked to employment relations and skills differentials. Manual workers and large-firm owners, for instance, would belong to different social classes. In this conceptualisation, social class is akin to occupational class, a relational notion in that the existence of, say, an 'employee' demands the existence of an 'employer'.

Another approach refers to 'socio-economic positions', of which there are two versions: one version adds hierarchical position and decision-making participation (top, upper, middle, or lower managerial, supervisory, and non-managerial) to occupational class as defined above (Wright 1996), and the other one adds occupational prestige (Galobardes et al. 2006a,b; Lynch and Kaplan 2000; Treiman 1977).

There are several measures of occupational class, some combining prestige and/or autonomy, including (among others):

- the Edwards' classification (Edwards 1917)
- the Duncan Socioeconomic Index (Duncan 1961)
- the Goldthorpe class schema (Goldthorpe and Hope 1974)
- the Hollingshead's occupational prestige scale (Hollingshead 2011)
- the Treiman's occupation prestige ranking (Treiman 1977)
- the Nam-Powers Socioeconomic Status Score (Nam and Powers 1983)
- the Stevens-Cho scheme (Stevens and Cho 1985)
- the Stevens-Hoisington (Stevens and Hoisington 1987)
- the Erikson, Goldthorpe, and Portocarero's schema (Erikson et al. 1979)[18]

- the Nakao-Treas score (Nakao and Treas 1994)
- the Hauser and Warren index (Hauser and Warren 1997)
- the Wright's Social Class scheme (Wright 1996)
- the Rose and Harrison's European schema (Rose and Harrison 2007)

These measures, with or without prestige, have three crucial pitfalls as proxies for social class. Firstly, they do not fully apply to individuals not in employment. Therefore, for our purposes, these measures may not be completely accurate indicators of social class of older people in retirement.[19] Secondly, occupations come and go due to changes in tastes, technology, and relocation of industrial processes[20] as well as their ascribed social prestige, so that they would not be informative in longitudinal studies that took a life-course approach. Thirdly, the main last job is an increasingly less valid indicator of social class in later life given the greater fluidity in contemporary labour markets with careers becoming more fragmented and the passage into retirement also becoming more unstructured (see Volume III). In this sense,

> …it is expected that the last recorded position in the labour market will show a weak discriminatory capacity to identify in a satisfactory manner the core characteristics of the ties each individual had with the labour market throughout the life course and the opportunities and disadvantages that were accumulated throughout
>
> (Lopes 2015, p. 57)

However, the main last job can be used provided an assumption is adopted: that there is continuity in the occupation prestige and class into retirement. And this is, by and large, the theoretical stance that, tacitly or explicitly, the authors who focus on social gradients in later life endorse. Its main empirical support comes from research findings in health-related studies.

Occupational class has been found to be statistically associated with a number of conditions in later life, from arthritis (Caban-Martinez et al. 2011) to flu vaccination (Damiani et al. 2007) to cognitive impairment (Li et al. 2002), and also with mortality (Wilkinson 1989) (see also Wilkinson and Pickett 2010). For example, using data from Great Britain, Arber and Ginn (1993) found that the previous main occupation of retired individuals was strongly associated with self-assessed health status and functional disability. Moreover, Breeze et al. (2001) found that occupational class at age of retirement was significantly associated with general health, mental health, physical performance, and disability (Marmot and Shipley 1996). Besides, in a study on demand for long-term care services among people aged 90 or over in Finland,

Enroth et al. (2018) found that occupational class at age of retirement was strongly associated with the probability of using private long-term care services. In addition, Christensen et al. (2014) reported a strong social gradient between a measure of socio-economic position based on the nature of and training required for the job, and three health-related indicators (chronic conditions, self-rated health, and mobility) among people aged 50–64 in Denmark, after controlling for chronological age and gender.

Some studies compared social class indicators based on occupational class with other social class markers such as education or income. Darin-Mattsson et al. (2017) reported similar findings among older people in Sweden using alternative indicators of social class: education level, income, occupational complexity, occupational class, and a composite index of these indicators. Moreover, Grundy and Holt (2001) examined the association between alternative indicators of social class and health outcomes among older people in Great Britain and recommended the use of occupationally defined social class and education coupled with a deprivation indicator.

However, Geyer et al. (2006) warned against using these indicators interchangeably: according to these authors, the indicators would measure different, though related, phenomena and would therefore operate via different causal mechanisms. Similarly, Chan and Goldthorpe (2007) drew on Max Weber's distinction between class and status, which these authors opine that has been virtually lost in the literature, and underline the need to maintain these concepts apart. According to Weber (1978), a social class is 'the totality of …class situations within which individual and generational mobility is easy and typical' [p. 302]. Class situation refers to the 'typical probability of procuring goods, gaining a position in life and finding inner satisfactions', stemming from 'the control over goods and skills and from their income-producing uses within a given economic order' [p. 302]. In turn, status refers to 'an effective claim to social esteem in terms of positive or negative privileges' [p. 305] based on lifestyle, education, or hereditary or occupational prestige. From this perspective, social stratification is multidimensional in nature: class determines the economic resources accumulated through employment because of its effect on opportunities and status influencing older people's consumption of cultural goods and services and personalised lifestyles. Chan and Goldthorpe suggested that class would have greater influence on the economic life chances of individuals than status.

Savage et al. (2005) developed the Capitals, Assets, and Resources (CARs) model in an attempt to combine three strands of the literature on social stratification: Pierre Bourdieu's theory of different types of capital (cultural and symbolic, as well as economic, social, and human), Erik Olin Wright's focus on assets, and John Goldthorpe's emphasis on resources. The CARs model proposes that individuals can accumulate certain forms of capital irrespective of their structural relationship with the means of production, and that they can eventually convert into specific advantages. The flip side of this process of capital accumulation is that inequalities are also cumulative and 'it is the potential of certain CARs to accumulate, store, and retain advantages that allow us to distinguish the most important causes of stratification' (Savage et al. 2005, p. 43). This model is based on the notion of accumulation; therefore it suggests a dynamic approach to stratification. In fact, Savage et al. (2015) argued that inequalities do not necessarily turn into classes, but that this transformation takes place 'when advantages endure over time in a way which extends beyond any specific transaction' [pp. 45–46].

Other areas of study have also found that social class is important. For example, longitudinal and event history analyses of retirement report that workers from lower social class occupations tend to retire earlier than their counterparts from higher social status, irrespective of work trajectories. The same applies to studies of political participation, which found a stable and significant association with income and wealth. The cultural turn has not provided with strong, persuasive counter-arguments to account for these disparities in later life outcomes. Furthermore, the rejection of social class as a valid theoretical construct would imply an off-hand rejection of the theories of cumulative advantage and disadvantage (Marmot and Bell 2016), of the hypotheses developed by continuity theory (Alen et al. 2017; Nimrod and Rotem 2012; Scherger et al. 2011), and, in general, of critical gerontological approaches with their emphasis on the importance of past labour market positions and occupational statuses for social standings in later life (see Part IV in this volume and Estes et al. (2003), Phillipson (2013)).

Whilst some sociologists and gerontologists contend that social class is still 'a pervasive form of hierarchy rooted in a person's wealth, education, and occupational prestige' (Piff et al. 2018, p. 5), not everyone is convinced—notwithstanding the mounting evidence that indicators of social class in later life have direct association with health outcomes, consumption patterns, long-care needs and demand structure, housing quality, and a myriad of other variables (e.g. managing money and cognitive abilities (Horvat et al. 2014)). In fact, some contemporary sociologists consider that the notion of social class is *demodé*, a conceptual relic from before industrial economies turned post-industrial, with their increasing affluence and individualisation of lifestyles. Class divisions have been eroded—or so these authors claim—with the demise of grand narratives and ideologies, the waning of the influence of trade unions and the disalignment of political parties to socio-occupational categories (see, for example, Pakulski and Waters (1996a,b), Clark and Lipset (1991)). One group of authors went so far as to decry the 'death of class'[21] or that social class is a 'persona non grata in ageing studies' (Formosa and Higgs 2015, p. 5). These authors do not deny the existence of inequalities, but argue that contemporary societies, though unequal, are classless and that, as a theoretical construct, class would not shed any light on the issue of the complex inequality pervasive in modern societies.

A different school of thought embraces a cultural view of social class, seen as

> …a coherent social and cultural existence, where members in similar class share a common lifestyle, educational background, kinship networks, consumption patterns, and beliefs.
>
> (Formosa 2014, p. 11)

The cultural turn on social class proposes that its complexities would extend beyond income and wealth particularly into the cultural realm, thus mapping onto consumption patterns. Contemporary social stratification would be less about economic inequality than about cultural markers of status and lifestyle. In fact, Gilleard et al. (2005) found (see also Gilleard and Higgs (2005, 2009)), using data for the United Kingdom, that birth cohort is more relevant for consumption patterns than class of origin:

> …people from British working class origins are no less likely, ceteris paribus, to participate in consumer culture than retired people from professional middle-class backgrounds but …people from cohorts born earlier in the century are less engaged than those born later.
>
> (Gilleard et al. 2005, p. S309)

In other words, an older person of working-class origin could not be told apart from an older person of professional background simply by observing their respective consumption patterns, but a 65-year-old person would be distinguished from an 85-year-old person from their lifestyle choices. Interesting, not only is the social class of origin not significant: neither is ageing: 'to view "age" rather than "cohort" as the engine driving third age consumerism risks mystifying rather than illuminating the structural processes underlying people's patterns of consumption pre and post working life' [p. S309]. That is, it is not that the 85-year-old person is older than the 65-year-old person what makes the difference, but that they belong to different birth cohorts. For many middle-aged individuals in developed countries, and a not negligible minority in the developing world, the lifestyles available through consumption represent 'mid-lifestyle' opportunities (Featherstone and Hepworth 1983). As opposed to 'middle-age' options, culturally followed by 'old age', the notion of mid-lifestyle has connotations of energy, vitality, and youthfulness that would be carried over well into later life, right into 'old-old age'. Mid-lifestyles imply, then, an extension of 'youth-full' predisposition, interests, and activities until they cannot be supported or pursued any longer for eventual declining health (Klein 2014). The adoption of mid-lifestyles is, in a sense, another anti-ageing strategy, but played out at the cultural plane rather than translated into the latest facial cream: it is an attempt to avoid the cultural associations of old age for as long as possible.

The classless society and the cultural views put excessive emphasis on agency. However, one thing is the will to adopt a particular lifestyle and a different thing is the capability of adopting it, which demands resources individuals may be lacking and, in certain cases, also overcoming cultural barriers existing either across a society or ascribed to, precisely, particular social classes. As Bauman (2004, p. 39) remarked: 'Desiring comes free, but to desire realistically, and so experience desire as a pleasurable state, requires resources'.

Skeggs (2005) noted that the declaration of the demise of social class has not been announced across the sociological board, but mainly by certain authors studying race, sexuality, culture, and feminism; those involved in work on education or health, for example, have continued applying social class as a variable with acceptable explanatory power. It seems that the field of social studies of ageing and old age, in cutting across various (if not all!) domains—as this textbook serves as (humble) evidence—is a naturally active hotbed of discussions between different approaches.

Before proceeding to the 'Review and Reflect' section, let me close this chapter with a general question. From the end of class society as a 'collectively experienced process of individualization within a post-traditional society of employees' (Beck and Beck-Gernsheim 2002, p. 36) to social class being 'the most powerful factor in determining aging and life course experiences' in the future (Settersten Jr and Trauten 2009, p. 459), the complete gamut of contending theoretical and conceptual options is on offer. Which one do you find more persuasive? Why?

Review and Reflect

1. *The story of proximate causes of inequality ...is a mixture of state interventions, capital income, and the labour market. As a result, a coherent story should make empirical predictions regarding the distribution of capital assets, the formation of a policy consensus, and labour market behaviour, whereas hypotheses that primarily focus on the latter are unsuited to provide a comprehensive account.*

    (Bogliacino and Maestri 2014, p. 37)

2. *...it is usually the case that elderly and disabled people can rely more strongly on less-stigmatizing benefits, than, for instance, unemployed people. In many countries widows are better protected by national benefit schemes than divorced women.*

    (Oorschot and Roosma 2017, p. 7)

3. Stiglitz (2012, p. 28) proposed the intergenerational transmission of wealth as a measure of equality of opportunities:

    If [a country] were really a land of opportunity, the life chances of success—of, say, winding up in the top 10 percent—of someone born to a poor or less educated family would be the same as those of someone born to a rich, well-educated, and well-connected family.

    Do you agree? How do you define 'life chances'? Do they depend on other variables than family wealth?

4. *...norms of reciprocity and of deservingness are important to support intergenerational redistribution, whereas the latter seems to be the relatively most important motivation. We can take this as a sign of intergenerational cohesion that is relevant against the background of accelerating demographic aging and resulting pressure on institutions of intergenerational redistribution.*

    (Prinzen 2016)

5. Savage et al. argued that social class and chronological age tend to be kept separate in most studies of stratification and concluded:

    The result of this unfortunate separation of class from age in much of our thinking is that generational divisions have instead been subsumed into an anxiety about 'declining social mobility,' which seeks to frame the issue of age as one of mobility between classes.

    (Savage et al. 2015, p. 176)

(continued)

> Why would an ageing population hamper upward social mobility? Prepare two responses: a) based on the life-cycle hypothesis and b) based on the CARs model.

Notes

1. See http://povertydata.worldbank.org/poverty/home/.
2. And given also that not everyone with a disability receives benefits.
3. An interesting trend in some developed countries is that partnering does not carry the same social expectation of caregiving as marriage, which would explain why cohabitation is finding increasing acceptance among older women compared to remarriage following dissolution or widowhood—see, for example, (Brown and Wright 2015; Noël-Miller 2011).
4. See Korenman and Remler (2013) for a dissenting view.
5. The 70 per cent cut-off level is used in combination with a deprivation index to obtain a measure of 'consistent poverty' (see below).
6. But not 'unreasonable', as Orshansky (1965, p. 4) defended the first poverty line for the United States of America (at 50 per cent).
7. See Villar (2017) for a detailed but accessible introduction.
8. See also Alkire et al. (2015).
9. In the United Kingdom, a household is defined as being in fuel poverty if it needs to spend more than 10 per cent of its income on fuel use.
10. For the alternative conceptualisations of 'place' in gerontological studies of social exclusion, see Moulaert et al. (2017) and Walsh (2017).
11. See www.cso.ie/releasespublications/documents/silc/current/silc.pdf.
12. For Great Britain, see Office for National Statistics. General Lifestyle Survey, 2011. Chapter 4. For the United States of America, US Census Bureau Historical Census of Housing Tables—Telephones.
13. For example, in the United States of America, 12 per cent of the population aged 55–64 had attained tertiary education in 1981; in 2016, the proportion increased to 42 per cent of all individuals aged 55–64 years old. *Source: OECD. Education at a Glance statistics.*
14. Severely criticised by, among others, Townsend (1979).
15. Assuming $\frac{G_{(i,t-1)}}{[(1-\tau) \cdot y_{(i,t-1)}]}$ is small.
16. Depreciation of capital stock: 6 per cent per year; technical progress: 1.4 per cent a year; all the individuals aged 65 or over are retired and all the population aged 20–64 are in paid employment; the ratio of the consumption level of 65

or over to that of 20–64 years old remains constant over time; and so on. See Sheiner et al. (2007).
17. Minimum age is also 90 years in Myanmar (see endnote 22).
18. Also known as the *CASRtIN* schema, after the Comparative Analysis of Social Mobility in Industrial Nations project (Hout and Hauser 1992).
19. Neither would they be appropriate for home-makers, the unemployed, and so on.
20. For example, in Bedfordshire, England, there were 17,316 milliners or hat makers in 1881, about 12 per cent of total population (ONS no date); the British Hat Guild closed in 2000.
21. A passing that has been ascribed to post-modern social traits such as liquid biographies, reflexivity, and risk.

References

Adato, Michelle, Michael R Carter, and Julian May (2006). "Exploring poverty traps and social exclusion in South Africa using qualitative and quantitative data". In: *The Journal of Development Studies* 42.2, pages 226–247.

Adena, Maja and Michal Myck (2014). "Poverty and transitions in health in later life". In: *Social Science & Medicine* 116, pages 202–210.

Aguirregabiria, Víctor (2006). "Sen-Shorrocks-Thon Index". In: *Encyclopedia of World Poverty*. Edited by Mehmet Odekon. Volume 3. Thousand Oaks, CA: United States of America: Sage Publications, pages 970–971.

Alen, Elisa, Nieves Losada, and Pablo De Carlos (2017). "Understanding tourist behaviour of senior citizens: lifecycle theory, continuity theory and a generational approach". In: *Ageing & Society* 37.7, pages 1338–1361.

Alkire, Sabina and James Foster (2007). *Counting and multidimensional poverty measures.* OPHI Working Paper Series 7. Oxford: United Kingdom.

Alkire, Sabina et al. (2015). *Multidimensional Poverty: Measurement and Analysis.* Oxford: United Kingdom: Oxford University Press.

American Psychological Association, Task Force on Socioeconomic Status (2007). *Report of the APA Task Force on Socioeconomic Status.* Report. Washington, D.C.: United States of America.

Araujo, M Caridad, Mariano Bosch, and Norbert Schady (2017). "Can cash transfers help households escape an inter-generational poverty trap?" In: *The Economics of Poverty Traps.* Edited by Christopher B Barrett, Michael R Carter, and Jean-Paul Chavas. Chicago, IL: United States of America: University of Chicago Press.

Arber, Sara and Jay Ginn (1993). "Gender and inequalities in health in later life". In: *Social Science & Medicine* 36.1, pages 33–46.

Aziz, Nusrate, Belayet Hossain, and Masum Emran (2018). "Role of income in intergenerational co-residence: Evidence from selected African and Asian countries". In: *Australasian Journal on Ageing* 37.2, E55–E60.
Barrett, Christopher B et al. (2006). "Welfare dynamics in rural Kenya and Madagascar". In: *The Journal of Development Studies* 42.2, pages 248–277.
Batana, Yélé, Maurizio Bussolo, and John Cockburn (2013). "Global extreme poverty rates for children, adults and the elderly". In: *Economics Letters* 120.3, pages 405–407.
Bauman, Zygmunt (2004). *Work, Consumerism And The New Poor.* Issues in society. Maidenhead: United Kingdom: Open University Press.
Bayaz Ozturk, Gulgun and Sean P Macdonald (2017). "Intertemporal Poverty among Older Americans". In: *Journal of Poverty* 21.4, pages 331–351.
Beck, Ulrich and Elisabeth Beck-Gernsheim (2002). *Individualization: Institutionalized Individualism and Its Social and Political Consequences.* London: United Kingdom: SAGE Publications.
Bernanke, Ben (2006). *The Coming Demographic Transition: Will We Treat Future Generations Fairly? Presented at the Washington Economic Club, Washington, DC, October 4, 2006.* Speech. Washington, DC: United States of America. URL: www.federalreserve.gov/newsevents/speech/bernanke20061004a.htm.
Bird, Kate (2013). "The intergenerational transmission of poverty: An overview". In: *Chronic Poverty. Concepts, Causes and Policy.* Edited by Andrew Shepherd and Julia Brunt. Rethinking International Development Series. Basingstoke: United Kingdom: Palgrave Macmillan, pages 60–84.
Blinder, Alan S (1974). *Toward an Economic Theory of Income Distribution.* Cambridge, MA: United States of America: The MIT Press.
Bloome, Deirdre (2014). "Income inequality and intergenerational income mobility in the United States". In: *Social Forces* 93.3, pages 1047–1080.
Böger, Tobias and Lutz Leisering (2017). *Social Citizenship for Older Persons? Measuring the Social Quality of Social Pensions in the Global South and Explaining Their Spread.* Discussion Paper 1703. Washington, DC: United States of America.
Bogliacino, Francesco and Virginia Maestri (2014). "Increasing Economic Inequalities?" In: *Changing Inequalities in Rich Countries. Analytical and Comparative Perspectives.* Edited by Wiemer Salvedra et al. Oxford: United Kingdom: Oxford University Press, pages 15–48.
Bonfatti, Andrea, Martina Celidoni, and Guglielmo Weber (2013). "Coping with risks during the Great Recession". In: *Active ageing and solidarity between generations in Europe: First results from SHARE after the economic crisis.* Edited by Axel Börsch-Supan et al. Berlin: Germany: Walter de Gruyter, pages 225–234.
Bossert, Walter, Satya R Chakravarty and Conchita dAmbrosio (2012). "Poverty and time" In: *The Journal of economic inequality* 10.2, pages 145–162.
Breeze, Elizabeth et al. (2001). "Do socioeconomic disadvantages persist into old age? Self-reported morbidity in a 29-year follow-up of the Whitehall Study". In: *American Journal of Public Health* 91.2, pages 277–283.

Bridges, Benjamin and Robert V Gesumaria (2013). "The supplemental poverty measure (SPM) and the aged: How and why the SPM and official poverty estimates differ". In: *Social Security Bulletin* 73.4, pages 49–69.
BritainThinks (2012). *Public attitudes to poverty and child poverty*. Report. London: United Kingdom.
Brody Elaine M. (2003). *Women in the Middle: Their Parent-Care Years, Second Edition*. Springer Series on Lifestyles and Issues in Aging. New York, NY: United States of America: Springer Publishing Company.
Brown, Susan L and Matthew R Wright (2015). "Older adults' attitudes toward cohabitation: Two decades of change". In: *Journals of Gerontology Series B: Psychological Sciences and Social Sciences* 71.4, pages 755–764.
Brunetti, Marianna, Elena Giarda, and Costanza Torricelli (2016). "Is financial fragility a matter of illiquidity? An appraisal for Italian households". In: *Review of Income and Wealth* 62.4, pages 628–649.
Buffel, Tine, Chris Phillipson, and Thomas Scharf (2013). "Experiences of neighbourhood exclusion and inclusion among older people living in deprived inner-city areas in Belgium and England". In: *Ageing & Society* 33.1, pages 89–109.
Buhmann, Brigitte et al. (1988). "Equivalence scales, well-being, inequality, and poverty: sensitivity estimates across ten countries using the Luxembourg Income Study (LIS) database". In: *Review of income and wealth* 34.2, pages 115–142.
Burchardt, Tania, Julian Le Grand, and David Piachaud (2002). "Degrees of exclusion: developing a dynamic, multidimensional measure". In: *Understanding Social Exclusion*. Edited by John Hills, Julian Le Grand, and David Piachaud. Oxford: United Kingdom: Oxford University Press, pages 30–43.
Burkhauser, Richard V, Timothy M Smeeding, and Joachim Merz (1996). "Relative inequality and poverty in Germany and the United States using alternative equivalence scales". In: *Review of Income and Wealth* 42.4, pages 381–400.
Busetta, Annalisa and Daria Mendola (2012). "The contribution to poverty persistence of children, adults, and the elderly: some empirical evidences from eleven European countries". In: *Population Association of America, 2012 Annual Meeting*. Population Association of America. San Francisco, CA: United States of America, pages 1–9.
Bussolo, Maurizio, Johannes Koettl, and Emily Sinnott (2015). *Golden aging: Prospects for healthy, active, and prosperous aging in Europe and Central Asia*. Washington, DC: United States of America: The World Bank.
Caban-Martinez, Alberto J et al. (2011). "Arthritis, occupational class, and the aging US workforce". In: *American Journal of Public Health* 101.9, pages 1729–1734.
Callander, Emily J and Deborah J Schofield (2015a). "Effect of asthma on falling into poverty: the overlooked costs of illness". In: *Annals of Allergy, Asthma & Immunology* 114.5, pages 374–378.
——(2015b). "Psychological distress and the increased risk of falling into poverty: a longitudinal study of Australian adults". In: *Social Psychiatry and Psychiatric Epidemiology* 50.10, pages 1547–1556.

———(2016). "Arthritis and the risk of falling into poverty: a survival analysis using Australian data". In: *Arthritis & Rheumatology* 68.1, pages 255–262.
———(2017). "Impact of multidimensional poverty on the self-efficacy of older people: Results from an Australian longitudinal study". In: *Geriatrics & Gerontology International* 17.2, pages 308–314.
Calvo, Cesar and Stefan Dercon (2009). "Chronic poverty and all that". In: *Poverty Dynamics: Interdisciplinary Perspectives*. Edited by Tony Addison, David Hulme, and Ravi Kanbur. Oxford: United Kingdom: Oxford University Press, pages 29–58/
Castañeda, Andrés et al. (2018). "A new profile of the global poor". In: *World Development* 101, pages 250–267.
Cavasso, Barbara and Guglielmo Weber (2013). "The effect of the great recession on the wealth and financial distress of 65+ Europeans". In: *Active ageing and solidarity between generations in Europe: First results from SHARE after the economic crisis*. Edited by Axel Börsch-Supan et al. Berlin: Germany: Walter de Gruyter, pages 27–36.
Chan, Tak Wing and John H Goldthorpe (2007). "Class and status: The conceptual distinction and its empirical relevance". In: *American Sociological Review* 72.4, pages 512–532.
Chavez, Koji et al. (2018). "Poverty among the Aged Population: The Role of out-of-Pocket Medical Expenditures and Annuitized Assets in Supplemental Poverty Measure Estimates". In: *Social Security Bulletin* 78.1, pages 47–75.
Chetty, Raj et al. (2014b). "Where is the land of opportunity? The geography of intergenerational mobility in the United States". In: *The Quarterly Journal of Economics* 129.4, pages 1553–1623.
Chien, Shaohua and Martin Ravallion (2001). "How did the world's poorest fare in the 1990s?" In: *Review of Income and Wealth* 47.3, pages 283–300.
Christelis, Dimitrios et al. (2009). "Income, wealth and financial fragility in Europe". In: *Journal of European Social Policy* 19.4, pages 359–376.
Christensen, Ulla et al. (2014). "Addressing social inequality in aging by the Danish occupational social class measurement". In: *Journal of Aging and Health* 26.1, pages 106–127.
Clark, Terry and Seymour Lipset (1991). "Are Social Classes Dying?" In: volume 6. 4, pages 397–410.
Commission Regulation (EU) No 112/2013 of 7 February 2013 implementing Regulation (EC) No 1177/2003 of the European Parliament and of the Council concerning Community statistics on income and living conditions (EU-SILC) as regards the 2014 list of target secondary variables on material deprivation (2013). European Commission Regulation 112/2013. Brussels: Belgium.
Corak, Miles (2013). "Income inequality equality of opportunity, and intergenerational mobility". In: *Journal of Economic Perspectives* 27.3, pages 79–102.
Coromaldi, Manuela and Mariangela Zoli (2012). "Deriving multidimensional poverty indicators: Methodological issues and an empirical analysis for Italy". In: *Social indicators research* 107.1, pages 37–54.

Council Decision of 19 December 1984 on specific Community action to combat poverty (85/8/EEC) (1984).

d'Addio, Anna Christina (2007). *Intergenerational transmission of disadvantage.* OECD Social, Employment and Migration Working Paper 52. Paris: France.

Dahrendorf, Ralf (1959). *Class and Class Conflict in Industrial Society.* Stanford, CA: United States of America: Stanford University Press.

Dalton, Patricio S, Sayantan Ghosal, and Anandi Mani (2016). "Poverty and aspirations failure". In: *The Economic Journal* 126.590, pages 165–188.

Damiani, Gianfranco et al. (2007). "The impact of socioeconomic level on influenza vaccination among Italian adults and elderly: a cross-sectional study". In: *Preventive Medicine* 45.5, pages 373–379.

Darin-Mattsson, Alexander, Stefan Fors, and Ingemar Kåreholt (2017). "Different indicators of socioeconomic status and their relative importance as determinants of health in old age". In: *International Journal for Equity in Health* 16.1. https://doi.org/2010.1186/s12939-017-0670-3.

Deaton, Angus and Salman Zaidi (2002). *Guidelines for constructing consumption aggregates for welfare analysis.* LSMS (Living Standards Measurement Study) Working Paper 135. Washington, DC: United States of America.

Department, Social Protection (2014). *Social protection for older persons: Key policy trends and statistics.* Social protection Policy Paper. Geneva: Switzerland.

Dixon, John (2012). "On being Poor-by-Choice: A Philosophical Critique of the Neoliberal Poverty Perspective". In: *Poverty & Public Policy* 4.2, pages 1–19.

Dixon, John, Kerry Carrier, and Rhys Dogan (2005). "On investigating the 'underclass': contending philosophical perspectives". In: *Social Policy and Society* 4.1, pages 21–30.

Doblas, Juan López, Mara del Pilar Daz Conde, and Mariano Sánchez Martnez (2014). "El rechazo de las mujeres mayores viudas a volverse a emparejar: cuestión de género y cambio social". In: *Poltica y Sociedad* 51.2, pages 507–532.

Duncan, Otis Dudley (1961). "A socioeconomic index for all occupations". In: *Occupations and Social Status.* Edited by Albert Reiss. Volume 1. New York, NY: United States of America: Free Press, pages 109–138.

Durlauf, Steven N and Ananth Seshadri (2018). "Understanding the Great Gatsby Curve". In: *NBER Macroeconomics Annual* 32.1, pages 333–393.

Edwards, Alba M (1917). "Socio-economic Groups of the United States". In: *American Statistical Association Quarterly* 15.118, pages 642–661.

Engelhardt, Gary and Jonathan Gruber (2006). "Social Security and the Evolution of Elderly Poverty". In: *Public Policy and the Income Distribution.* Edited by Alan J Auerbach, David Card, and John M Quigley. New York, NY: United States of America: Russell Sage Foundation, pages 259–287.

Enroth, Linda et al. (2018). "Does use of long-term care differ between occupational classes among the oldest old? Vitality 90+ Study". In: *European Journal of Ageing* 15.2, pages 143–153.

Erikson, Robert, John H Goldthorpe, and Lucienne Portocarero (1979). "Intergenerational class mobility in three Western European societies: England, France and Sweden". In: *The British Journal of Sociology* 30.4, pages 415–441.

Estes, Carroll L., Simon Biggs, and Chris Phillipson (2003). *Social Theory, Social Policy and Ageing. A Critical Introduction.* Open University Press.

Estivill, Jordi (2003). *Concepts and strategies for combating social exclusion: an overview.* Geneva: Switzerland: International Labour Organization.

Evans, Brooks and Robert Palacios (2015). *An Examination of Elderly Co-residence in the Developing World1.* Social Protection & Labor Policy Note 17. Washington, D.C.: United States of America: The World Bank.

Feagin, Joe R (1975). *Subordinating the Poor: Welfare and American Beliefs.* Englewood Cliffs, NJ: United States of America: Prentice-Hall.

Featherstone, Mike and Mike Hepworth (1983). "The Midlifestyle of 'George and Lynne': Notes on a Popular Strip". In: *Theory, Culture & Society* 1.3, pages 85–92.

Ferreira, Francisco HG et al. (2015). *A global count of the extreme poor in 2012. Data issues, methodology and initial results.* Policy Research Working Paper 7432. Washington, DC: United States of America.

Formosa, Marvin (2014). "Social Class Structure and Identity in Later Life". In: *Research on Ageing and Social Policy* 2.1, pages 2–27.

Formosa, Marvin and Paul Higgs, editors (2015). *Social class in later life: Power, identity and lifestyle.* Bristol: United Kingdom: Policy Press.

Foster, James E (2009). "A Class of Chronic Poverty Measures". In: *Poverty Dynamics: Interdisciplinary Perspectives.* Edited by Tony Addison, David Hulme, and Ravi Kanbur. Oxford: United Kingdom: Oxford University Press, pages 59–76.

Foster, James, Joel Greer, and Eric Thorbecke (1984). "A class of decomposable poverty measures". In: *Econometrica* 52.3, pages 761–765.

Fox, Liana E and José Pacas (2018). *Deconstructing Poverty Rates among the 65 and Older Population: Why Has Poverty Increased Since 2015?* SEHSD Working Paper 13. Washington, DC: United States of America: Social, Economic and Housing Statistics Division - U.S. Census Bureau.

Fox, Liana et al. (2015). "Waging war on poverty: Poverty trends using a historical supplemental poverty measure". In: *Journal of Policy Analysis and Management* 34.3, pages 567–592.

Galea, Sandro et al. (2011). "Estimated deaths attributable to social factors in the United States". In: *American Journal of Public Health* 101.8, pages 1456–1465.

Galobardes, Bruna et al. (2006a). "Indicators of socioeconomic position (part 1)". In: *Journal of Epidemiology & Community Health* 60.1, pages 7–12.

—— (2006b). "Indicators of socioeconomic position (part 2)". In: *Journal of Epidemiology & Community Health* 60.2, pages 95–101.

Garner, Thesia I and Kathleen S Short (2003). "Personal assessments of minimum income and expenses: what do they tell us about minimum living' thresholds and equivalence scales?" In: *Inequality, Welfare and Poverty: Theory and Measurement.* Edited by Yoram Amiel and John A Bishop. Volume 9. Research on Economic

Inequality. Bingley: United Kingdom: Emerald Group Publishing Limited, pages 191–243.

Geyer, Siegfried et al. (2006). "Education, income, and occupational class cannot be used interchangeably in social epidemiology Empirical evidence against a common practice". In: *Journal of Epidemiology & Community Health* 60.9, pages 804–810.

Ghilarducci, Teresa (2004). *What You Need to Know about the Economics of Growing Old (but Were Afraid to Ask): A Provocative Reference Guide to the Economics of Aging*. Notre Dame, IN: United States of America: University of Notre Dame Press.

Gilleard, Chris and Paul Higgs (2005). *Contexts of Ageing: Class, Cohort and Community*. Cambridge: United Kingdom: Polity Press.

—— (2009). "The third age: field, habitus or identity". In: *Consumption and generational change: The rise of consumer lifestyles*. Edited by Ian Jones, Paul Higgs, and David J Ekerdt. New Brunswick, NJ: United States of America: Transaction Publishers, pages 23–36.

Gilleard, Chris et al. (2005). "Class, cohort, and consumption: The British experience of the third age". In: *The Journals of Gerontology Series B: Psychological Sciences and Social Sciences* 60.6, S305–S310.

Goedhart, Theo et al. (1977). "The poverty line: concept and measurement". In: *Journal of Human Resources* 12.4, pages 503–520.

Goldthorpe, John and Keith Hope (1974). *The Social Grading of Occupations: A New Approach and Scale*. Oxford: United Kingdom: Clarendon Press.

Gradín, Carlos, Coral Del Ro, and Olga Cantó (2012). "Measuring poverty accounting for time". In: *Review of Income and Wealth* 58.2, pages 330–354.

Greene, William H (2003). *Econometric Analysis*. 5th edition. Upper Saddle River, NJ: United States of America: Prentice-Hall.

Grundy Emily and Gemma Holt (2001). "The socioeconomic status of older adults: How should we measure it in studies of health inequalities?" In: *Journal of Epidemiology and Community Health* 55.12, pages 895–904.

Guio, Anne-Catherine et al. (2017). *Revising the EU material deprivation variables*. Statistical Working Papers. Luxembourg: Luxembourg.

Harper, Caroline (2004). *Child Ill Health and Mortality How Can We Prevent the Preventable*. CHIP Policy Briefing 7. London: United Kingdom: Childhood Poverty Research and Policy Centre.

Hauser, Robert M and John Robert Warren (1997). "Socioeconomic indexes for occupations: A review, update, and critique". In: *Sociological Methodology* 27.1, pages 177–298.

Haveman, Robert and Barbara Wolfe (1995). "The determinants of children's attainment: a review of methods and findings". In: *Journal of Economic Literature* 33, pages 1829–1878.

Hollingshead, August B (2011). "Four factor index of social status". In: *Yale Journal of Sociology* 8, pages 21–51.

Horvat, Pia et al. (2014). "Life Course Socioeconomic Position and Mid-Late Life Cognitive Function in Eastern Europe". In: *The Journals of Gerontology Series B: Psychological Sciences and Social Sciences* 69.3, pages 470–481.

Hout, Michael and Robert M Hauser (1992). "Symmetry and hierarchy in social mobility: a methodological analysis of the CASMIN model of class mobility". In: *European Sociological Review* 8.3, pages 239–266.

Hoy, Michael and Buhong Zheng (2011). "Measuring lifetime poverty". In: *Journal of Economic Theory* 146.6, pages 2544–2562.

Hulme, David, and Andy McKay (2013). "Identifying and Measuring Chronic Poverty: Beyond Monetary Measures?.". In: *The many dimensions of poverty*, pp. 187–214. London: United Kingdom. Palgrave Macmillan

Hunt, Matthew O and Heather E Bullock (2016). "Ideologies and beliefs about poverty". In: *The Oxford Handbook of the Social Science of Poverty*. Edited by David Brady and Linda M Burton. Oxford: United Kingdom: Oxford University Press, pages 93–116.

Iceland, John (2012). *Poverty in America: A Handbook*. Berkeley: United States of America: University of California Press.

Jäntti, Markus and Sheldon Danziger (2000). "Income poverty in advanced countries". In: *Handbook of Income Distribution*. Edited by Anthony B Atkinson and François Bourguignon. Volume 1. Amsterdam: The Netherlands: Elsevier, pages 309–378.

Jehoel-Gijsbers, Gerda and Cok Vrooman (2008). *Social exclusion of the elderly: A comparative study of EU member states*. ENEPRI Research Report 57. Brussels: Belgium.

Kamara, J.; Allieu-Kekura, I.; and Kanu, A. (2017). Sierra Leone 2015 Population and Housing Census. Thematic Report on Elderly Population. Statistics Sierra Leone. Freetown, Sierra Leone.

Kakwani, Nanak, and Kalanidhi Subbarao (2005). *Ageing and poverty in Africa and the role of social pensions*. Social Protection Discussion Paper Series N 0521. Social Protection Unit. The World Bank. Washington, DC: United States of America.

Kallio, Johanna and Mikko Niemelä (2014). "Who Blames the Poor? Multilevel evidence of support for and determinants of individualistic explanation of poverty in Europe". In: *European Societies* 16.1, pages 112–135.

Kanbur, Ravi and Diganta Mukherjee (2007). "Premature mortality and poverty measurement". In: *Bulletin of Economic Research* 59.4, pages 339–359.

Klein, Daniel (2014). *Travels with Epicurus: A Journey to a Greek Island in Search of a Fulfilled Life*. London: United Kingdom: Penguin Publishing Group.

Kluegel, James R and Eliot R Smith (1981). "Beliefs about stratification". In: *Annual Review of Sociology* 7.1, pages 29–56.

Korenman, Sanders and Dahlia Remler (2013). *Rethinking elderly poverty: time for a health inclusive poverty measure?* NBER Working Paper 18900. Cambridge, MA: United States of America: National Bureau of Economic Research.

Kotecha, Mehul, Sue Arthur, and Steven Coutinho (2013). *Understanding the relationship between pensioner poverty and material deprivation*. DWP Research Report 827. London: United Kingdom: Department for Work and Pensions.

Krueger, Alan B (2012). *The rise and consequences of inequality*. Presentation made to the Center for American Progress. URL: http://www.%20americanprogress. %20org/events/2012/01/12/17181/the-rise-and-consequences-of-inequality.

Kwan, Crystal and Christine A Walsh (2018). "Old age poverty: a scoping review of the literature". In: *Cogent Social Sciences*. URL: https://doi.org/10.1080/23311886. 2018.1478479.

Lefebvre, Mathieu, Pierre Pestieau, and Grégory Ponthière (2013). "Measuring poverty without the mortality paradox". In: *Social Choice and Welfare* 40.1, pages 285–316.

— (2014). "Mortalité différentielle et pauvreté par âge". In: *Revue française d'économie* 29.4, pages 173–196.

— (2018). "FGT Old-Age Poverty Measures and the Mortality Paradox: Theory and Evidence". In: *Review of Income and Wealth* 64.2, pages 428–458.

Lelli, Sara (2005). "Using functionings to estimate equivalence scales". In: *Review of Income and Wealth* 51.2, pages 255–284.

Lewis, Oscar (1961). *The Children of Sánchez: Autobiography of a Mexican Family*. New York: United States of America: Random House.

— (1966). "The culture of poverty". In: *Scientific American* 215.4, pages 19–25.

Lewis, Oscar et al. (1966). *La vida: A Puerto Rican family in the culture of poverty-San Juan and New York*. Volume 13. New York: United States of America: Random House.

Li, Chung-Yi, Shwu Chong Wu, and Fung-Chang Sung (2002). "Lifetime principal occupation and risk of cognitive impairment among the elderly". In: *Industrial Health* 40.1, pages 7–13.

Lokshin, Michael (2004). *Robustness of Subjective Welfare Analysis in a Poor Developing Country: Madagascar 2001*. World Bank Policy Research Working Paper 3191. New York, NY: United States of America.

Lopes, Alexandra (2015). "Measuring social class in later life". In: *Social class in later life. Power, identity and lifestyle*. Edited by Marvis Formosa and Paul Higgs. Bristol: United Kingdom: Policy Press, pages 53–71.

Lynch, John and George Kaplan (2000). "Socioeconomic position". In: *Social epidemiology*. Edited by Lisa F Barkman and Ichiro Kawachi. New York, NY: United States of America: Oxford University Press, pages 13–35.

MacDonald, Bonnie-Jeanne, Doug Andrews, and Robert L Brown (2010). "The Canadian Elder Standard - pricing the cost of basic needs for the Canadian elderly". In: *Canadian Journal on Aging/La Revue canadienne du vieillissement* 29.1, pages 39–56.

Marchand, Joseph and Timothy M Smeeding (2016). "Poverty and Aging". In: *Handbook of the Economics of Population Aging*. Edited by John Piggott and Alan Woodland. Volume 1B. Handbooks in Economics. Amsterdam: The Netherlands: Elsevier, pages 905–950.

Marmot, Michael G and Martin J Shipley (1996). "Do socioeconomic differences in mortality persist after retirement? 25 year follow up of civil servants from the first Whitehall study". In: *British Medical Journal* 313.7066, pages 1177–1180.

Marmot, Michael and Ruth Bell (2016). "Social inequalities in health: a proper concern of epidemiology". In: *Annals of Epidemiology* 26.4, pages 238–240.

Mendola, Daria, Annalisa Busetta, and Arnstein Aassve (2009). "What keeps young adults in permanent poverty? A comparative analysis using ECHP". In: *Social Science Research* 38.4, pages 840–857.

Mendola, Daria, Annalisa Busetta, and Anna Maria Milito (2011). "Combining the intensity and sequencing of the poverty experience: a class of longitudinal poverty indices". In: *Journal of the Royal Statistical Society: Series A (Statistics in Society)* 174.4, pages 953–973.

Moen, Phyllis (2013). "New Directions in the Sociology of Aging". In: edited by Linda J Waite and Thomas J Plewes. Washington, DC: United States of America: National Academies Press, pages 175–216.

Morciano, Marcello, Ruth Hancock, and Stephen Pudney (2015). "Disability costs and equivalence scales in the older population in Great Britain". In: *Review of Income and Wealth* 61.3, pages 494–514.

Morduch, Jonathan (2005). "Poverty measures". In: *Handbook on poverty statistics: concepts, methods and policy use*. New York: United States of America: United Nations Statistics Division, pages 52–84.

Moulaert, Thibauld, Anna Wanka, and Matthias Drilling (2017). *Mapping the Relations Between Age, Space, and Exclusion*. Knowledge Synthesis 5.

Mowafi, M and M Khawaja (2005). "Poverty". In: *Journal of Epidemiology and Community Health* 59, pages 260–264.

Mutchler, Jan E et al. (2015). "The elder economic security standard index^{TM}: A new indicator for evaluating economic security in later life". In: *Social Indicators Research* 120.1, pages 97–116.

Nakao, Keiko and Judith Treas (1994). "Updating occupational prestige and socioeconomic scores: How the new measures measure up". In: *Sociological Methodology* 24, pages 1–72.

Nam, Charles B and Mary G Powers (1983). *The socioeconomic approach to status measurement*. Houston, TX: United States of America: Cap and Gown Press.

Naschold, Felix (2013). "Welfare Dynamics in Pakistan and Ethiopia - Does the estimation method matter?" In: *The Journal of Development Studies* 49.7, pages 936–954.

Nimrod, Galit and Arie Rotem (2012). "An exploration of the innovation theory of successful ageing among older tourists". In: *Ageing & Society* 32.3, pages 379–404.

Noël-Miller, Claire M (2011). "Partner caregiving in older cohabiting couples". In: *Journals of Gerontology Series B: Psychological Sciences and Social Sciences* 66.3, pages 341–353.

OECD (2017). *Pensions at a Glance 2017. OECD and G20 Indicators*. Technical report. Paris: France.

Ogg, Jim (2005). "Social exclusion and insecurity among older Europeans: the influence of welfare regimes". In: *Ageing & Society* 25.1, pages 69–90.

ONS (no date). *200 years of the Census in ...Bedfordshire*. Statistical Monitor. Titchfield: United Kingdom.

Oorschot, Wim van and Femke Roosma (2017). "The Social Legitimacy of Targeted Welfare and Welfare Deservingness". In: *The Social Legitimacy of Targeted Welfare. Attitudes to Welfare Deservingness*. Edited by Wim van Oorschot et al. Cheltenham: United Kingdom: Edward Elgar Publishing, pages 3–33.

Orshansky Mollie (1965). "Counting the poor: Another look at the poverty profile". In: *Social Security Bulletin* 28, pages 3–29.

Osberg, Lars (2001). *Poverty Among Senior Citizens: A Canadian Success Story in International Perspective*. Luxembourg Income Study Working Paper 274. Luxembourg: Luxembourg: LIS Cross-National Data Center.

Osberg, Lars and Kuan Xu (2000). "International comparisons of poverty intensity: Index decomposition and bootstrap inference". In: *Journal of Human Resources* 35.1, pages 51–81.

Pakulski, Jan and Malcolm Waters (1996a). *The Death of Class*. London: United Kingdom: Sage.

—— (1996b). "The reshaping and dissolution of social class in advanced society". In: *Theory & Society* 25.5, pages 667–691.

Phillipson, Chris (2013). *Ageing*. Polity Press.

Piachaud, David (1981). "Peter Townsend and the Holy Grail". In: *New Society* 10, pages 418–420.

Piff, Paul K, Michael W Kraus, and Dacher Keltner (2018). "Unpacking the inequality paradox: The psychological roots of inequality and social class". In: *Advances in Experimental Social Psychology*. Volume 57. London: United Kingdom: Elsevier, pages 53–124.

Piketty, Thomas (2014). *Capital in the Twenty-First Century*. Cambridge, MA: Harvard University Press.

Ponthière, Grégory (2017). *Économie du vieillissement*. Collection Repéres. Paris: France: La Découverte.

Pradhan, Menno and Martin Ravallion (2000). "Measuring poverty using qualitative perceptions of consumption adequacy". In: *Review of Economics and Statistics* 82.3, pages 462–471.

Prinzen, Katrin (2016). "Attitudes toward intergenerational redistribution in the Welfare State". In: *Social Demography. Forschung an der Schnittstelle von Soziologie und Demografie*. Edited by Karsten Hank and Michaela Kreyenfeld. Kölner Zeitschrift für Soziologie und Sozialpsychologie. Wiesbaden: Germany: Springer, pages 349–370.

Proctor, Bernadette D, Jessica L Semega, and Melissa A Kollar (2016). *Income and Poverty in the United States: 2015*. Current Population Reports P60-256. Suitland, MD: United States of America.

Ravallion, Martin, Shaohua Chen, and Prem Sangraula (2009). "Dollar a day revisited". In: *The World Bank Economic Review* 23.2, pages 163–184.

Ravallion, Martin, Gaurav Datt, and Dominique Van De Walle (1991). "Quantifying absolute poverty in the developing world". In: *Review of Income and Wealth* 37.4, pages 345–361.

Renwick, Trudi and Liana E Fox (no date). *The Supplemental Poverty Measure: 2015.* Current Population Report P60-258. Washington, DC: United States of America.

Rissanen, Sari and Satu Ylinen (2014). "Elderly poverty: risks and experiences a literature review". In: *Nordic Social Work Research* 4.2, pages 144–157.

Rodgers, Joan R and John L Rodgers (1993). "Chronic poverty in the United States". In: *Journal of Human Resources* 28.1, pages 25–54.

—— (2010). "Chronic and transitory poverty over the life cycle". In: *Australian Journal of Labour Economics* 13.2, pages 117–136.

Rose, David and Eric Harrison (2007). "The European socio-economic classification: a new social class schema for comparative European research". In: *European Societies* 9.3, pages 459–490.

Ruggles, Patricia (1992). "Measuring poverty". In: *Focus* 14.1, pages 1–9.

Russell, Laura Henze, Ellen A Bruce, and Judith Conahan (2006). *The WOW-GI National Elder Economic Security Standard: A methodology to determine economic security for elders.* Technical report. Boston, MA and Washington, DC: United States of America.

Sachs, Jeffrey (2005). *The End of Poverty. How we can make it happen in our lifetime.* London: United Kingdom: Penguin Books.

Saunders, Peter and Timothy M. Smeeding (1998). *How Do the Elderly Fare Cross-Nationally? Evidence from the Luxembourg Income Study Project.* Discussion Papers. University of New South Wales, Social Policy Research Centre.

Savage, Mike, Alan Warde, and Fiona Devine (2005). "Capitals, assets, and resources: some critical issues". In: *The British Journal of Sociology* 56.1, pages 31–47.

Savage, Mike et al. (2015). *Social Class in the 21st Century.* London: United Kingdom: Pelican Books.

Scharf, Thomas, Chris Phillipson, and Allison E Smith (2005). "Social exclusion of older people in deprived urban communities of England". In: *European Journal of Ageing* 2.2, pages 76–87.

Scharlach, Andrew E and Amanda J Lehning (2013). "Ageing-friendly communities and social inclusion in the United States of America". In: *Ageing & Society* 33.1, pages 110–136.

Scherger, Simone, James Nazroo, and Paul Higgs (2011). "Leisure activities and retirement: do structures of inequality change in old age?" In: *Ageing & Society* 31.1, pages 146–172.

Schwartzman, Simon (no date). *The Statistical Measurement of Poverty.* Summary report of the "Expert Group on Poverty Statistics" (Rio Group) Rio de Janeiro, 13–15 May, 1998. URL: https://unstats.un.org/unsd/statcom/doc99/rio.pdf.

Settersten Jr, Richard A and Molly E Trauten (2009). "The new terrain of old age: Hallmarks, freedoms, and risks". In: *Handbook of Theories of Aging.* Edited by Vern L. Bengtson et al. 2nd. Volume 2. New York, NY: United States of America, pages 455–470.

Settersten, Richard A and Lynn Gannon (no date). "Structure, agency, and the space between: on the challenges and contradictions of a blended view of the life course".

In: *Towards an Interdisciplinary Perspective on the Life Course*. Edited by René Levy et al. Volume 10. Advances in Life Course Research. Amsterdam: The Netherlands: Elsevier, pages 35–55.

Sheiner, Louise, Daniel Sichel, Lawrence Slifman, et al. (2007). *A primer on the macroeconomic implications of population aging*. Staff Working Paper 2007-01. Washington, DC: United States of America.

Shepherd, Andrew (2013). "An Evolving Framework for Understanding and Explaining Chronic Poverty". In: *Chronic Poverty. Concepts, Causes and Policy*. Edited by Andrew Shepherd and Julia Brunt. Rethinking International Development Series. Basingstoke: United Kingdom: Palgrave Macmillan, pages 7–37.

Short, Kathleen (2011). *The Research Supplemental Poverty Measure: 2010*. Current Population Report P60-241. Washington, DC: United States of America.

Short, Kathleen and Thesia I Garner (2002). "Experimental poverty measures: Accounting for medical expenditures". In: *Monthly Labor Review* 125, pages 3–13.

Shröder-Butterfill, Elisabeth and Ruly Marianti (2006). "A framework for understanding old-age vulnerabilities". In: *Ageing Society* 26.1, pages 9–35.

Skeggs, Beverley (2005). "The re-branding of class: Propertising culture". In: *Rethinking class: Culture, identities and lifestyle*. Basingstoke: United Kingdom: Palgrave Macmillan, pages 46–68.

Smeeding, Timothy M. (1999). *Social Security Reform: Improving Benefit Adequacy and Economic Security for Women*. Center for Policy Research Policy Briefs 16. Center for Policy Research, Maxwell School, Syracuse University.

Smeeding, Timothy and Susanna Sandstrom (2004). *Poverty and Income Maintenance in Old Age: A Cross-National View of Low Income Older Women*. Working Papers, Center for Retirement Research at Boston College. Center for Retirement Research.

Smeeding, Timothy et al. (2008). *Older Women's Income and Wealth Packages in Cross-National Perspective*. Working Papers, Center for Retirement Research at Boston College. Center for Retirement Research.

Solon, Gary (2004). "A model of intergenerational mobility variation over time and place". In: *Generational income mobility in North America and Europe*. Edited by Miles Corak. Cambridge: United Kingdom: Cambridge University Press, pages 38–47.

Spicker, Paul, Sonia Álvarez Leguizamón, and David Gordon (2013). *Poverty: An International Glossary*. International Studies in Poverty Research. London: United Kingdom: Zed Books.

Stevens, Gillian and Joo Hyun Cho (1985). "Socioeconomic indexes and the new 1980 census occupational classification scheme". In: *Social Science Research* 14.2, pages 142–168.

Stevens, Gillian and Elizabeth Hoisington (1987). "Occupational Prestige and the 1980 U.S Labor Force". In: *Social Science Research* 16, pages 74–105.

Stewart, Mark (2009). "The estimation of pensioner equivalence scales using subjective data". In: *Review of Income and Wealth* 55.4, pages 907–929.

Stiglitz, Joseph E (2012). *The Price of Inequality*. New York, NY: United States of America: W. W. Norton.

Stiglitz, Joseph, Amartya Sen, and Jean Fitoussi (2009). *Report by the commission on the measurement of economic performance and social progress*. Technical report. Paris: France: Commission on the measurement of economic performance and social progress.

Suppa, Nicolai (2015). "Towards a multidimensional poverty index for Germany". In: *Empirica*, pages 1–29. https://doi.org/2010.1007/s10663-017-9385-3.

The World Bank (2015). *The State of Social Safety Nets 2015*. Technical report. Washington, DC: United States of America.

Townsend, Peter (1979). *Poverty in the United Kingdom. A Survey of Household Resources and Standards of Living*. Harmondsworth: United Kingdom: Penguin Books.

—— (1981). "The structured dependency of the elderly: a creation of social policy in the twentieth century". In: *Ageing & Society* 1.1, pages 5–28.

Treiman, Donald J (1977). *Occupational prestige in comparative perspective*. Quantitative studies in social relations series. New York, NY: United States of America: Academic Press.

Turner, Kelly and Amanda J Lehning (2007). "Psychological Theories of Poverty". In: *Journal of Human Behavior in the Social Environment* 16.1–2, pages 7–72.

UK, Age (2018). *Poverty in later life*. Technical report. London: United Kingdom.

UNDP (2016). *Human Development Report 2016. Human Development for Everyone*. New York, NY: United States of America: United Nations Development Programme.

UNFPA—HelpAge International (2012). *Ageing in the Twenty-First Century: A Celebration and A Challenge*. New York, NY: United States of America and London: United Kingdom.

United Nations (1996). *Report of the World Summit for Social Development, Copenhagen 6–12 March 1995*. A/CONF. 166/9. New York: United States of America.

United Nations (2015). World Ageing Report 2015. Department of Economic and Social Affairs. Population Division. United Nations. New York, NY. United States of America.

Van de Water, Paul, Arloc Sherman, and Kathy A Ruffing (2013). *Social Security keeps 22 million Americans out of poverty: A state-by-state analysis*. Technical report. Washington, DC: United States of America: Center for Budget and Policy Priorities.

Veall, Michael R. (2007). *Which Canadian Seniors Are Below the Low-Income Measure? Quantitative Studies in Economics and Population Research Reports*. Technical report. McMaster University.

Veit-Wilson, John (2002). "States of welfare: A response to Charles Atherton". In: *Social Policy & Administration* 36.3, pages 312–317.

Vespa, Jonathan (2012). "Union formation in later life: Economic determinants of cohabitation and remarriage among older adults". In: *Demography* 49.3, pages 1103–1125.

Villar, Antonio (2017). *Lectures on Inequality, Poverty and Welfare* Volume 685. Lecture Notes in Economics and Mathematical Systems. Cham: Switzerland: Springer International Publishing AG.

Walker, Alan (2006). "Reexamining the political economy of aging: Understanding the structure/agency tension". In: *Aging, globalization and inequality: The new critical gerontology.* New York, NY: United States of America: Baywood, pages 59–79.

Wallace, Steven P, D Imelda Padilla-Frausto, and Susan E Smith (2013). "Economic Need among Older Latinos: Applying the Elder Economic Security Standard^{TM} Index". In: *Journal of cross-cultural gerontology* 28.3, pages 239–250.

Walsh, Kieran (2017). "Constructions of old-age social exclusion". In: *Geographical gerontology: Perspectives, concepts, approaches.* Edited by Mark W Skinner, Gavin J Andrews, and Malcolm P Cutchin. Routledge, pages 252–266.

Walsh, Kieran, Thomas Scharf, and Norah Keating (2017). "Social exclusion of older persons: a scoping review and conceptual framework". In: *European Journal of Ageing* 14.1, pages 81–98.

Watts, Harold W (1968). *An economic definition of poverty.* Discussion Paper 5–68. Madison, WI: United States of America.

Weber, Max (1978). *Economy and Society: An Outline of Interpretive Sociology.* Berkley, CA: United States of America: University of California Press.

Wilkinson, Richard G (1989). "Class mortality differentials, income distribution and trends in poverty 1921–1981". In: *Journal of Social Policy* 18.3, pages 307–335.

Wilkinson, Richard G and Kate E Pickett (2010). *The Spirit Level: Why Equality is Better for Everyone.* London: United Kingdom: Penguin Books.

Wimer, Christopher et al. (2016). "Progress on poverty? New estimates of historical trends using an anchored supplemental poverty measure". In: *Demography* 53.4, pages 1207–1218.

Wood, Claudia et al. (2012). *Poverty in Perspective.* London: United Kingdom: Demos.

Wright, Erik Olin (1996). *Class Counts: Comparative Studies In Class Analysis.* New York, NY: United States of America: Cambridge University Press.

Wu, Zheng, Christoph M Schimmele, and Nadia Ouellet (2014). "Repartnering after widowhood". In: *Journals of Gerontology Series B: Psychological Sciences and Social Sciences* 70.3, pages 496–507.

Xu, Kuan, Lars Osberg, et al. (2001). "How to Decompose Sen-Shorrocks-Thon Poverty Index: A Practitioners Guide". In: *Journal of Income Distribution* 10.1-2, pages 77–94.

Zaidi, Asghar and Tania Burchardt (2005). "Comparing incomes when needs differ: equivalization for the extra costs of disability in the UK". In: *Review of Income and Wealth* 51.1, pages 89–114.

6

Some Questions of Intergenerational Economics

> **Overview**
> This chapter brings together some developments in intergenerational economics and policy. Topics include intergenerational transfers and mobility, equity, solidarity, conflict and ambivalence, and the discussions around justice between generations. The chapter closes with a presentation of indices of intergenerational fairness.

Not only do questions of transmission of poverty and inequality incorporate an intergenerational dimension: in several other places along this textbook, we have encountered intergenerational issues. The overlapping generations and the dynastic models, as well as the national transfer accounts, deal with economic relations between generations. Questions of altruism and bequests touch upon intergenerational contacts, and so do considerations of informal care by family members (often provided by adult children to older parents) or childcare provided by grandparents to their grandchildren. Moreover, social security is centred upon discussions of implicit social compacts between generations and the projected 'burden' on future generations. Similarly, some of the macroeconomic implications of population ageing and, more generally, the apocalyptic projections of the allegedly negative consequences of ageing populations have a clear intergenerational flavour.[1] Other areas where issues of intergenerational economics flare up are the discussions about the existence of a 'lump' of labour supply caused by older workers, the analyses of family-run businesses, decision-making in multigenerational households, and so on.

Therefore it is appropriate to devote a short chapter to put together some of the key theoretical insights on intergenerational economic relations and policy.

The concept of 'generation' (see Volume I, Chap. 2) may refer to kinship relationships, consecutive birth cohorts, or a group of birth cohorts sharing and being influenced by historical, cultural, economic, political, and social circumstances. Intergenerational economics usually deals with the first two definitions, although in certain topics (e.g. attitudes to work, consumer behaviour) the third acceptation of the term is also used. Generation may also refer to the position of individuals in the ranked descent within a biosocial family of procreation and succession (Bengtson and Allen 1996, p. 481). Besides, the term may reflect microsocial roles and interactions between age groups within families (Walker 1996b, p. 6). We must not fall into a 'cohort trap' by assuming either that relations between members of successive birth cohorts are characteristic of all intergenerational relations involving those cohorts, or that membership of a particular cohort is necessarily more influential than other factors in determining the nature of intergenerational relations (Walker 1996b, p. 7).

According to Fragnière et al. (2014), intergenerational conflict is the cultural, social, or economic opposition between generations, a conflict that emanates from value differences or divergences regarding the interests and objectives of each generation—that is, a conflict that has a cultural origin. Improbable sources as they may be, two traditional proverbs with a clear intergenerational economic focus provide some initial insight. There is a Chinese saying that can be loosely translated as 'bring up children who can look after you in your older life'. According to Goh (2011, p. 5), this ideal of children as old age insurance is shifting to children as providers of emotional and psychological companionship but with no economic demands—an interesting departure. The second proverb is the Yiddish saying '[w]hen a father gives to his son, both laugh; when a son gives to his father, both cry'[2] (Stone 2006, p. 142). We can see in the first case a social expectation that younger generations will look after and support their parents, whereas in the second instance, the same support and care may entail a feeling of shame for both parties. This ambivalence runs through most of the developments in questions of intergenerational economics. It is related to cultural changes in social roles, expectations, and aspirations.

It is important not to lose sight of the influences of cultural forces on people's thinking, views, and concepts about life and the world—in a (German) word, on their *Weltanschauung*. This applies to all areas of intellectual endeavour, it goes without saying, but is salient in discussions on intergenerational issues. For example, Hareven (1994, p. 447) noted that whilst '[i]n the late nineteenth century children continued to stay in the parental home or moved back and forth in order to meet the needs of their family of orientation by taking care of aging parents or, in some cases, of younger siblings', nowadays 'young adult children reside with their parents in order to meet their own needs, because of their inability to develop an independent work career or to find affordable housing'. Or, consider for example, the following points raised by the US social anthropologist Margaret Mead:

- …in class societies in which there is a high expectation of mobility, problems of generation conflict are endemic (Mead 1970, p. 41)
- Within two decades, 1940–1960, events occurred that have irrevocably altered men's relationships to other men and to the natural world. The invention of the computer, the successful splitting of the atom and the invention of fission and fusion bombs, the discovery of the biochemistry of the living cell, the exploration of the planet's surface, the extreme acceleration of population growth and the recognition of the certainty of catastrophe if it continues, the breakdown in the organisation of cities, the destruction of the natural environment, the linking up of all parts of the world by means of jet flights and television, the preparations for the building of satellites and the first steps into space, the newly realised possibilities of unlimited energy and synthetic raw material and, in the more advanced countries, the transformation of man's age-old problems of production into problems of distribution and consumption –all these have brought about a drastic, irreversible division between the generations. (Mead 1970, p. 61)

Here Mead (1970) opines that economic outcomes (expected or realised) are the main driver behind intergenerational clashes. But to what extent is mobility expectation not the result of the confluence between a particular level of economic development and a set of cultural norms? And to what extent are the events listed above not the result of the interplay between cultural forces and economic development? Hence, my call is for all the readers to be culturally aware, and also aware of the cultural influences under which authors and scholars are and may have been.

One telling aspect to take into account is the viewpoint of the commentator, the policy maker, and the social scientist. Consider this assertion: '[i]n most discussions of the generation gap, the alienation of the young is emphasised, while the alienation of their elders may be wholly overlooked' (Mead 1970, p. 78). Of course, this may simply reflect an alleged bias 'easily' explained by the chronological age group to which most of the individuals who influence public opinion and make policy and even research at any one time belong. However, the opposite slant can also be found in part of the literature. The point is that it is of great significance to be able to separate the scientific wheat from the ideological and cultural chaff—an exercise that goes beyond empirical analyses, either quantitative or qualitative.

One cultural influence, which—according to most scholars—is closer to a myth than to historical evidence, is the romantic idea that in the past there was no intergenerational strife. We could be forgiven if we concluded, on reading the excerpt from Mead above, that intergenerational relations went pear-shaped between 1940 and 1960. In reality, Mead is underlining the transformations in those relations that took place in that period and which, in her view, were deep and irrevocable and mostly deleterious to intergenerational relations, but she is not affirming that before the invention of the computer and the other events, different generations used to live in harmony. OK, you say, perhaps not during the twentieth century, but if we went further back in history would we not be able to see a rosier picture? I am afraid we would not. Samuelson (1958, p. 473) surmised that in 'cold and selfish competitive markets', the young would not 'teleologically' respect older people. This negligence goes beyond 'cold and selfish' free markets. A general theme going through the literature is that since remote times parents and adults in general had to 'work at it' to gain any favours in later life from their siblings. As Reinhold (1970) explained, even in the patriarchal and gerontological societies of the ancient cultures of the Near East, older people as a matter of social policy indoctrinated the youth with obedience to and respect for their elders because they did not take it for granted that generational balance would necessarily follow due to economic dependence from the youth. The Hittites in 1400–1200 BC developed myths about intergenerational power struggles which were adopted later by the Greeks; it is not otiose to reflect that the intergenerational transmission of these myths was 'a matter of social policy'.

However, what is intergenerational policy? Lüscher et al. (2010, p. 114) defined it as 'an expression of the current discursive ethical negotiations about "intergenerational justice" and welfare (state) institutions redistributing resources between generations'. These authors suggested two types of intergenerational policy, descriptive and programmatic. Descriptive intergenerational policy 'includes all efforts to institutionalize individual and collective relations between the generations in the private and public spheres'. Programmatic intergenerational policy comprises 'establishing societal conditions that allow the creation of private and public intergenerational relations in present and in future, in a way that guarantees the development of a responsible and community-oriented personality on the one hand and of societal progress on the other'.

On studying applied quantitative research on intergenerational issues, it is important to take into account that empirically three main statistical approaches have been used, which usually provide heterogeneous results (Costanza et al. 2017):

- age groups are compared in one or more periods of time, but not *over* time
- cross-temporal meta-analysis, where individuals from one age group are compared to individuals from the same age group in previous periods (e.g. the studies of self-esteem among college students in the United States of America in between 1988 and 2008 (Gentile et al. 2010) and in Australia between 1978 and 2014 (Hamamura and Septarini 2017))
- cross-classified hierarchical linear models, where changes in the units individuals belong to over time (residence, school, etc.) are taken into account (Leckie 2014)

Moreover, Guillemard (1996), among others, criticised the use of cross-sectional data to study differences between 'generations' (the first of the three approaches listed above), when all these data can indicate is the distribution of a variable of interest across age groups.

6.1 Intergenerational Transfers

Masson (2007) noted that the economic analysis of intergenerational flows narrowly focuses on the sustainability and optimality of macroeconomic growth, that is, on the allocation and use of non-renewable resources and on the accumulation of capital. However, there are four main types of flows between generations: financial and material, co-residence, and care and support of children and of older adults (Arber 2013), and all these linkages have direct and indirect economic implications.

Some of the exchanges between members of different generations take place between people who are alive, but not always. Bequests, for example, frequently link a member of an older generation leaving, on dying, property and other valuables to members of, usually, a younger generation. In addition, inter-vivos intergenerational transfers, by definition, take place between living members of different generations. Apart from financial transfers, other inter-vivos flows include educational expenses, co-housing, caregiving, and so on. In this regard, it is worth considering the model proposed by Arrondel and Masson (2001), which distinguishes between three types of inter-vivos transfers depending on their primary objective—education, assistance, or transmission—and on the period in the life cycle of the beneficiaries they take place:

- Investments in children's human capital, which are received early in their life cycles
- Financial assistance (e.g. access to credit, down-payments for property acquisition), received during the early part of agents' working lives
- Wealth gifts interpreted as anticipated inheritance and substitutes for bequests, usually received later in the beneficiary's life cycle

To this list, we could add unpaid caregiving.

Flows can go from a member of a younger generation to a member of an older generation (upward transfers) and vice versa (downward transfers). Shanas (1967) reported findings about upward and downward intergenerational financial and in-kind transfers from extensive interviews of 2500 people aged 65 or over carried out in the United Kingdom, Denmark, and the United States of America. This author concluded that the direction of the flow and the nature and magnitude transferred depended on the country and the social class of the older person. In the United Kingdom and United States of America, older people of white-collar occupational background reported giving more help to their adult children than the help they received from them. In contrast, in the United Kingdom, older people from blue collar occupational background received more from their children than what they gave them. In Denmark, in turn, the amount of help received from or given to adult children was much lower than in the United States of America or the United Kingdom.

Regarding the timing of the transmissions, we need to distinguish between the life-cycle stage in which the intergenerational transfers take place and when recipients benefit from them, as there may be a gap between these two processes. One example of a delay between the period a donor makes the transfer and the timing of the resulting benefit for the recipient are the intergenerational transfers of human capital (Blinder 1976; Ishikawa 1975): they tend to materialise much earlier in the lives of the beneficiaries than financial bequests, but their beneficiaries benefit from them much later than when the donors (usually, the parents) made the transfer. Transfers that take the form of human capital investments are usually received before an agent starts her working life, but she actually begins to benefit from the investment later in her career when the human capital investment is reflected in her wages. From the point of view of the donor, instead of spreading savings over the income-generating part of her life cycle, the economic agent has to concentrate these transfers in a shorter period of time. Given usual age-earnings profiles (see Chap. 2 in Volume III) and fertility patterns, these transfers tend to take place before earnings reach their peak. This timing delay would have no implications at all should the agent had perfect access to capital markets when she decided to start a, or enlarge her, family and could borrow an amount equivalent to the human capital transfers she plans to make. However, if there exist borrowing restrictions or other imperfections in the capital markets, then the agent might not be able to access to loans large enough for the desired human capital transfers to materialise.

Sometimes, intergenerational policy gets embroiled in discussions regarding welfare state transfers and benefits. Whilst it is true that most intergenerational policies involve transfers between generations (Klimczuk 2013), much of the redistributive effort by welfare states, as Hills (1996) explained, actually take place across the own life cycles of individuals, rather than between different individuals.

Echoing the objection by Arber about the narrowness of focus in economic analysis of intergenerational issues on fiscal implications of upward transfers of resources to older people, these authors urge that intergenerational linkages and transfers of time as well as money within families (beyond the nuclear unit) should be also taken into account. Moreover, contemporary societal changes and trends have not reduced the level of intergenerational linkages: 'smaller families, serial marriages and geographical mobility have changed the relations between generations but have not undermined mutuality and support' (Johnson 2005, p. 569).

Finally, Gál et al. (2018) combined data on public and private transfers, and unpaid household labour from ten European countries and carried out an extended National Transfer Accounts exercise (see Chap. 1, Volume II). Their findings show that children receive more than twice as many per capita resources as older persons in general, although older people do get more in terms of net public spending and transfers. According to these authors, focusing exclusively on public resources is 'in danger of looking for a lost car key only where the streetlight shines' resulting in the 'apparent pro-elderly welfare state bias' when in fact, 'European societies, as societies, transfer more than twice as many resources on average to each child as to each older person' [p. 952].

6.2 Intergenerational Mobility

'Mobility' is a concept that appears in various sections along this textbook, although denoting different things. For example, in the context of disability and long-term caregiving, mobility refers to the ability to walk for a certain distance, get up from a chair after a given period of time, stoop, kneel, crouch, pick up a coin, and so on. Geographical mobility refers to migration, internal and external. Production factor mobility (e.g. labour mobility or capital mobility) may also refer to migration, but sometimes this concept is used in relation with movements of factors across industrial sectors. Mobility is also applied to changes in education attainment, social class, occupation, or income or wealth brackets of birth along an individual's life course—individuals in countries or welfare state regimes, and so on, with higher social mobility show lower probability of remaining, after a given number of years, in the same education level, social class, and so on, of birth. This is known as intragenerational mobility. Income mobility, for example, refers to 'the movements of a given individual through the distribution of income over time' (OECD 2018, p. 64).

There is another use of mobility, related to the previous acceptation, which is the focus of this section: intergenerational mobility. The OECD defines intergenerational mobility as 'the extent to which some key characteristics and outcomes of individuals differ from those of their parents' (OECD 2006, p. 74). Intergenerational mobility (and its opposite, intergenerational persistence) refers to the degree to which an individual is in the same social position as their parents. Higher intergenerational mobility reflects a higher degree or probability of moving up (or down) the social, occupational, income, or wealth ladders compared to the respective parental position.

To measure of intergenerational mobility, data are required from at least two generations, usually from same-sex dyads (son-father, daughter-mother), although some studies use households as the observational units. The variable of interest may be their respective highest educational attainment, occupation, income, wealth, and so on. Depending on which topic is under study, the intervening variables also differ. For example, a study of intergenerational transmission of wealth may consider the inheritance tax regime or the role of bequests, whilst a study of the intergenerational transmission of educational outcomes may consider a measure of grandparent-grandchild ties (say, frequency of contact). Some variables of interest are discrete (occupation), whereas other variables are continuous (e.g. income), which means that different statistical approaches are used—although the latter are sometimes grouped by quantiles such as quintiles or deciles. Another methodological aspect to take into account is that some studies of intergenerational mobility based on income or wealth data look into changes in levels, whereas other studies focus on changes in rankings (Table 6.1).

Imagine we get data of income levels from fathers and their sons. One approach to measure intergenerational mobility consists in running a regression model of sons' levels of income against their fathers' level of income.[3]

6 Some Questions of Intergenerational Economics

Table 6.1 Intergenerational mobility matrices

	Father's income quintile				
	1	2	3	4	5
(a) No intergenerational mobility					
Son's income quintile 1	0	0	0	0	0
2	0	1	0	0	0
3	0	0	0	0	0
4	0	0	0	0	0
5	0	0	0	0	0
(b) Upward intergenerational mobility					
Son's income quintile 1	0	0	0	0	0
2	0	0	0	0	0
3	0	1	0	0	0
4	0	0	0	0	0
5	0	0	0	0	0

Using logarithms, we get:

$$\log Y_s = \alpha + \beta \cdot \log Y_f + \varepsilon \tag{6.1}$$

where ε is the stochastic error measure, and β is the intergenerational elasticity, from which a measure of intergenerational mobility can be obtained: $1 - \beta$. One problem with this approach is that unless the permanent or the life-cycle incomes are used, the regression estimates will be contaminated by a 'life-cycle bias' if the earnings of fathers and sons are measured at a different ages (as it is the case if data from the same year are used). Using the same ages for fathers and sons, rather than data from the same period, attenuates this bias, but the regression approach still faces the problem that earnings at particular ages may be more representative of lifetime earnings than at other ages. Earnings over the whole life course of the fathers and the sons would be needed. One way to reduce these difficulties is to run a regression on an extended specification that includes the age of the fathers and the sons, and their squared values, such as:

$$\log Y_s = \alpha + \beta_1 \cdot \log Y_f + \beta_2 \cdot \text{age}_s + \beta_3 \cdot \text{age}_s^2 + \beta_4 \cdot \text{age}_f + \beta_5 \cdot \text{age}_s^f + \varepsilon \tag{6.2}$$

A different approach consists in the estimation of intergenerational transition matrices. To fix ideas, let's concentrate on income quintiles (where quintile 1 corresponds to the lowest 20 per cent income levels and quintile 5 groups the top 20 per cent levels of income). For each father-son dyad, at a given age (say, when both the father and the son were 45 years old), the quintile that corresponds to the incomes of each individual is recorded. A matrix is prepared for each dyad, with a value equal to 1 for the cell that represents the father's and son's income quintiles, and equal to 0 for the other cells. Once the data for all the father-son dyads are tabulated, a population matrix is obtained with the observed frequencies for each cell. This matrix is known as the intergenerational mobility matrix. Table 6.2 shows the intergenerational mobility matrix of the occupational categories of fathers' and sons' aged between 30 and 60 years in Norway for the period 1960–1980 (Modalsli 2017, Table A5). Individuals were classified into four occupational categories, from white collar (category with the highest occupational status) to unskilled manual worker (lowest):

In general, there are $i = 1, 2, \ldots, I$ rows and $j = 1, 2, \ldots, I$ columns.[4] The sum of each row is given by $f_{i+} = \sum_{j=1}^{j=I} f_{i,j}$, and the sum of each column by $f_{+j} = \sum_{i=1}^{i=I} f_{i,j}$. The total or grand sum is equal to $f_{++} = \sum_{i=1}^{i=I} \sum_{j=1}^{j=I} f_{i,j}$. With these three aggregates, the marginal distributions by row and column are obtained: f_{i+}/f_{++} and f_{+j}/f_{++}, respectively. The intergenerational transition matrix can be used to estimate the intergenerational transition probability matrix. Continuing the example above, we need to estimate the probability that a son is classified in category j given that his father was classified in category i. This probability is equal to the ratio between the cell $f_{i,j}$ and the four possible categories open to the son: $p_{i,j} = \frac{f_{i,j}}{\sum_{j=1}^{j=I} f_{i,j}}$. In situations of low intergenerational mobility, the social status of a son/daughter is more strongly dependent on the social status of his father/mother than in periods or countries with higher intergenerational mobility. Table 6.2 presents the resulting intergenerational transition probability matrix:

Table 6.2 Male occupational intergenerational mobility matrix, Norway 1960–1980

		Father's occupation				
		White collar	Farmer	Skilled manual	Unskilled manual	Column sum
Son's occupation	White collar	152,363	14,264	119,788	13,433	299,848
	Farmer	1259	5983	2417	615	10,274
	Skilled manual	39,538	11,253	91,062	9365	151,218
	Unskilled manual	11,817	3029	24,416	4839	44,101
Row column		204,977	34,529	237,683	28,252	505,441

Source: Modalsli (2017, Table A5)

Table 6.3 Father-son intergenerational transition probability matrix, Norway 1960–1980

		Father's occupation			
		White collar	Farmer	Skilled manual	Unskilled manual
Son's occupation	White collar	0.508	0.048	0.399	0.045
	Farmer	0.123	0.582	0.235	0.060
	Skilled manual	0.261	0.074	0.602	0.062
	Unskilled manual	0.268	0.069	0.554	0.110

Source: Own estimations based on Modalsli (2017, Table A5)

Table 6.3 shows that there was a relatively low probability that a son born to a father employed in an unskilled manual job remained in the same occupational category. In turn, almost 40 per cent of sons whose fathers were employed in skilled manual jobs were white-collar workers.

Based on transition matrices, various measures of mobility have been proposed (Richey and Rosburg 2018). One consideration is that the 'social ladders' have grown in length over time in most countries (e.g. the distance between minimum and maximum income levels has expanded—see Chetty (2016, Fig. 1.2)). Therefore, social positions may reflect different social standings over time. Some authors have criticised the use of quintiles (and other transformations of the income distribution into discrete categories) to compute these matrices and proposed more finely divided categories (e.g. percentiles). However, using percentiles places more weight on small, materially imperceptible, changes (say, a move between a father's category in the sixty-seventh percentile and a son's income belonging to the sixty-sixth percentile). For continuous variables of interest such as income, another popular method is the rank-rank specification, which consists in regressing the rankings of parental income against other parents in their birth cohort and the rankings of their children's incomes relative to other children in their cohort (Chetty et al. 2014b).

One of the areas of interest in this field is the impact on the aggregate (or social) welfare function of differences in intergenerational mobility. If we assume a continuous variable of interest (say, income) and that the utility functions of all individuals of one same generation can be aggregated into a generational utility function by means of adding them up,[5] then the total welfare function is the sum of the utility functions of each generation in the population. To continue with the father-son dyads, and assuming that they are all alive, the effect of changes in intergenerational mobility on the social welfare function can be represented by:

$$\Delta W = \int_0^{a_f} \int_0^{a_s} U(s,f) \cdot \Delta f(s,f) dx dy \qquad (6.3)$$

where Δ represents a difference or change under the assumption that the utility of each generation increases or at least does not diminish with upward variations in their respective variable of interest (say, that the utility of sons is a positive function of their own income), that is, if we assume that $\frac{\partial U}{\partial s} \geq 0$ and $\frac{\partial U}{\partial f} \geq 0$, and also that the social welfare function is subject to diminishing marginal utility (i.e. <0), then social welfare increases if $\Delta f(s,f) < 0$ (Atkinson and Bourguignon 1982).[6]

The empirical literature on the intergenerational mobility of incomes (see Black and Devereux (2011) for a survey) suggests that the intergenerational elasticity of earnings varies across countries:

- It is higher in the United States of America (around 0.6) than in Nordic European countries.
- Relatively high levels (0.4–0.5) have been reported for Italy (Mocetti 2007; Piraino 2007; Roccisano et al. 2013) and France (Lefranc 2011; Lefranc and Trannoy 2005) and Spain (Cervini-Plá 2015).
- Leigh (2007) found much lower intergenerational elasticity coefficients in Australia (0.2–0.3), although Mendolia and Siminski (2016) reported an intergenerational earnings elasticity of 0.35 for that country.
- Low persistence was reported for Singapore, with intergenerational elasticity of income at about 0.23–0.28 (Ng 2007).
- In contrast, high coefficients are found in many emerging economies:
  - Piraino (2015) reported elasticities of between 0.62 and 0.68 for South Africa.
  - Nunez and Miranda (2010) estimated elasticities of between 0.57 and 0.74 in Chile.

Elasticities tend to differ by gender and marital status of children:

- In Japan, Ueda (2009) found that intergenerational elasticity of earnings for married sons was between 0.41 and 0.46, but only around 0.30–0.38 for married daughters, and lower than 0.30 for single daughters.

- In contrast, in Argentina, Jiménez (2012) found almost complete persistence (0.937) in labour earnings intergenerational elasticity between mothers and daughters, the results for fathers and sons was much lower: 0.642.
- Estimates for South Korea, a country with much lower intergenerational persistence than South Africa or Argentina, Ueda (2013) also reported lower elasticities for sons (0.25) than daughters (0.35).
- In Taiwan, Kan et al. (2015) estimated that the father–son income elasticity is 0.18 whilst the elasticity of incomes of mothers and daughters is 0.54.

Intergenerational persistence of income may result from institutional mechanisms that prevent downward mobility or from institutional penalisations of upward mobility. Raitano and Vona (2015) argue that the former are in place in Spain and Italy, whilst the latter are present in the United Kingdom.

Using wealth instead of income is favoured by some economists as wealth is a proxy for lifetime income. Adermon et al. (2018) reported intergenerational wealth elasticities of around 0.3 and 0.4 in Sweden. As mentioned, inheritance tax and bequests are pertinent explanatory variables for the intergenerational transmission of wealth. In a study of children and parents aged 45–50 years from 2003 to 2013 in Denmark, Boserup et al. (2016) found that bequests can explain as much as 26 per cent of the average wealth over three years following a parent's death. Black et al. (2015) looked into intergenerational transmission of wealth between natural parents and their children and between adoptive parents and their adopted children in Sweden. These authors found higher persistence among adoptive parents, from which they inferred that transmission of wealth across generations is mainly due to environmental factors within the households.

6.2.1 Multigenerational Mobility

Studies of intergenerational mobility do not circumscribe to two contiguous generations, but may extend to three or more generations in which case the topic is known as 'multigenerational mobility'. Becker and Nigel (1986) opined that it takes three generations for any advantages or disadvantages within a family to disappear. In contrast, Clark (2014) argued that it takes several generations and hundreds of years for individuals born in families with incomes or wealth far above or below the mean to get close to that level.

The starting point of the economic models of multigenerational mobility (see Solon (2014) for a review) is a relationship between three adjacent generations, not two as in the previous chapters. A household's budget constraint reflects the parents' decision to either consume for their own or to invest in their children's human capital. If we denote the parents as generation $t-1$ and their children as generation t, the parents' budget constraint becomes:

$$Y_{i,t-1} = C_{i,t-1} + I_{i,t-1} \tag{6.4}$$

where Y, C, I denote, respectively, income, consumption, and human capital investment. Human capital investment is assumed to impact on the children's human capital positively:

$$h_{i,t} = \theta \cdot \log I_{i,t-1} + e_{i,t} \tag{6.5}$$

where $\theta > 0$ is the marginal product or addition to a child's human capital derived from their parent's investment, and e corresponds to the child's endowment of human capital independent of their parent's investment. Although assumed to be independent of parental investments on their children's human capital, this level of endowment is assumed to be related to parental human capital stock. Therefore we can express it thus (Solon 2004):

$$e_{i,t} = \delta + \lambda \cdot e_{i,t-1} + v_{i,t} \tag{6.6}$$

with v as the stochastic error. In time series analysis, Eq. (6.6) is known as a first order autoregressive process: the value of a variable in period t is a function of its value in period $t-1$. Multigenerational models extend the order of the autoregressive process. For example, including grandparents, it becomes a second order process:

$$e_{i,t} = \delta + \lambda_1 \cdot e_{i,t-1} + \lambda_2 \cdot e_{i,t-2} + v_{i,t} \tag{6.7}$$

where $t-2$ corresponds to the second generation from t, that is, t's grandparents.

Empirical approaches of multigenerational mobility focus almost exclusively on men. They test the assumption that the influence of one generation onto the next one diminishes at a geometric rate so that it converges to zero: whatever the social status of your ancestors was in, say, the seventeenth century, it has little to no effect on your current social status. Moreover, many researchers propose that we do not need to look that far back in time: even grandparental status or income would have little predictive power on the position of their grandchildren—a finding closer to Becker and Nigel' intuition than Clark's hypothesis. In other words, the assumption is that the coefficient λ_2 in Eq. (6.7) is not significant.

However, according to Solon (2018) in his review of the literature of multigenerational mobility, there is no theoretical basis for such assumption. Besides, recent research has cast doubt on previous findings: it seems that grandparents, at least, do matter (i.e. $\lambda_2 > 0$). Another empirical approach incorporates the relative importance of demographic change on multigenerational transmission mechanisms: Mare and Song (2014) developed a model that takes into account the effects of population renewal and cohort size; see also Mare (2014). For example, a 'social reproduction model' of men involving four generations can be expressed thus:

$$S_{i|jkl,c} = F_{jkl,c} \cdot m_{jkl,c} \cdot r_{jkl,c} \cdot p_{i|jkl,c} \tag{6.8}$$

where $S_{i|jkl,c}$ represents the number of men in the offspring generation in a given social position i (say, the second lowest income decile) and group c (say, ethnic or religious group), whose fathers were in social position j, grandfathers were in social position k, and great-grandfathers were in social position l. F denotes the number of men in the paternal generation in position j with parents and grandparents in positions k and l, respectively. m stands for the marriage term: the number of wives per man in the paternal generation in each group. r is the fertility term: the expected number of sons born to each woman in the paternal generation by group. p is the positional mobility term: the probability of survival of a son born in each group by parental, grandparental, and great-grandparental social positions.

Evidence is being gathered supportive of the hypothesis that there exists a net grandparental effect on mobility:

- Chan and Boliver (2013) looked into data for three generations in England from three different birth cohorts (1946, 1958, and 1970). These authors found a strong persistence in occupational class between grandparents and their grandchildren: once the parents' occupation is taken into account, the odds that grandchildren enter a professional–managerial occupation instead of an unskilled manual job are two and a half times bigger if their grandparents were in a professional–managerial occupation.
- In a study of income mobility across three generations in rural areas of China using data from 2002, Zeng and Xie (2014) found a significantly higher effect of the level of educational attainment of co-resident grandparents on their grandchildren's educational attainment compared to the effect of the level of incomes of deceased or non-co-resident grandparents.

- Hertel and Groh-Samberg (2014) looked into class mobility in the United States of America and Germany using data from grandparents born around 1923 (in the United States of America) and 1916 (Germany) and grandchildren born in 1972 and 1978, respectively.
- Using Swedish data on first and second cousins, Hällsten (2014) found that inequality in grandparental wealth, occupational prestige, and schooling performance persist across at least four generations. Who your grandparents are or were is significant for your own social position, even in egalitarian country such as Sweden.
- There is strong multigenerational persistence in occupational aspiration, educational attainment, and social class in the United Kingdom, particularly among men, according to (Zhang and Li 2018) who studied longitudinal data from three generations of both men and women between 1991 and 2017.
- Using data from the United States of America between 1910 and 2013, Ferrie et al. (2016) studied educational mobility across four generations for both men and women using a regression approach and found a strong influence of grandparents on their grandchildren. These authors concluded that restricting the studies to parent-child dyads underestimates intergenerational persistence by around 25 per cent.

Direct effects of grandparents on their grandchildren may have to do with several transmission mechanisms, including inheritance of accumulated financial and non-financial wealth, access to social networks, and even biological (epigenetic; see Chap. 4 in Volume I) channels. In addition to these direct effects, there also seems to be additional indirect grandparental effects: for example, in a study based on the United States of America, Fomby et al. (2014) reported that the age at which grandparents had their children influences their grandchildren's mathematical achievement.

Other than a direct influence on their grandchildren, two additional explanations have been propounded to account for a significant multigenerational effect from grandparents: group influences and measurement error (Solon 2018). Group effects relate to race, religion, or any other source of discrimination or privilege that is transmitted from generation to generation in a given social group: a significant second (and higher autoregression order) coefficient could merely reflect the omission of a variable such as ethnicity, if ethnic belonging somehow affects social status, either positively or negatively, over time. The measurement error explanation is more prosaic: if the variable of interest is measured with an error in each generation, the multigenerational coefficient would be contaminated by a positive bias, indicating either a relationship when there is none or a stronger relationship across generations than the true association.

You guessed it right: the empirical evidence is still mixed and the point is far from settled. For example, in contrast to Zeng and Xie's findings for rural China, Bol and Kalmijn (2016) failed to find any significant grandparental effect on grandchildren's schooling in the Netherlands, even of co-residing grandparents. In a systematic review of the literature on grandparent effects on educational outcomes of their grandchildren, Anderson et al. (2018) reported that out of sixty-nine studies since 2000, forty found significant grandparent effects but added that 'the extant evidence gives a remarkably incoherent picture as to whether grandparental resources are associated with the educational outcomes of their grandchildren independently of the characteristics of the parental generation' [p. 135]. These authors conclude that more frequent contact and co-residence do not seem to be influential on grandchildren's educational outcomes, that the gender of the grandparent is not significant either, but that the effects of grandparents on their grandchildren's outcomes are stronger when the parents have lower educational qualifications (i.e. the 'grandparent compensation hypothesis'). Many potential intervening variables are seldom included in empirical studies. For example:

- The gerontological literature has developed the concept of 'grand cultures', that is, the 'overall identifiable ways of interacting between grandparents and grandchildren that are consistent within and across generations' (Kemp 2007, p. 864). Two main cultures are distinguished: a culture of affinity, where grandparents are highly engaged in and relevant to the lives of their grandchildren, and a culture of dissociation, with much weaker grandparent–grandchild ties and therefore influence.
- Another variable of possible relevance is the timing of the transition into grandparenthood. As Hank et al. (2018) argue, not only the year of birth of the youngest grandchild (which is usually recorded in household surveys) is important, but also the age of the oldest grandchild giving different implications for caregiving and contact.
- The health status of the grandparents during their grandchildren's formative years has also been found to be effectual (Margolis and Wright 2017) but not always factored in.
- Finally, the importance of incorporating data from the four grandparents of a grandchild is hardly ever considered (Sheppard and Monden 2018).

6.3 Justice Between Generations

In contemporary Western societies, the debate about intergenerational justice is ultimately framed under the concept of equality of opportunity. Utilitarian approaches, particularly the consequentialist and welfarist versions, focus on equality of welfare or utility. Issues of intergenerational mobility in turn are part of public debates on equality of opportunity. Three main conceptions of equality of opportunity are meritocracy, non-discrimination, and the levelling of the playing field principle (Cavanagh 2002; Mason 2006).[7]

- Meritocracy establishes a moral claim to positions (say, a job) or pay based on qualifications.
- Non-discrimination refers to equal chances for all individuals with equal relevant characteristics and to non-relevant attributes not affecting opportunities at all.
- The playing field metaphor refers to interventions to bring all individuals to the same level regarding relevant characteristics or attributes.

To illustrate the difference between non-discrimination and the playing field, we can consider that whilst non-discrimination protects people against being denied access to public services due to their gender or chronological age, for example, the levelled playing field approach insists on the removal of barriers to access to public services or on the dissemination of information about available services among groups with low rates of engagement and so on. Both approaches recognise that ultimately individual choice and responsibility do play a role in outcomes. As Roemer (1998a, p. 2) asserts:

> Common to all these views, however, is the precept that the equal-opportunity principle, at some point, holds the individual accountable for the achievement of the advantage in question, whether that advantage be a level of educational achievement, health, employment status, income, or the economist's utility or welfare.

Equality of opportunity is tied with the notion of merit, which must be distinguished from desert. Desert is a moral claim based on past doings and achievements; merit, in turn, is forward-looking.[8] So is equality of opportunity in general: meritocracy and the levelled playing field principle aim to establish fairness in outsets, in the starting points—equality of opportunity is *not* opposed to unequal outcomes but to unequal initial positions.

6　Some Questions of Intergenerational Economics

Some authors object to the application of equality of opportunity as a guide to assess the fairness of intergenerational policies because young people have more opportunities ahead of them than older people. Two answers to this objection are the 'prudential lifespan account' approach and the 'fair innings' argument.

6.3.1　Prudential Lifespan

Equality legislation protects certain areas or characteristics against discrimination, including disability, gender, race, religion, or sexual orientation, among others. In many countries, chronological age is one of these characteristics under protection. However, there is a fundamental difference regarding age: we are all ageing, we are all changing categories. A gay man is not expected to become heterosexual nor a Caucasian woman is expected to turn into a Melanesian, but a 20-year-old man in, for example, England is expected to live for another 60.04 years and a 20-year-old woman for another 63.55 years.[9] This seemingly basic distinction[10] allowed Daniels to answer to how to distribute public goods over a lifespan: the prudential lifespan account approach (Daniels 1988, 2008). The prudential lifespan account states that treating people differently by age is not morally objectionable provided everyone is treated differently over their lifespan[11]:

> We must not look at the problem as one of justice between distinct groups in competition with each other, between working adults who pay high premiums and the elderly who consume so many services. Rather, we must see that each group represents a stage of our lives. We must view the prudent allocation of resources through the stages of life as our guide to justice between groups.
>
> (Daniels 1988, p. 45)

As mentioned, this approach starts by distinguishing age groups from birth cohorts: age groups do not age; birth cohorts do. The birth cohort of, say, 1963 turned 50 in 2013 and will reach 80 in 2043. In contrast, the 50–64 age group will always be the 50–64 age group, formed of different people as they move onto older ages, of course, but there will always be a group of people aged 50 to 64. The prudential lifespan account posits that a 'whole lives' perspective may demonstrate that under certain circumstances treating individuals differently because of their chronological age is not morally objectionable. Inequality across individuals of different ages, that is, across different age groups, in one moment of time (i.e. the synchronic view) is, according to Daniels, mistaken:

Table 6.4 Wealth stock of two generations over time

	A	B
T1	3	
T2	5	
T3	1	3
T4		5
T5		1

Source: *Table is illustrative, prepared with mock data*

a whole lives, diachronic approach is the proper moral guide. It is important to highlight that prudence is not exclusively about maximisation over the life course, although maximisation over the life course is a prudent behaviour. The following example provides a first approximation to the prudential lifespan approach.[12]

Table 6.4 shows the wealth stock (as an illustration to fix the idea; any other variable of interest would do) of two generations, A and B, which live for three periods and overlap in T3.

Each generation has the same wealth trajectory, with a stock of three units in their first period, five in the second, and one in the last period of their lives. When they overlap, the wealth of generation A amounts to one and that of generation B is equal to three. Is this an unfair situation? Should wealth be distributed from generation A to B? Daniels opine it should not, because considering the whole lives of both generations, there is no unfairness to be seen; in fact, a transfer from one generation to the other at any point would introduce inequality over their lifetimes. Simultaneous segment inequality is not morally objectionable if over the whole lifespan there is a fair distribution, a fair share of the variable of interest. However, such distribution as shown in Table 6.4 may not be *prudent*. The approach assumes that each individual lives the whole of their lifespan. It may not be prudent to reach the third period with such little wealth. It would be prudent, although, to design institutions that transfer resources from one generation to the other to improve the conditions of each generation in their last period of their lives.

Even maintaining the assumption that each generation lives for the same length of time, if the generations are not of equal size, fairness may not be assured under this approach. Population ageing from below implies that younger generations are smaller than older generations. In Daniels (2008), this author addresses this difficulty, but cannot offer a solution:

Prolonged societal aging poses a real-world challenge to the feasibility of my proposed solutions to the age group and birth cohort problems. The challenge can best be met by developing strategies that stabilize population levels, something we do not know how to do at the level of policy. If we cannot meet that challenge, then integrated solutions to the age group and birth cohort problems face destabilization.

(Daniels 2008, p. 493)

6.3.2 Fair Innings

Developed in health economics by Williams (1997) as a justification for age-based rationing of certain services, the 'fair innings' argument postulates that every individual is entitled to a 'normal' span of life (or healthy life in the context of health economics) and it is morally objectionable that this length is curtailed by any institution, whereas it is not objectionable to ration services or goods to people who have lived beyond a given number of years. Needless to say, what accounts for a normal or 'reasonable' length of life is subject to debate (life expectancy at birth is often invoked), but Williams points out that in common parlance, premature or early deaths refer to those occurring when individuals are in their 20s or 60s, but not in their 90s; similarly we do not say that a worker has retired 'early' if she stopped paid employment in her 90s. Therefore, even though agreement on the exact cut-off point may not be reached, the argument calls for restraints in transfers and allocations to older people beyond a certain age, which in this view would be morally permissible (see Bognar (2015) for a justification based on prioritarian arguments).

6.4 Equity, Solidarity, Conflict, and Ambivalence

In discussions about justice between generations, a number of concepts are used, which is worth exploring. However, it is not only a matter of terminology: Williamson and Watts-Roy (1999) argued that there are two opposing narratives or 'interpretive packages', rhetorically framing the issue of justice between generations in the public sphere in developed countries: the generational equity frame and the generational interdependence frame.

Gamson and Modigliani (1989, pp. 1–2)[13] explained:

> An archivist might catalog the metaphors, catchphrases, visual images, moral appeals, and other symbolic devices that characterize [a] discourse. The catalog

would be organized, of course, since the elements are clusters; we encounter them not as individual items but as interpretive *packages*...

...one can view policy issues as, in part, a symbolic context over which interpretation will prevail.

These two narratives under which generational justice is framed vie for gaining influence on and acceptance among policy makers and the public discourse. Therefore, the delimitation of terms sometimes used loosely by commentators should also examine the wider catalogue of their respective components. Such an exercise goes beyond the objectives of this textbook, but in what follows, I present a glimpse into the discussion. Perhaps we could bear in mind, alongside Phillipson (2015a), that the mix of liberal social policies, economic growth, and the promotion of social and generational ties that were put in place in the twentieth century to accommodate ageing populations and extending longevity are not enough to provide security in later life in the current century. This author argues that a re-conceptualisation of the role of the State (and the Welfare State) and new forms of intergenerational solidarity are needed.

6.4.1 Generational Equity

The generational equity frame advocates for cuts in public spending on older people and increasing spending on children and young adults. In terms of justice between generations, the argument is that today's taxpayers would not receive the same level of benefits as today's older people and therefore the former suffer a double whammy: they are paying more taxes now but will not get more benefits in the future compared to what they are funding today. The main ethical argument on which this frame is based is that each generation should fend for itself. This interpretive package postulates values such as individualism, freedom, and self-reliance. Some rhetorical narratives spinning from this frame include the fiscal burden on future generations imposed by current levels of consumption (see Part I in Volume II) and the worrying presence of elements of gerontocracy (see Part IV in this volume) in political systems and policy decision-making.

Generation equity refers to the concept that 'different generations should be treated in similar ways and should have similar opportunities' and 'links the economic costs of the aged to those of the young, with the argument that the young are being deprived of opportunities for well-being because of excessive allocation of resources to the old.' (Marshall et al. 1993, p. 119). Similarly, as Héran (2007, p. XIV) expounds, generational inequity

> …arises when a generation's hopes of advancement, based on the advancement obtained by previous generations, are disappointed. It becomes particularly acute when the new generation feels it has been disinherited by its elders, whether deliberately or through negligence.

Generational inequity, then, is about unequal lifetime opportunities because of belonging to a given cohort. One example is the incisive analyses of intergenerational disparities in contemporary France by French sociologist Louis Chauvel, who noted that in France there has been a 'generational fracture' as a result of the economic slowdown in the mid-1970s, a rift between the generations born before and after 1955—the latter were starting the employment phase in their life cycles when the economy was being hit by the slowdown (Chauvel 2006, 2010). In this author's view, older generations acted as 'insiders' in the labour markets preventing younger people (the 'outsiders') from getting into jobs. High youth unemployment caused a scarring effect[14] that resulted in:

- increasing pay gaps between generations: the difference between average earnings of the 30–35 and 50–55 age groups was around 15 per cent in 1977; by 2005, it had reached 40 per cent
- stunted life chances for younger age groups
- lack of socialisation and political mobilisation

Chauvel proposes two concepts as guides to understand contemporary generational strife: the distance between upper and lower classes ('moyennisation' in French) and the upward social aspirations. The middle classes would be increasingly detached from the upper echelons and in a more precarious and state: downward social mobility is more likely than upward social mobility. With social aspirations still alive, the gulf between aspirations and reality is growing:

> Today's generational transmission problem comes from a lack of correspondence between the values and ideas that the new generation receives (individual freedom, self achievement, valorization of leisure, etc.) and the realities it will face (centrality of market, heteronomy, scarcity, lack of valuable jobs, boredom, etc.)
>
> (Chauvel 2006, pp. 159–160)

Similarly, a report by the UK Intergenerational Commission announced that generational progress is 'a promise under threat', a source of pessimism and anxieties among younger groups about employment, housing, living standards, and retirement (Commission 2018).

Intergenerational inequity and unfairness is vividly portrayed in the powerful symbolic device that younger generations will be the first not to be better off than their parents. In the United Kingdom, Hood and Joyce (2013) compared incomes and wealth of individuals born between the 1940s and the 1970s and concluded that the 1960s and 1970s cohorts have not experienced the high rates of income growth when they were 30–50 years old as the previous generation did, but have more unequal incomes in early adulthood. People born in the 1970s are finding it more difficult to buy a house, are accumulating less pension wealth, and are more reliant on inheritances (which are also more unequally distributed in favour of the already better off) than preceding cohorts. Portes (2014) considered that this sense of foreboding is misplaced. This author looked into the data for the period 1996–1997 to 2011–2012 and found that inequality between retired and non-retired households was smaller than inequality within each group, with improving signs (catching up) for average income of retirees (a finding in support of the generational interdependence interpretive package). Whilst this positive trend in retired households is mostly financed by taxes and other transfers from working-age households, the author correctly points out:

> those currently of working age will eventually become pensioners; the net impact on them will depend on the extent to which the policy is in fact sustainable and sustained
>
> (Portes 2014, p. F7)

Regarding the findings by Hood and Joyce (2013), among others, France and Roberts (2015), Goodwin and O'Connor (2009) and Portes (2014) argued that intragenerational inequity remains much greater than intergenerational inequity, and that its impact is growing. For example, Portes (2014, p. F10) stated:

> The circumstances of your birth and early life (who your parents are, education, gender, ability, effort and just plain luck) still matter far more – indeed, more so than before – than when you were born

As can be seen, not everyone agrees with the intergenerational conflict approach. Conflict is one of the tropes that the generational equity frame propounds and is a component of the dystopian narrative that puts forth apocalyptic views of population ageing (see the Introduction to Part I, in Volume I). According to Bristow (2016), the 'lucky, affluent, large, selfish, and reckless' Baby Boomer generation (Bristow 2015) is used as a scapegoat and has been socially constructed as the 'problem', especially the 'economic problem' behind the fiscal impact ('burden') of pension and healthcare systems in ageing societies (see also Curryer et al. 2018).

According to Attias-Donfut and Arber (2000), the generational equity frame confuses between (see also Attias-Donfut and Wolff 1997):

- the synchronic allocation of public spending between people of different age groups at any one time
- the 'questionable' principle of equivalence in living standards between different generations
- the right of 'just returns' for any one cohort on the contributions they have made

Attias-Donfut and Arber criticise the use of synchronic data much in the same sense as Daniels in his prudential lifespan approach (see above). Besides, they find questionable the principle of equivalent intergenerational living standards because people in different eras, even within one same country, are 'completely separated in terms of consumption patterns, the environment, and acquired customs and habits, etc.' [p. 7]. Finally, regarding the 'just returns' argument, these authors point out that it is not only public transfers what matter in this social accounting exercise, but private transfers too. This brings us to the second interpretive frame: the generational interdependence.

6.4.2 Generational Interdependence and Solidarity

The generational interdependence frame emphasises the presence of intergenerational linkages and intragenerational inequality. The emphasis is less on whether consumption levels are sustainable over time and more on transfers and supply of public and private goods and services between generations. In terms of justice between generations, this argument does not view child poverty as the consequence of affluence among older people and insists, on the contrary, that high levels of poverty in both groups can coexist. The main ethical argument on which this frame is based is that justice between generations cannot be isolated from gender, ethnicity, and other cleavages through which inequality seeps in. This interpretive package postulates values

Table 6.5 Conceptual dimensions of intergenerational solidarity

Dimension	Related to
Associational	The quality and frequency of interactions
Affectual	Positive sentiments
Consensual	Agreement on values and beliefs
Functional	Support and resources given and received
Normative	Commitment to family roles
Structural	Variables such as family size and geographical proximity

Source: *Bengtson and Roberts (1991)*

such as solidarity and community obligations towards people in need. Some rhetorical narratives spinning from this frame include redistribution to reduce inequality, vulnerability, and need (Table 6.5).

One concept close to the generational interdependence frame is intergenerational solidarity. Intergenerational solidarity has been incorporated as a central policy objective by many intergovernmental organisations. The United Nations recognised the need to 'strengthen solidarity among generations' in the Political Declaration and the Madrid International Plan of Action on Ageing of 1992. It identified intergenerational solidarity as one of the issues included in its first Priority Direction, setting out the objective to strengthen 'solidarity through equity and reciprocity between generations' through the following actions (United Nations 2002, Issue 5):

- Promote understanding of ageing through public education as an issue of concern to the entire society.
- Consider reviewing existing policies to ensure that they foster solidarity between generations and thus promoting social cohesion.
- Develop initiatives aimed at promoting mutual, productive exchange between the generations, focusing on older persons as a societal resource.
- Maximize opportunities for maintaining and improving intergenerational relations in local communities, inter alia, by facilitating meetings for all age groups and avoiding generational segregation.
- Consider the need to address the specific situation of the generation of people who have to care, simultaneously, for their parents, their own children, and their grandchildren.
- Promote and strengthen solidarity among generations and mutual support as a key element for social development.
- Initiate research on the advantages and disadvantages of different living arrangements for older persons, including familial co-residence and independent living in different cultures and settings.

The Treaty of the European Union, signed in 1992, states that the Union shall promote 'solidarity between generations' (*Consolidated version of the Treaty on European Union* 2012, Article 3, para 3). Moreover the European Union has officially proclaimed 29 April as the 'European Day of Solidarity between Generations'.

But, what is intergenerational solidarity? It has been defined as 'an intentional connection between two or more persons of different age groups' that 'reflects personal wishes and material goals, emotional bonds and rational justifications, altruism and self-interest, caregiving and care receiving' (Cruz-Saco 2010, p. 9).

One approach to intergenerational solidarity draws upon social exchange and family developmental theories and focuses on the dynamics within families and the relations between their members. Its main concept of interest is the degree of cohesiveness within families. This approach is based on a theoretical model that posits that there is a spectrum of intergenerational relations from solidarity to conflict. Six conceptual dimensions have been identified:

As Bengtson and Oyama (2010) pointed out, economic studies have mainly focused on the functional dimension and included structural and associational aspects as intervening explanatory variables. Other approaches extend the dimension of intergenerational solidarity to care settings, communities, and nations. In these other settings, solidarity may be related either to interaction (or consensus) between generations (OECD 2011). One such model is the 'convoy' of social relations (Kahn and Antonucci 1980)—see Volume II, Chap. 7—, according to which families provide protection, support, and care depending on contextual variables (such as ethnicity, chronological ages of members, or social class), structural variables (geographical proximity and family size and composition), and individual variables (e.g. psychological).

Two aspects of intergenerational solidarity can be identified: a distributional and a relational aspect (Cruz-Saco 2010). The distributional element manifests in transfers of resources between members of different generations, including income, wealth, caregiving, bequests, or gifts, with net flows going downwards, from older to younger generations. In gerontology, the 'flow reversal' hypothesis posits that older parents are net providers of their adult children until a certain advanced age beyond which they become net receivers. Most empirical studies have rejected this hypothesis with regard to financial flows: older parents remain net providers all their lives albeit the net balance diminishes substantially at very old ages (Albertini et al. 2007; Litwin et al. 2008; Mudrazija 2014).[15] The distributional element of intergenerational solidarity also includes, at a macro level, pension benefits in pay-as-you-go systems (see Chap. 4, in Volume III) and other transfers to older generations funded by taxes levied on younger generations. The relational aspect includes education, norms, values, trust, and political, cultural, and civic engagement which are transmitted from one generation to another.

Inasmuch as not everyone agrees with the negative bias of the rhetoric of the generational equity frame, it is also true that not everyone agrees with the positive bias of the intergenerational solidarity approach either. The main objection addressed to the latter is that it idealises family relationships and equates any negative aspect as a lack of solidarity (Lowenstein 2005).

Moreover, not everyone agrees that intergenerational solidarity is the opposite to intergenerational conflict. Lüscher et al. (2010) opined that solidarity is the idealistic pole in the intergenerational rhetoric opposed to conflict and threat at the other end (identified with the intergenerational equity frame). However, Bengtson and Oyama (2010) argued that high solidarity and conflict can coexist. Out of this realisation, some gerontologists developed the concept of intergenerational 'ambivalence' (Lüscher and Pillemer 1998; Pillemer et al. 2007).[16] Lüscher and Pillemer (1998) defined intergenerational ambivalence as the 'contradictions in relationships between parents and adult offspring that cannot be reconciled and explained that ambivalence takes place at two levels: institutional (resources and requirements from statuses, roles, and norms) and subjective (cognitions, emotions, and motivations).

Bengtson and Oyama (2010, p. 46) explains that a number of reasons may lead to *less* intergenerational conflict in the future but another group of reasons could lead to *more*. Hope for less conflict stems from:

- changes in social structures and cultural values that reflect and allow for more effective responses to demographic change
- strong intergenerational solidarity within families
- strong norms of reciprocity between generations
- changes in the roles of older people and the meaning of old age towards more positive conceptualisations

In turn, Bengtson and Oyama highlight three trends that do not bode well for future levels of intergenerational solidarity:

- increasing old-age dependency ratios
- increasing perception of generational inequity (see next section)
- increasing ageism

Véron et al. leave the question about the future of intergenerational solidarity open and opine that it will depend 'on the way that family changes, transformations of the labour market and population ageing interact' (Véron et al. 2007, p. 7).

6.4.3 Generational Contract

One of the main messages of the generational equity frame is that, on average, younger generations are worse off than their parents at the same age and will be worse off over their lifetimes. This message leads to opinions about broken generational 'contracts'—'the most important and also the most contentious dimension of contemporary welfare systems', according to Albertini et al. (2007, p. 319).

A generational contract is an implicit agreement or compact between generations for mutual support. The welfare state can be seen as the institutional mechanism to implement this contract, with transfers from taxpayers to pensioners and children, as pensioners transferred part of the income they earned when they were economically active and children will do likewise when they begin to work. Willets (2010) proposed a typology of social contracts between successive generations, centred on the generation currently of working age:

- Direct exchange: if working-age people care for their young, the latter will care for the former when they reach later life.
- Replication of own behaviour for own benefit: if working-age people care for their older parents, children will learn this behaviour and replicate it in due course.
- Replication of own behaviour for benefit of future generation: if working-age people care for their children, they will learn this behaviour and replicate it with their own children.
- Replication of behaviour from which one has benefited: grandparents look after their grandchildren, so working-age individuals learn this behaviour and will care for their own grandchildren in the future.

This learning and replication of behaviours regulate transfers across generations, which according to Komp and Tilburg (2010) is a key function of the implicit intergenerational contract. These authors identified two conditions for sustainable intergenerational contracts:

- that the demands on the working-age groups to support the young and the older persons are not so excessive that the middle age becomes a precarious and vulnerable period in the life course
- that the size of future generations does not substantially decrease, which would jeopardise the prospects of currently middle-age individuals of being supported in later life

Komp and Tilburg noted that population ageing poses a serious challenge to both conditions and, consequently, to the future of the intergenerational contract, but remarked that the pressure is not uniform and that gender and social class determine the type and amount of upward and downward transfers as well as the level of support for different types of policies of intergenerational distribution. As other policy issues around ageing, intergenerational policy should not be designed in isolation of other potential causes of inequality, such as ethnicity and gender. As Catchen (1989, p. 21) warned in a US context: 'Reducing current federal benefit levels to the elderly would do little to alleviate the problems experienced by younger generations, and would cause severe hardship for many older people. Women and minorities would be especially hard hit by such reductions since they are more likely to be poor than white males'. (In these arguments, the imprint of the generational interdependence frame can be clearly identified.)

6.4.4 Intergenerational Ambivalence

Williamson and Watts-Roy (1999) pointed out that arguments in favour of the generational equity frame tend to be presented in the mass media and grey literature, whilst the arguments supportive of the generational interdependence frame tend to be published in academic journals. From this disparity in how each view is channelled and from the values each package promotes, the authors concluded that—at least in the United States of America—the generation equity frame has had the upper hand in the quest to influence public opinion. However, there is a third, middle-of-the-road approach: intergenerational ambivalence.

Masson (2007) proposed that intergenerational relations studies should embrace the notion of ambivalence stemming from a 'generation dilemma' due to the impossibility of long-term intra-family contracts. There is no market mechanism that forces a working adult to, for example, look after her frail father because he looked after her when she was a child. Parents do invest in their children and specifically in shaping their preferences so that they will look after them if frailty and impairment ever knock at their door, but there is no assurance that they 'will be there for them'. Moreover, whatever parents may have invested is irreversible.

Enter altruism as a solution or the State as a guarantor and representative of future generations. These are, theoretically, two ways out of the generation dilemma. However, these solutions do not answer the practical questions of justice between generations and the viability of the transfer mechanisms. Masson proposal incorporates the indirect reciprocities of transfers between generations, such as the non-monetary returns on investment in education on the one hand and pension rights on the other—in other words, issues of fair debts and claims close to the generational interdependence frame. Generational accounting (see Chap. 1 in Volume II), in turn, focuses exclusively on the fairness of claims ignoring the fairness of inheritance and returns of public investments made by the currently older generation (and therefore it is preferred by supporters of the generational equity approach). This duality, Masson adds, needs to be analysed at the same time, as part and parcel of both intergenerational conflict and cooperation: they are not balanced out; on the contrary, their presence should be *emphasised*: for example, 'many "selfish" seniors, accused of "taking the lion's share" of public resources, look after their grandchildren' (Masson 2007, p. 97).

Masson warns that tensions may arise at many levels: when delimiting age groups and stages in the life cycle, when discussing levels of commitment to future generations, or the extension of longevity with its increasing claims on resources. He understands that the structure of a representative contemporary family is not a three-generational unit formed by a child,[17] working-age adult, and older pensioner (his terminology) but a four-generational unit formed by a child, a young working adult, a middle-aged working adult, and older pensioner. In this alternative configuration, the only upward transfers are those from the middle-aged adults to their older parents. This author considers that this structure helps to better understand the modern roles of women throughout their life cycles: choices between maternity and career, between career and support for older parents, the 'sandwich' generation (see Chap. 6, Volume II), and so on. Finally, Masson also considers that the configuration is better suited to study political economy games (Volume IV, Chap. 10): a coalition of young and middle-aged working adults against older people, or a coalition of older people and the younger working adults against the middle aged (i.e. against the Baby Boomers)—see Chap. 10 in this volume.

6.5 Indices of Intergenerational Fairness

In ecology and related disciplines, there are various indices of sustainability and intergenerational fairness. However, these constructs—though useful in their field—only deal with the current and future state of the environment. On the other hand, there have been attempts at measuring fiscal and other economic aspects of intergenerational justice, and Gál and Monostori (2017) proposed a taxonomy of indicators of intergenerational fairness and economic sustainability broken down into cross-sectional and long-term horizons with an exclusive economic content.

Both approaches lack the necessary integral view of fairness across generations. There are two indices that attempt to measure justice between generations from an interdisciplinary perspective more attuned to the study of economics and ageing: the Intergenerational Fairness Index and the Intergenerational Justice Index.

According to Laub and Hagist (2017), both indices provide comprehensive insight into intergenerational justice; however, they have also been subject to criticism, as we will see.

6.5.1 The Intergenerational Fairness Index

This index is composed of indicators from nine policy areas (Leach and Hanton 2012):

- Unemployment
- Housing
- Pensions
- Government debt
- Participation in democracy
- Health
- Income
- Environmental impact
- Education

Its European version (Leach et al. 2016) is composed of the same nine areas plus:

- Population structure
- Tertiary education (proportion of people aged 25–34 holding a university degree or equivalent)
- Expenditure on R&D as a proportion of GDP
- Poverty and social exclusion (proportion of younger people with incomes below 60 per cent the median)

The measure combines two types of data from these areas: a measure of advantage or disadvantage that younger people experience against the average and a measure of advantage or disadvantage that unborn generations will experience as a result of current policy decisions.

The index assumes that only area in which current decisions affect younger and unborn generations is housing: it is assumed that housing costs have an impact on younger generations, whereas construction will impact the generations yet to be born.

All the other areas affect one generation or the other. Younger groups are affected by unemployment and income (as these areas are measured with the youth unemployment rate and the ratio of income from younger workers and the average). Democratic participation is operationalised with the average age of local councillors (i.e. local representatives). The only indicator to measure the health domain is the use of selected services by people aged under 60. Three indicators are used to measure education, all related to the current younger cohorts: public spending on education as a proportion of GDP, average tuition fee liability of students in higher education, and the proportion of students leaving secondary education with high marks.

Environmental impact (measured by green gas emissions and concentration of carbon dioxide in the atmosphere), pensions (measured with state pensions and unfunded public sector occupational pensions), and government debt (public debt per head) are assumed exclusively to affect future generations.

The index is calculated as the unweighted arithmetic average of the annual changes in the indicators for each area.

Shaw (2018) criticised, in particular, the inclusion of public debt per head in the index. The rationale given by the authors of the Intergenerational Fairness Index is that as public debt must be serviced in the future, its level affects later generations. Shaw argues that such procedure does not take into account the purposes for which the debt was incurred, the distribution between generations of the public spending the debt funded, and the generational consequences of the array of possible policies that could be implemented to service it.

In addition, Boston (2017) questioned the choice of indicators and the adoption of equal weights (remember that if an index is unweighted, it means that each of its components is given the same importance).

Finally, Rowlingson et al. (2017, p. 6), whilst acknowledging the index is 'interesting', opined that it is not 'helpful to suggest that this is an index of fairness between generations when indicators are only gathered in relation to one generation'.

6.5.2 Intergenerational Justice Index

The Intergenerational Justice Index (IJI) was initially designed to compare international justice across twenty-nine developed countries and therefore offers a macro-level picture. It considers four dimensions (Vanhuysse 2013):

- Sustainability
- Child poverty
- Public sector debt per child
- Pro-elderly public spending bias

Each dimension is operationalised by one indicator. Except for sustainability, the indicators for the other dimensions are straightforward although the biased explanation of their rationale (particularly of the fiscal policy indicators) is less so. Sustainability is measured by the ecological footprint created by generations currently alive. The ecological footprint is an estimation of the surface of land and water (in hectares) required by a country to produce all the goods consumed by its inhabitants (Rees 1992). Child poverty rates are used to indicate early-life starting conditions. Public sector debt per child (i.e. per person aged 14 or under) was chosen to assess the fiscal 'burden' upon currently young generations. Finally, the ratio between public spending on people aged 65 or over and public spending on people aged under 65 years is proposed as an indicator of 'overall pro-elderly bias'.

Unlike the Intergenerational Fairness Index, the IJI uses different weights for each dimension, determined according to the 'benefit-of-the-doubt' method (Melyn and Moesen 1991): the weights are obtained from a comparative exercise of how each country ranks along each dimension after normalising the indicators according to the formula

$$Xn_i = \frac{X_{max} - X_i}{X_{max} - X_{min}} \qquad (6.9)$$

The rationale behind this weighting method is that if a country scores highest in one particular dimension, this reveals, under democratically elected governments, the preferences of the electorate, and that consequently the weights should reflect these preferences. The three best performing countries in terms of intergenerational justice according to this method are Estonia, South Korea, and Israel. At the other end ranked Italy, Greece, and Canada.

Review and Reflect
1. The intergenerational equity thesis...is merely a politically expedient use of demographic change to conceal, on the one hand, the falling welfare surplus and, on the other, welfare restructuring. Walker (1996a, p. 24)
2. Comment on the following two excerpts from the same paper on multigenerational mobility (Mare 2011):

   - *Most social science population researchers are interested in intergenerational processes based on shorter time spans for people who are more directly observable. Yet most social mobility research avoids this issue altogether by conditioning on the distributions of parents' and offspring's traits (e.g., income or educational attainment) and describing these associations. This tradition of work has served well for answering the narrow question, "Who gets ahead?" ...But it is inadequate for analyzing the population question of how a socioeconomic distribution in one generation gets transformed into a distribution in later generations. And it is even inadequate for such causal questions as, What would happen to the next generation if we made it easier for girls (the potential mothers of the next generation) to stay in school [p. 15]*

   - *Our models of social mobility have a strong mid-twentieth century American middle- and working-class bias. That is, they emphasize the pivotal role of formal schooling in transmitting the advantages conferred by parents on offspring and inducing new variation in the socioeconomic positions of the next generation that is independent of those of the previous generation...such an orientation limits our ability to see other types of intergenerational mobility and immobility patterns that may dominate the highest and lowest segments of social hierarchies and to envision that the relative sizes of elite and underclass populations may change over time. We should be open to pluralistic models of mobility that regard populations as containing mixtures of two-generational and multigenerational modes of socioeconomic persistence. [p. 20]*

3. In relation to the prudential lifespan account, consider the following objection by McKerlie:

(continued)

> *For most of us, the middle part of life will have the most influence on the success and failure of our lives...the prudential choice will favor middle age and short-change the elderly in an unfair way. But it is not clear how powerful the objection is. It shows that prudence would distribute resources unequally over the different temporal stages of a life. However, it also provides a reason for the inequality. The goals that we pursue in the middle part of our lives are the most important for the value of our complete lives. We might think that, if the inequalities have this basis, there is no reason to consider them objectionable. Furthermore, the synchronic inequality between age groups that this factor would generate might not run very deep.*
>
> *(McKerlie 2006, p. 45)*

Do you agree with this author in that the 'goals that we pursue in the middle part of our lives are the most important for the value of our complete lives'?

4. Consider the following assertion:

   > *...the potential for distributional conflicts among generations certainly exists and is fuelled by the current challenges of public finance and changing demography.*
   >
   > *(Lowenstein 2010, p. 58)*

   Do you agree? If so, do you think that economists have played an active role in fuelling intergenerational conflict? How could they also become part of the answer towards increasing solidarity between generations?

5. According to Bengtson and Oyama, the argument by advocates of generational 'equity' can be summarised thus: in order to combat poverty in later life and as a result of effective political lobbying, older people enjoy better economic status and receive increasing proportions of public funds, whilst resources to other age groups are decreasing. This author stated that 'this perspective relies largely upon projected demographic trends and theoretical conceptualizations that have not been empirically demonstrated by public opinion polls and other research' and that the approach merely consists in 'a "symbolic battle" created and disseminated by the mass media and political interests' (Bengtson and Oyama 2010, p. 40).

   Discuss.

6. According to Shaw (2018, p. 13), 'Choosing between conflictual and solidaristic framings of intergenerational fairness is ...more than a descriptive exercise: these are alternative causal stories that open up (or close off) different policy options'.

   Do you agree?

Notes

1. See, for example, Beckett (2011), Howker and Malik (2010), and Willets (2010).
2. A Spanish version of the same saying replaces 'gives' with 'helps'.
3. To circumvent some data limitations, a technique known as the two-sample two-stage least squares (TSTSLS) estimator is frequently used. See Jerrim et al. (2016).
4. I follow the notation in Xie and Killewald (2013).
5. The assumption of additive separability.
6. For a detailed description of this and other measures of mobility, see Jäntti and Jenkins (2015).
7. See Roemer (1998a), who uses the levelling metaphor, although he does not cover meritocracy.
8. Segall (2013).
9. Source: *National Life Tables, England, 1980–1982 to 2015–2017. Period expectation of life based on data for the years 2015–2017.* Office for National Statistics. London: United Kingdom.
10. Which Daniels points out it was not obvious when he first proposed this approach—see Daniels (2008, p. 478).
11. For critical expositions of the prudential lifespan account approach, see Lazenby (2011) and McKerlie (2006, ch. 3).
12. This example is a modified version of Daniels (2008, p. 481).
13. See also Gamson (1988).
14. See Chap. 1, Volume III.
15. Moreover, the probability of receiving or giving varies significantly across countries (Roll and Litwin 2013).
16. See also the special issue of the *Journal of Marriage and Family* on ambivalence in intergenerational relationships (Volume 64, Issue 3, August 2002).
17. A configuration used in Masson (1999).

References

Adermon, Adrian, Mikael Lindahl, and Daniel Waldenström (2018). "Intergenerational wealth mobility and the role of inheritance: Evidence from multiple generations". In: *The Economic Journal* 128.612, F482–F513.

Albertini, Marco, Martin Kohli, and Claudia Vogel (2007). "Intergenerational transfers of time and money in European families: common Patterns different regimes?" In: *Journal of European Social Policy* 17.4, pages 319–334.

Anderson, Lewis R, Paula Sheppard, and Christiaan WS Monden (2018). "Grandparent effects on educational outcomes: a systematic review". In: *Sociological Science* 5, pages 114–142.
Arber, Sara (2013). "Gender, Marital Status and Intergenerational Relations in a Changing World" In: *Global Ageing in the Twenty-First Century. Challenges, Opportunities and Implications*. Edited by Susan McDaniel and Zachary Zimmer. Farnham: United Kingdom: Ashgate, pages 215–234.
Arrondel, Luc and André Masson (2001). "Family Transfers Involving Three Generations". In: *The Scandinavian Journal of Economics* 103.3, pages 415–443.
Atkinson, Anthony B and François Bourguignon (1982). "The comparison of multi-dimensioned distributions of economic status". In: *Review of Economic Studies* 49.2.
Attias-Donfut, Claudine and Sara Arber (2000). "Equity and Solidarity Across The Generations". In: *The myth of generational conflict: the family and state in ageing societies*. Edited by Sara Arber and Claudine Attias-Donfut. London: United Kingdom: Routledge, pages 1–21.
Attias-Donfut, Claudine and François-Charles Wolff (1997). "Transferts publics et privés entre générations: incidences sur les inégalités sociales". In: *Retraite et Société* 20, pages 20–39.
Becker, Gary S and Nigel Tomes (1986). "Human capital and the rise and fall of families". In: *Journal of Labor Economics* 4.3, pages 1–39.
Beckett, Francis (2011). *What Did the Baby Boomers Ever Do For Us?* London: United Kingdom: Biteback Publishing.
Bengtson, Vern L and Katherine R Allen (1996). "The Life Course Perspective Applied to Families Over Time". In: *Sourcebook of Family Theories and Methods. A contextual approach*. Edited by Pauline Boss et al. New York, NY: United States of America: Springer, pages 56–80.
Bengtson, Vern L and Petrice S Oyama (2010). "Intergenerational Solidarity and Conflict: What Does It Mean and What Are the Big Issues?" In: *Intergenerational Solidarity. Strengthening Economic and Social Ties*. Edited by María Amparo Cruz-Saco and Sergei Zelenev. New York, NY: United States of America: Palgrave Macmillan, pages 35–52.
Bengtson, Vern L and Robert EL Roberts (1991). "Intergenerational solidarity in aging families: An example of formal theory construction". In: *Journal of Marriage and the Family* 53.4, pages 856–870.
Black, Sandra E and Paul J Devereux (2011). "Recent developments in intergenerational mobility". In: *Handbook of Labor Economics*. Edited by Orley C Ashenfelter and David Card. Volume 4B. Amsterdam: The Netherlands: Elsevier/North Holland, pages 1487–1541.
Black, Sandra E et al. (2015). *Poor little rich kids? The determinants of the intergenerational transmission of wealth*. NBER Working Paper 21409. Cambridge, MA: United States of America: National Bureau of Economic Research.
Blinder, Alan S (1976). "Intergenerational transfers and life cycle consumption". In: *The American Economic Review* 6 (2), pages 87–93.

Bognar, Greg (2015). "Fair innings". In: *Bioethics* 29.4, pages 251–261.
Bol, Thijs and Matthijs Kalmijn (2016). "Grandparents' resources and grandchildren's schooling: Does grandparental involvement moderate the grandparent effect?" In: *Social Science Research* 55, pages 155–170.
Boserup, Simon H, Wojciech Kopczuk, and Claus T Kreiner (2016). "The role of bequests in shaping wealth inequality: evidence from Danish wealth records". In: *American Economic Review* 106.5, pages 656–61.
Boston, Jonathan (2017). *Governing for the Future: Designing Democratic Institutions for a Better Tomorrow*. Volume 25. Public Policy and Governance. Bingley: United Kingdom: Emerald Group Publishing Limited.
Bristow, Jennie (2015). *Baby Boomers and Generational Conflict*. Basingstoke: United Kingdom: Palgrave Macmillan.
— (2016). "The making of Boomergeddon': the construction of the Baby Boomer generation as a social problem in Britain". In: *The British Journal of Sociology* 67.4, pages 575–591.
Catchen, Harvey (1989). "Generational Equity Issues of Gender and Race". In: *Women & Health* 14.3-4, pages 21–38.
Cavanagh, Matt (2002). *Against Equality of Opportunity*. Oxford: United Kingdom: Oxford University Press.
Cervini-Plá, Mara (2015). "Intergenerational Earnings and Income Mobility in Spain". In: *Review of Income and Wealth* 61.4, pages 812–828.
Chan, Tak Wing and Vikki Boliver (2013). "The grandparents effect in social mobility: Evidence from British birth cohort studies". In: *American Sociological Review* 78.4, pages 662–678.
Chauvel, Louis (2006). "Social Generations, Life Chances and Welfare Regime Sustainability". In: *Changing France: the politics that markets make*. Edited by Pepper D Culpepper, Peter A Hall, and Bruno Palier. Basingstoke: United Kingdom: Palgrave Macmillan, pages 150–175.
— (2010). *Le destin des générations, structure sociale et cohortes en France du XX e siècle aux années 2010*. Paris: France: Presses Universitaires de France.
Chetty Raj (2016). "Socioeconomic mobility in the United States: New evidence and policy lessons". In: *Shared Prosperity in America's Communities*. Edited by Susan M Wachter and Lei Ding. Philadelphia, PA: United States of America: University of Pennsylvania Press, pages 7–19.
Chetty Raj et al. (2014b). "Where is the land of opportunity? The geography of intergenerational mobility in the United States". In: *The Quarterly Journal of Economics* 129.4, pages 1553–1623.
Clark, Gregory (2014). *The Son Also Rises: Surnames and the History of Social Mobility*. Volume 49. The Princeton Economic History of the Western World. New Jersey NJ: United States of America: Princeton University Press.
Commission, Intergenerational (2018). *A New Generational Contract. The final report of the Intergenerational Commission*. London: United Kingdom: Resolution Foundation.

Consolidated version of the Treaty on European Union (2012). Brussels: Belgium.

Costanza, David P et al. (2017). "A review of analytical methods used to study generational differences: Strengths and limitations". In: *Work, Aging and Retirement* 3.2, pages 149–165.

Cruz-Saco, María Amparo (2010). "Intergenerational Solidarity". In: *Intergenerational Solidarity. Strengthening Economic and Social Ties*. Edited by María Amparo Cruz-Saco and Sergei Zelenev. New York, NY: United States of America: Palgrave Macmillan, pages 9–34.

Curryer, Cassie, Sue Malta, and Michael Fine (2018). "Contesting Boomageddon? Identity politics and economy in the global milieu". In: *Journal of Sociology* 54.2, pages 159–166.

Daniels, Norman (1988). *Am I my parents' keeper? An essay on justice between the young and the old*. London: Oxford University Press.

—— (2008). "Justice between adjacent generations: further thoughts". In: *Journal of Political Philosophy* 16.4, pages 475–494.

Ferrie, Joseph, Catherine Massey, and Jonathan Rothbaum (2016). *Do grandparents and great-grandparents matter? Multigenerational mobility in the US, 1910-2013*. NBER Working Paper 22635. Cambridge, MA: United States of America.

Fomby, Paula, Patrick M Krueger, and Nicole M Wagner (2014). "Age at childbearing over two generations and grandchildren's cognitive achievement". In: *Research in Social Stratification and Mobility* 35, pages 71–88.

Fragnière, Jean-Pierre, François Höpflinger, and Valérie Hugentobler (2014). "Glossaire pour l'étude des relations entre les générations". In: URL: http://www.jpfragniere.ch/textes/pdf/C52-Glossaire_generations.pdf.

France, Alan and Steven Roberts (2015). "The problem of social generations: a critique of the new emerging orthodoxy in youth studies". In: *Journal of Youth Studies* 18.2, pages 215–230.

Gál, Róbert I and Judit Monostori (2017). "Economic sustainability and intergenerational fairness: a new taxonomy of indicators". In: *Intergenerational Justice Review* 11.2, pages 77–86.

Gál, Róbert Iván, Pieter Vanhuysse, and Lili Vargha (2018). "Pro-elderly welfare states within child-oriented societies". In: *Journal of European Public Policy* 25.6, pages 944–958.

Gamson, William A (1988). "The 1987 distinguished lecture: A constructionist approach to mass media and public opinion". In: *Symbolic interaction* 11.2, pages 161–174.

Gamson, William A and Andre Modigliani (1989). "Media discourse and public opinion on nuclear power: A constructionist approach". In: *American Journal of Sociology* 95.1, pages 1–37.

Gentile, Brittany, Jean M Twenge, and W Keith Campbell (2010). "Birth cohort differences in self-esteem, 1988–2008: A cross-temporal meta-analysis." In: *Review of General Psychology* 14.3, pages 261–268.

Goh, Esther (2011). *China's One-Child Policy and Multiple Caregiving: Raising Little Suns in Xiamen*. Routledge Contemporary China Series. Taylor & Francis.

Goodwin, John and Henrietta O'Connor (2009). "Youth and generation: In the midst of an adult world". In: *Handbook of Youth and Young Adulthood. New Perspectives and Agendas*. Edited by Andy Furlong. New York, NY: United States of America: Routledge, pages 38–46.

Guillemard, Anne-Marie (1996). "Equity between generations in aging societies: The problem of assessing public policies". In: *Aging and generational relations over the life course: A Historical and Cross-cultural Perspective*. Edited by Tamara Hareven. Berlin: Germany: Walter de Gruyter, pages 208–224.

Hällsten, Martin (2014). "Inequality across three and four generations in egalitarian Sweden: 1st and 2nd cousin correlations in socio-economic outcomes". In: *Research in Social Stratification and Mobility* 35, pages 19–33.

Hamamura, Takeshi and Berlian Gressy Septarini (2017). "Culture and Self-Esteem Over Time: A Cross-Temporal Meta-Analysis Among Australians, 1978–2014". In: *Social Psychological and Personality Science* 8.8, pages 904–909.

Hank, Karsten et al. (2018). "What do we know about grandparents? Insights from current quantitative data and identification of future data needs". In: *European Journal of Ageing* 15.3, pages 1–11.

Hareven, Tamara (1994). "Aging and Generational Relations: A historical and life course perspective". In: *American Review of Sociology* 20, pages 437–461.

Héran, François (2007). "Preface". In: *Ages, Generations and the Social Contract: The Demographic Challenges Facing the Welfare State*. Edited by Jacques Véron, Sophie Pennec, and Jacques Légaré. Springer, Dordrecht: The Netherlands, pages 301–321.

Hertel, Florian R and Olaf Groh-Samberg (2014). "Class mobility across three generations in the US and Germany". In: *Research in Social Stratification and Mobility* 35, pages 35–52.

Hills, John (1996). "Does Britain have a welfare generation?". In: *The New Generational Contract. Intergenerational relations, old age and welfare*. Edited by Alan Walker. London: United Kingdom: University College London Press, pages 56–80.

Hood, Andrew and Robert Joyce (2013). *The Economic Circumstances of Cohorts Born between the 1940s and the 1970s*. IFS Report 89. London: United Kingdom.

Howker, Ed and Shiv Malik (2010). *Jilted Generation: How Britain has Bankrupted its Youth*. London: United Kingdom: Icon Books Limited.

Ishikawa, Tsuneo (1975). "Family Structures and Family Values in the Theory of Income Distribution". In: *Journal of Political Economy* 83.5, pages 987–1008.

Jäntti, Markus and Stephen P Jenkins (2015). "Income Mobility". In: *Handbook of Income Distribution*. Edited by Anthony B Atkinson and François Bourguignon. Volume 2. Amsterdam: The Netherlands: Elsevier, pages 808–935.

Jerrim, John, Álvaro Choi, and Rosa Simancas (2016). "Two-Sample Two-Stage Least Squares (TSTSLS) estimates of earnings mobility&58; how consistent are they&63". In: *Survey Research Methods*. Volume 10. 2, pages 85–101.

Jiménez, Maribel (2012). "Movilidad o persistencia intergeneracional del ingreso en la Argentina? Una aproximación emprica". In: *Revista de Economa Poltica de Buenos Aires* 9,10, pages 91–143.
Johnson, Malcolm L (2005). "The Social Construction of Old Age as a Problem". In: *The Cambridge Handbook of Age and Ageing*. Edited by Malcolm L Johnson. Cambridge: United Kingdom: Cambridge University Press, pages 563–571.
Kahn, Robert L and Toni C Antonucci (1980). "Convoys over the life course: Attachment, roles, and social support". In: *Life-span development and behavior*. Edited by Paul B Baltes and Orville G Brim. Volume 3. Cambridge, MA: United States of America: Academic Press.
Kan, Kamhon, I-Hsin Li, and Ruei-Hua Wang (2015). "Intergenerational income mobility in Taiwan: Evidence from TS2SLS and structural quantile regression". In: *The BE Journal of Economic Analysis & Policy* 15.1, pages 257–284. URL: https://doi.org/10.1515/bejeap-2013-0008.
Kemp, Candace L (2007). "Grandparentgrandchild ties: Reflections on continuity and change across three generations". In: *Journal of Family Issues* 28.7, pages 855–881.
Klimczuk, Andrzej (2013). "Analysis of Intergenerational Policy Models". In: *AD ALTA: Journal of Interdisciplinary Research*, pages 66–69.
Komp, Kathrin and Theo van Tilburg (2010). "Ageing societies and the welfare state: where the inter-generational contract is not breached". In: *International Journal of Ageing and Later Life* 5.1, pages 7–11.
Laub, Natalie and Christian Hagist (2017). "Pension and Intergenerational Balance - A case study of Norway Poland and Germany using Generational Accounting". In: *Intergenerational Justice Review* 11.2, pages 64–77.
Lazenby, Hugh (2011). "Is age special? Justice, complete lives and the prudential lifespan account". In: *Journal of Applied Philosophy* 28.4, pages 327–340.
Leach, Jeremy and Angus Hanton (2012). *Intergenerational Fairness Index. Measuring Changes in Intergenerational Fairness in the United Kingdom*. Technical report. London: United Kingdom.
Leach, Jeremy et al. (2016). *European Intergenerational Fairness Index. A crisis for the young*. Technical report. London: United Kingdom.
Leckie, George (2014). "Cross-Classified Hierarchical Linear Modeling". In: *Encyclopedia of Quality of Life and Well-Being Research*. Edited by Alex C Michalos. Dordrecht: The Netherlands: Springer, pages 1359–1363.
Lefranc, Arnaud (2011). *Educational expansion, earnings compression and changes in intergenerational economic mobility: Evidence from French cohorts, 1931–1976*. THEMA Working Paper 11. Cergy: France.
Lefranc, Arnaud and Alain Trannoy (2005). "Intergenerational earnings mobility in France: Is France more mobile than the US?" In: *Annales d'Economie et de Statistique* 78, pages 57–77.
Leigh, Andrew (2007). "Intergenerational mobility in Australia". In: *The BE Journal of Economic Analysis & Policy* 7.2. URL: https://doi.org/10.2202/1935-1682.1781.

Litwin, Howard et al. (2008). "The balance of intergenerational exchange: Correlates of net transfers in Germany and Israel". In: *European Journal of Ageing* 5.2, pages 91–102.

Lowenstein, Ariela (2005). "Global Ageing and Challenges to Families". In: *The Cambridge Handbook of Age and Ageing*. Edited by Malcolm L Johnson. Cambridge: United Kingdom: Cambridge University Press, pages 403–412.

—— (2010). "Determinants of the Complex Interchange among Generations: Collaboration and Conflict". In: *Intergenerational Solidarity. Strengthening Economic and Social Ties*. Edited by María Amparo Cruz-Saco and Sergei Zelenev. New York, NY: United States of America: Palgrave Macmillan, pages 53–80.

Lüscher, Kurt and Karl Pillemer (1998). "Intergenerational ambivalence: A new approach to the study of parent-child relations in later life". In: *Journal of Marriage and the Family*, pages 413–425.

Lüscher, Kurt et al. (2010). *Generations, intergenerational relationships, generational policy: A trilingual compendium*. Technical report. Bern: Switzerland.

Mare, Robert D (2011). "A multigenerational view of inequality". In: *Demography* 48.1, pages 1–23.

—— (2014). "Multigenerational aspects of social stratification: Issues for further research". In: *Research in Social Stratification and Mobility* 35, pages 121–128.

Mare, Robert D and Xi Song (2014). *Social mobility in multiple generations*. UCLA CCPR Population Working Paper 014. Los Angeles, CA: United States of America: California Center for Population Research, University of California, Los Angeles.

Margolis, Rachel and Laura Wright (2017). "Healthy Grandparenthood: How Long Is It, and How Has It Changed?" In: *Demography* 54.6, pages 2073–2099.

Marshall, Victor W, Fay Lomax Cook, and Joanne Gard Marshall (1993). "Conflict over intergenerational equity: Rhetoric and reality in a comparative context". In: *The Changing Contract across Generations*. Edited by Vern L Bengtson and W Andrew Achenbaum. New York, NY: United States of America: Walter de Gruyter, pages 119–140.

Mason, Andrew (2006). *Levelling the Playing Field: The Idea of Equal Opportunity and its Place in Egalitarian Thought: The Idea of Equal Opportunity and its Place in Egalitarian Thought*. Oxford Political Theory. Oxford: United Kingdom: Oxford University Press.

Masson, André (1999). "Quelle solidarité intergénérationnelle?" In: *Revue Française d'Économie* 14.1, pages 27–90.

—— (2007). "Economics of the Intergenerational Debate: Normative, Accounting and Political Viewpoints". In: *Ages, Generations and the Social Contract: The Demographic Challenges Facing the Welfare State*. Edited by Jacques Véron, Sophie Pennec, and Jacques Légaré. Springer, Dordrecht: The Netherlands, pages 61–104.

McKerlie, Dennis, editor (2006). Oxford: United Kingdom: Oxford University Press.

Mead, Margaret (1970). *Culture and Commitment. A Study of the Generation Gap*. London: United Kingdom: The Bodley Head.

Melyn, Wim and Willem Moesen (1991). *Towards a synthetic indicator of macroeconomic performance: unequal weighting when limited information is available*. Public Economics Research Papers 17. Leuven: Belgium.

Mendolia, Silvia and Peter Siminski (2016). "New estimates of intergenerational mobility in Australia". In: *Economic Record* 92.298, pages 361–373.

Mocetti, Sauro (2007). "Intergenerational earnings mobility in Italy". In: *The BE Journal of Economic Analysis & Policy* 7.2. URL: https://doi.org/10.2202/1935-1682.1794.

Modalsli, Jørgen (2017). "Intergenerational mobility in Norway 1865–2011" In: *The Scandinavian Journal of Economics* 119.1, pages 34–71.

Mudrazija, Stipica (2014). "The balance of intergenerational family transfers: A life-cycle perspective". In: *European Journal of Ageing* 11.3, pages 249–259.

Ng, Irene (2007). "Intergenerational Income Mobility in Singapore". In: *The BE Journal of Economic Analysis & Policy* 7.2. URL: https://doi.org/10.2202/1935-1682.1713.

Nunez, Javier I and Leslie Miranda (2010). "Intergenerational income mobility in a less-developed, high-inequality context: The case of Chile". In: *The BE Journal of Economic Analysis & Policy* 10.1. URL: https://doi.org/10.2202/1935-1682.2339.

OECD (2006). *Society at a Glance 2006*. Paris: France.

—— (2011). *Session 3. Paying for the Past, Providing for the Future: Intergenerational Solidarity*. Background Document. OECD Ministerial Meeting on Social Policy. Paris, 2–3 May 2011. Paris: France.

—— (2018). *A Broken Social Elevator? How to Promote Social Mobility*. Paris: France.

Phillipson, Chris (2015a). "The Political Economy of Longevity: Developing New Forms of Solidarity for Later Life". In: *The Sociological Quarterly* 56.1, pages 80–100.

Pillemer, Karl et al. (2007). "Capturing the complexity of intergenerational relations: Exploring ambivalence within later-life families". In: *Journal of Social Issues* 63.4, pages 775–791.

Piraino, Patrizio (2007). "Comparable estimates of intergenerational income mobility in Italy". In: *The B.E. Journal of Economic Analysis and Policy* 7 (2). URL: https://doi.org/10.2202/1935-1682.1711.

—— (2015). "Intergenerational earnings mobility and equality of opportunity in South Africa". In: *World Development* 67, pages 396–405.

Portes, Jonathan (2014). "Intergenerational and Intragenerational Equity". In: *National Institute Economic Review* 227.1, F4–F11.

Raitano, Michele and Francesco Vona (2015). "Measuring the link between intergenerational occupational mobility and earnings: evidence from eight European countries". In: *The Journal of Economic Inequality* 13.1, pages 83–102.

Rees, William E (1992). "Ecological footprints and appropriated carrying capacity: what urban economics leaves out". In: *Environment and Urbanization* 4.2, pages 121–130.

Reinhold, Meyer (1970). "The generation gap in antiquity". In: *Proceedings*. New York, NY: American Philosophical Society pages 347–363.

Richey, Jeremiah and Alicia Rosburg (2018). "Decomposing economic mobility transition matrices". In: *Journal of Applied Econometrics* 33.1, pages 91–108.

Roccisano, Federica et al. (2013). "On intergenerational mobility in Italy: what a difficult future for the young". In: *Review of Applied Socio-Economic Research* 6.2, pages 203–216.

Roemer, John E (1998a). *Equality of Opportunity*. Cambridge, MA: United States of America: Harvard University Press.

Roll, Anat and Howard Litwin (2013). "The exchange of support and financial assistance: differences in exchange patterns and their implications for ageing well". In: *Active Ageing and Solidarity Between Generations in Europe. First Results from SHARE After the Economic Crisis*. Edited by Axel Börsch-Supan et al. Berlin: Germany: Walter De Gruyter.

Rowlingson, Karen, Ricky Joseph, and Louise Overton (2017). *Inter-generational Financial Giving and Inequality: Give and Take in 21st Century Families*. Palgrave Macmillan Studies in Family and Intimate Life. London: United Kingdom: Palgrave Macmillan.

Samuelson, Paul A (1958). "An exact consumption-loan model of interest with or without the social contrivance of money". In: *The Journal of Political Economy* 66.6, pages 467–482.

Segall, Shlomi (2013). *Equality and Opportunity*. Oxford: United Kingdom: Oxford University Press.

Shanas, Ethel (1967). "Family Help Patterns and Social Class in Three Countries". In: *Journal of Marriage and Family* 29.2, pages 257–266.

Shaw, Kate Alexander (2018). *Baby Boomers versus Millennials: rhetorical conflicts and interest-construction in the new politics of intergenerational fairness*. Technical report. Sheffield: United Kingdom.

Sheppard, Paula and Christiaan Monden (2018). "The additive advantage of having educated grandfathers for children's education: evidence from a cross-national sample in Europe". In: *European Sociological Review* 34.4, pages 365–380.

Solon, Gary (2004). "A model of intergenerational mobility variation over time and place". In: *Generational income mobility in North America and Europe*. Edited by Miles Corak. Cambridge: United Kingdom: Cambridge University Press, pages 38–47.

——— (2014). "Theoretical models of inequality transmission across multiple generations". In: *Research in Social Stratification and Mobility* 35, pages 13–18.

——— (2018). "What do we know so far about multigenerational mobility?". In: *The Economic Journal* 128.612, F340–F352.

Stone, Jon R (2006). *The Routledge Book of World Proverbs*. New York: United States of America: Routledge.

Ueda, Atsuko (2009). "Intergenerational Mobility of Earnings and Income in Japan". In: *The BE Journal of Economic Analysis & Policy* 9.1. URL: https://doi.org/10.2202/1935-1682.2203.

—— (2013). "Intergenerational mobility of earnings in South Korea". In: *Journal of Asian Economies* 27, pages 33–41.

United Nations (2002). *Political Declaration and Madrid International Plan of Action on Ageing. Second World Assembly on Aging, 8–12 April 2002*. New York, NY: United States of America: United Nations.

Vanhuysse, Pieter (2013). "Measuring Intergenerational Justice–Toward a Synthetic Index for OECD Countries". In: *Intergenerational Justice in Aging Societies* page 10.

Véron, Jacques, Sophie Pennec, and Jacques Légaré, editors (2007). *Ages, Generations and the Social Contract: The Demographic Challenges Facing the Welfare State*. Springer, Dordrecht: The Netherlands.

Walker, Alan (1996a). "Intergenerational relations and the provision of welfare". In: *The New Generational Contract. Intergenerational relations, old age and welfare*. Edited by Alan Walker. London: United Kingdom: University College London Press, pages 10–36.

—— (1996b). "Introduction: the new generational contract". In: *The New Generational Contract. Intergenerational relations, old age and welfare*. Edited by Alan Walker. London: United Kingdom: University College London Press, pages 1–9.

Willets, David (2010). *The Pinch: how the Baby Boomers Took Their Children's Future And Why They Should Give it Back*. London: United Kingdom: Atlantic Books.

Williams, Alan (1997). "Intergenerational equity: an exploration of the 'fair innings' argument". In: *Health Economics* 6.2, pages 117–132.

Williamson, John B and Diane Watts-Roy (1999). "Framing the generational equity debate". In: *The Generational Equity Debate*. Edited by John B Williamson, Diane Watts-Roy and Eric R Kingson. New York, NY: United States of America: Columbia University Press, pages 3–37.

Xie, Yu and Alexandra Killewald (2013). "Intergenerational occupational mobility in Great Britain and the United States since 1850: Comment". In: *American Economic Review* 103.5, pages 2003–2020.

Zeng, Zhen and Yu Xie (2014). "The effects of grandparents on children's schooling: Evidence from rural China". In: *Demography* 51.2, pages 599–617.

Zhang, Min and Yaojun Li (2018). "Family fortunes: The persisting grandparents' effects in contemporary British society". In: *Social Science Research* URL: https://doi.org/10.1016/j.ssresearch.2018.08.010.

7

Ageing, House Prices, and Economic Crises

Overview

This chapter looks into the associations between individual and population ageing and housing issues, including the discussions about the possibilities of equity release to reduce the risk of poverty in later life and as a source of funding for long-term care. Other topics include residential mobility in later life, housing poverty, and the links between housing and pension saving.

7.1 Introduction

A chapter on housing and ageing? Well, yes. 'Finally, housing is a hot topic for economics', as Smith et al. (2010, p. 1) announced, ' is not just another consumption good from which consumers derive utility. It is not even just another durable consumption good: housing wealth accounts for around 70 per cent of total household wealth (excluding pension wealth) in Australia[1]; 63 per cent in the Euro countries (Bank 2018, Table 1) and 59 per cent in the United Kingdom (National Statistics 2015, Table 2.1)—in the United States of America, also excluding pension accounts, it represents around 42 per cent of total household wealth (Wolff 2017, Table 3).[2] In addition, housing is an asset that can be used as collateral to access credit, thus affecting savings. Moreover, housing equity can be used as a pension (Doling and Elsinga 2012), and as a contingency asset or fund, as a 'safety net' or insurance against future risk especially, during retirement (Skinner 1996; Smith and Searle 2010). Furthermore, there is some evidence that as housing wealth increases, government spending on older people diminishes (De Deken et al. 2012; Doling and Horsewood 2011), which adds another dimension to the link between ageing and housing. On top of this, housing tenure, adequacy, and affordability impacts on older people's health, well-being, independence, ability to age in place, and social inclusion (Morris 2016). Besides, in many countries housing markets specifically for older people are being developed as a response to population ageing.

Housing also has psychological, social, and temporal dimensions that are unique compared to other durable consumption goods and which have to be incorporated in the studies of the economics of housing in later life (Roy et al. 2018). It is crucial to realise that human dwellings have cultural values and representations attached to them and that, at an intimate personal level, people develop a relationship with the houses they live in, so much so that dwellings become their 'homes' (Frank 2002), or, as Milligan (2009) highlighted, that they provide a protected place or haven, a preconscious sense of setting for familiar daily routines and a site for the embodiment of self-expression and identity.

According to many economic historians, all the episodes of financial crises following a weakening or loosening of the monetary policy and a large influx of capital from abroad were preceded by a run-up in house prices (Kaminsky and Reinhart 1999; Reinhart and Rogoff 2008). The 2008 crisis was no exception (Ahrend et al. 2008; Duca et al. 2010; Taylor 2009). Surely Leamer (2007) overdid it when he stated that 'housing is the business cycle'—though only just. It would certainly be far-fetched to ascribe *sole* causal responsibility for the house price bubble and burst to population ageing. However, some authors did find a strong association between an increasing proportion of older people in a population and increasing prices in the housing markets. For example, Lisack et al. (2017) estimated that population ageing could explain around three quarters of the 50 per cent rise in real house prices between 1970 and 2009 in developed countries. Besides, Nishimura (2011) suggested that changes in the old-age dependency ratios in developed countries preceded the onset of the financial crisis in most of them. Beltratti and Morana (2010), in turn, estimated that shocks in the housing markets (particularly in the United States of America) accounted for 20 per cent of macroeconomic fluctuations in developed countries between 1980 and 2007, a higher contribution than that of stock market shocks.

7.2 Residential Mobility in Later Life

This section looks into the decisions by older people to change residence within a country as well as some of their wider effects—see Volume I Sect. 3 for the economics of migration in later life.

7.2.1 Ageing, Moving, and House Prices

Why would the ageing of the population in a country influence the prices in the housing market? Four mechanisms have been identified, three of which operate via the demand side of the housing market—a population size effect, an age structure effect, and an investment demand channel—whilst the fourth one affects the supply side of the market (Hiller and Lerbs 2016):

- The population size effect is, possibly, the easiest to understand: given the relative price inelasticity of the supply of housing (the supply is not highly responsive to changes in prices as it takes time to find the right location, obtain the planning permissions, and build the properties), changes in the number of people living in an area—that shift the demand for housing—may affect house prices. Population ageing leads to a slower rate of population growth (or to a negative population growth rate), so it would put downward pressure on house prices via a population size effect *ceteris paribus*.
- The age structure effect is related to the varying demand for housing along an agent's life cycle. The life-cycle hypothesis points to various variables with influence on the demand for housing that exhibit change over the lives of economic agents: from family formation and household size—that is, the 'family life-cycle' effect (Clark and Onaka 1983; Doling 1976)—to job-related or retirement-related mobility; from income level and security to accumulated wealth towards putting down upfront payments for buying the first home, and so on. The age structure effect would suggest an inverted U-shaped age profile of the demand for housing.
- The investment demand channel is related to the role of housing as a saving vehicle and a source of income during retirement. As during retirement, part of the housing stock owned by retirees would be sold, population ageing would place additional downward pressure on house prices.
- The supply side effects of population ageing in the housing markets are related to changes in land availability and in planning legislation to adjust the markets to demographic changes. The effects of population ageing,

however, tend to be marginal compared to its demand side implications although supply side effects may drive a wedge between residential mobility intentions and decisions (see below).

The 'asset meltdown' hypothesis posits that the ageing of the 'Baby Boomers' generation would lead to a fall in the prices of financial assets in developed countries (see Chap. 9 in Volume III). As a matter of fact, this hypothesis was first put forward by Mankiw and Weil (1989) in the context of housing in the United States of America during the 1970s and the 1980s. These authors noticed that the demand for housing had followed a clear pattern in relationship with the chronological age of home buyers, with an inverted U-shape peaking at around age 40. Using population projections by age, they forecast that house prices would fall by 47 per cent in real terms between 1987 and 2007. Tongue-in-cheek, the Danish physicist Niels Bohr stated that 'prediction is very difficult, especially if it's about the future', and I accept that economists have a poor track record, in best of times, when it comes to forecasting, but the reduction by 47 per cent of real house prices in twenty years was way off the mark even for the standards of the profession: according to US Bureau of Labor Statistics, the Consumer Price Index for Housing *increased* by 83 per cent during the period.[3] Among the explanations as to why this prediction never materialised was that Mankiw and Weil had failed to distinguish between an age and a cohort effect underlying the shape in the data they had studied and based their predictions upon: older people back then had, on average, lower human capital and lower lifetime income than then younger people whose demand was being predicted, so that the demand for housing over the whole lives of the previous generation had been lower than that of the older members of subsequent cohorts (Green and Hendershott 1996; Pitkin and Myers 1994). The age profile with its inverted U shape was merely apparent: it masked a cohort effect. Green and Lee (2016) confirmed this explanation using data for the United States of America between 1990 and 2014: the fact that younger birth cohorts, on average, are better educated and earn higher income than previous cohorts—coupled with the large size of the 'Millennial generation' (i.e. people born after 1982)—would increase the demand for housing, thus preventing a meltdown in prices.

It goes without saying that not every economist is convinced about this cohort effect, and that—more generally—the jury is still out on whether population ageing affects house prices. Some authors have found a negative relationship whilst others have reported a positive association. To illustrate:

- Takáts (2012) estimated that between 1970 and 2009, population ageing was responsible for an *increase* in house prices by around 30 percentage points on average across twenty-two developed countries and projected that demographic forces would reduce house prices by around 80 percentage points between 2010 and 2050 in those countries. However, the author predicted that despite being economically and statistically significant, this reduction would not be large enough to represent the price meltdown of colossal proportions that Mankiw and Weil predicted.
- Similarly, Lisack et al. (2017)—using a multi-country OLG model—estimated that population ageing explained three quarters of the 50 per cent *increase* in real house prices between 1970 and 2009 in developed countries, mainly through its lowering effect on interest rates, but
- Jäger and Schmidt (2017) found that population ageing reduced house prices in thirteen developed countries between 1950 and 2012. These authors reported that longer life expectancy was not associated with variations in house prices and that the only demographic indicator that would affect prices in the housing markets was the age distribution of the population: a higher proportion of people aged 60 or older would depress house prices whilst a greater share of people aged 30–34 years would be associated with a 2.5 per cent increase in real house prices.
- Park et al. (2017) reported an inverse relationship between ageing and house prices using data for Korea between 1990 and 2014 and forecast that as a result of the projected demographic change, house prices would fall by more than 20 per cent by 2030.
- However, Yang (2009) studied data for the United States of America between 1984 and 2000 and reported that the life-cycle profile of housing demand increases with chronological age until around age 60 to stabilise thereafter: this author failed to find an inverted U shape; once the age-profile reached its peak, it would flatten out until very late in the life cycle when it would go down but not in a great proportion.

A number of considerations and caveats are in order.

First, the dissociation between intentions and actions is starkest at very old ages: Abramsson and Andersson (2016) investigated data from a survey carried out in Sweden in 2013 and reported that the oldest old exhibited the lowest intentions to move but the highest residential mobility rates of all age groups studied. The authors explained the paradox in terms of adjustments to changing needs—see also Hasu (2018) for a similar discrepancy between intentions and decisions among older people in Finland.

Besides, whilst most non-institutionalised older people own the houses they live in, a non-negligible percentage either pay a mortgage or rent. Retirement decisions are influenced by housing tenure: it is more difficult to 'afford' retiring if there is a monthly mortgage or a rent to be paid. Moreover, not being an owner outright (i.e. without an outstanding mortgage) of a property adds an element of risk regarding the place of abode in later life: if for any reason (and later life may bring some of them) an individual cannot continue earning an income, the chances of facing eviction are higher if she needs to pay a mortgage or a rent. On the other hand, downsizing may be easier for an older person who does not own a house—and therefore does not need to sell it before moving to a smaller property. However, the risks remain, of course, as she does not own the smaller property she moves in.

In addition, references to the housing 'market', though correct at a general level, mask great geographical and product heterogeneity. Hence some authors opine that referring to a variety of housing 'markets' would be more appropriate; otherwise, the studies might suffer from aggregation bias. The differences across housing 'markets' lie in many factors: for example, the elasticity of supply may differ due to building regulations, and so on, but the crucial aspect in our case is to do with the selective demand for types of houses and regions among older people: do older people prefer one housing type over others and one type of location over the rest? This is the approach taken by Hiller and Lerbs (2016), who looked into the markets for different housing options (including, for example, condominiums and single-family homes) in eighty-seven German cities between 1995 and 2014. The authors reported that population ageing had negative effects on the prices of condominiums and single-family homes, but a positive impact on rental prices in urban areas. The increased demand for rented city apartments was not driven by the older people's desire to 'downsize' from, say, bigger suburban houses; it came as a result of capital investment decisions: older buyers expected that these flats would provide them with additional income in retirement—see also Attanasio et al. (2012) for evidence for the United Kingdom that the price elasticity of demand for flats and houses differ among older people. In addition, Andersson et al. (2018) reported different a self-image among older people living in non-urban middle-class areas and in inner cities in Sweden: for the former, being a homeowner—among other aspects—is a central tenet of their self-image whilst it is much less so for urban dwellers.

Gerontologists have proposed two models of residential mobility in later life: the life-cycle and the stress threshold models. The life-cycle model of residential mobility points to changes in life stages as main drivers to move house such as marriage, having children, changing jobs, widowhood, or infirmity (Rossi 1955). The stress threshold model of residential mobility points to a mismatch between preferences and the utility derived from not only the actual and perceived amenities of the property but also its surroundings, neighbourhood, location, and so on: a bigger mismatch would reduce the relative satisfaction with the residential location and cause enough stress to precipitate a move (Fokkema et al. 1996; Speare 1974). Hansen and Georg Gottschalk (2006) analysed motivations to move and actual moves among older people in the United States of America between 1997 and 2002 and found that although these two models are supplementary—certain transitions in later life lead to dissatisfaction and stress, which increase the probability of moving residence—the stress threshold model would explain mobility considerations better. Interestingly, the authors suggested that the threshold model would also fit better the data on actual mobility decisions, except that several barriers to move that older people face (from the inadequacy of affordable alternatives to a preference not to go through the 'ordeal' of moving house) would have weakened the statistical link.

Furthermore, the 'housing regime' prevalent in a country must be taken into account. The housing regime refers to how the housing system is organised, which depends on the wider dominant ideology, the balance of power, and the political structures (Kemeny 1992, 1995). Higher rates of home ownership, for example, would be aligned with a greater influence of a pro-privatisation ideology in which welfare provision would be given a residual role as a safety net. Housing regimes 'determine which social groups have access to home ownership, at which age, for which price, and to which extent they experience capital gains and losses' (Wind et al. 2017, p. 627). Home ownership rates increased substantially across European countries, except Sweden, between 1980 and 2010. In contrast, Sweden exhibits the highest mobility rates among older people across developed countries.

Finally, the life-cycle hypothesis predicts that older people would bank on the properties they own to finance consumption—that is, that they would reduce housing equity as they age. With a number of financial products, older people can release equity without having to leave their homes. I discuss this in the following subsection.

7.2.2 Housing-Related Financial Products

Housing wealth may be a source of income in later life. Other than selling and moving into cheaper accommodation, renting second homes, or sub-renting rooms in the main residence to lodgers, for housing wealth to be

Table 7.1 Sources of housing income

		Income		
			Full	Reduced
Income from equity (cash)	Not dissaving	Zero	1. Continue to live in home	2. Let out part of home to lodger
	Dissaving	Reduced	3. Home equity loans and lines of credit	4. Sell home and move into smaller/cheaper house (i.e. downsize)
		Full	5. Reverse mortgage	6. Sell home. Move into rented accommodation

translated into an income flow, older people need access to particular financial instruments. Table 7.1—based on Doling and Elsinga (2013, Table 4.1)—presents a classification of sources of housing income:

Houses can be understood as accumulated savings, so drawing on these assets by moving into a smaller and cheaper owned property (i.e. downsizing; option 4) or into rented accommodation (option 6) would be rational responses in later life to help smooth lifetime consumption. However, Venti and Wise (1990a, 1991) showed that mobility among older people was low during the period 1969–1979 in the United States of America—similar results were reported in Venti and Wise (2004) for the years 1993–1998.[4] These studies suggested that older people would sell up and reduce equity, but only following bereavement or to move into residential or nursing caregiving institutions. Concepts such as life events, transitions, or turning points—see Volume I, Chap. 3—would have greater explanatory power than simply economic life-cycle considerations.

The decision not to move may not respond to older people's preferences also due to constraints originating in financial market imperfections, confining older people to houses where they do not want to live in any longer and to illiquid assets they would be willing to use to finance consumption—that is, the almost stereotypical 'house-rich, cash-poor' older person.

Therefore a number of financial products have been designed for older people who prefer to remain in their homes for as long as possible that generate either a regular stream of income or a one-off amount of money out of the value of a property—financial instruments that allow older people to continue living in their homes while dissaving part or all of their housing wealth (options 3 and 5). The most popular of these equity conversion instruments are known as 'reverse mortgages' (option 5)—also as 'home equity conversion mortgages' or 'reverse annuity mortgages'. Other instruments include 'home equity loans' (HEL) and 'home equity lines of credit' (HELOC) (option 3). HELs consist of a loan the home owner receives as a lump-sum amount against the equity in her house as collateral, but that generates regular—usually monthly—interest payments. HELOCs are open-ended credit facilities up to pre-agreed amounts to be tapped into at any time over a given period. The home owner does not pay any interest unless she borrows some amount and only on that amount, known in the industry as the 'takedown'—this practice is popularly referred to as using the house as an ATM. Agarwal et al. (2006) compared borrowers who chose equity loans and lines of credit between 1994 and 2001 in the United States of America and found that the former had, on average, lower credit quality scores, higher initial loan-to-value ratios, and lower access to credit than borrowers who took HELOCs.[5]

A reverse mortgage is a loan against a property owned by the debtor. A reverse mortgage would release equity without the older person having to leave her home. There are three main differences with a typical mortgage (known also as a 'forward' mortgage to distinguish from a reverse one): the owner of the property is the recipient of the loan, the loan is not paid off in regular instalments but becomes repayable when the debtor passes away or moves permanently, and the borrower's future flow of income is not taken into consideration. How the borrower receives the money varies according to the particular instrument (e.g. via a lump-sum amount, regular payments for a fixed period, regular payments until the end of the contract—i.e. until the death of the borrower or a permanent change of residence).

Lenders of reverse mortgages are not totally risk-free of eviction and foreclosure. Even though technically borrowers would not incur in arrears on the mortgage, lenders are exposed to a default risk: in 2014, 12 per cent of reverse mortgage borrowers in the United States of America (about 78,000 older people) were in default on their property taxes or insurance, for example—see (Moulton et al. 2015).

Reverse mortgages would increase an older homeowner's utility: they would lower any financial constraints, they would permit higher consumption in later life and better consumption smoothing over the life cycle, and they would allow older people to stay put in their homes for as long as they decide to. One crucial consideration is the presence of a bequest motive: the higher an older homeowner's motivation to bequeath her house, the less desirable a reverse mortgage becomes as an option (Chiang and Tsai 2016).

Potentially, the market for housing equity is huge: estimates for 2017 put older people's total housing equity at €6.6 trillion in the United States of

America (Spanko 2018) and for 2013 at €8 trillion in France, Germany, Italy, Spain, and the United Kingdom combined (Haurin and Moulton 2017)—and it is part of the process of globalisation of mortgage markets following the financial markets liberalisation that started in the 1980s (Lowe et al. 2012). However, older people do not access this equity as expected—the average older person's portfolio is unbalanced with a greater than optimal proportion of total wealth held as housing wealth—in fact, according to Hanewald et al. (2016), reverse mortgages provide higher utility gains for older people than home reversion plans, annuities, and long-term care insurance; see also Li and Chand (2018).

Studies in the 1990s in the United States of America concluded that the demand for reverse mortgages was low among older people (Mayer and Simons 1994; Merrill et al. 1994; Venti and Wise 1991). After carrying out in-depth interviews to older home owners, Leviton (2002) cited the following factors behind the decision not to move in later life: a strong attachment to the home and neighbourhood, a preference towards frugality despite considerable debt or financial instability, secrecy, and privacy, and a desire to leave the house as a legacy. The take-up of reverse mortgages was still low in 2011, reaching merely 2.1 per cent of eligible households in the United States of America (Nakajima and Telyukova 2017). Current demand is still below predictions in most countries: in Australia, the industry has declined between 2013 and 2018 (IBISWorld 2017), and in Spain, following low demand and adverse judicial rulings, most banks withdrew from the market, which has dwindled severely since 2010 and is now catered for by specialised financial entities.

Nakajima and Telyukova (2017) estimated the impact of the 2008 financial crisis on the future demand for reverse mortgages. Based on period effects that suggest that the crisis would affect the most people who were in their 20s and 30s at the time because of the impact that long spells of unemployment in those chronological ages may have on future careers, the authors distinguished between short- and long-term wealth effects of the crisis and then short- and long-term effects on the reverse mortgage market. They predicted that the demand would fall in the short run among higher income individuals and would increase among low-income households—and this increase will be higher with chronological age. This positive shift in demand for reverse mortgage will respond to additional adverse wealth effects which would drive older people on low incomes towards buying these financial instruments. In the long term, the demand for reverse mortgages is projected to grow across the income and age distributions.

Some researchers suggested that educational attainment would be a significant explanatory factor behind the decision to buy a reverse mortgage or not:

in a study in Hong Kong, Chou et al. (2006) found that the willingness to buy such a product is higher among well-educated older people who hold shares, bonds, and other financial assets than among older people on lower income and with lower educational attainment. Two possible explanations are that the former group may be more familiar with financial instruments than the latter group, and that less educated older people who would be more influenced by traditional Chinese values (filial piety, leaving a legacy, protection against old age) and would not be as comfortable with novel financial products. Costa-Font et al. (2010) also found that educational attainment was positively associated with willingness to buy a reverse mortgage instrument among older people in Spain. In contrast, Fornero et al. (2016) responded that high literacy is associated with *lower* demand for reverse mortgages among older people in Italy because the instruments would be seen as a 'last resort' vehicle to finance later life: risk aversion would reduce its demand, and greater uncertainty about future income would increase it. In a similar vein, Chiang and Tsai (2016) found that higher retirement income is negatively associated with the demand for reverse mortgage products. Education attainment is also associated with familiarity with the financial instruments: Davidoff et al. (2017) found that, even after almost thirty years of being introduced in the market and despite high awareness among older people, the knowledge of contractual terms remained low and the misunderstandings were high, which explained the low levels of demand for these products.

Another point worth considering is the existence of a well-developed market for traditional mortgages. Considering that in many countries most older people own the houses they reside in, developments and regulations of the market for residential mortgages would not be of major interest for this chronological age group, other than from a life-course perspective as, for example, favouring or hampering borrowing towards buying a house earlier in the life cycle. However, Angelini et al. (2011) identified a strong correlation between mortgage market development and the proportion of older people undergoing financial distress. The reason, the authors surmised, lay in the link between how developed the market for traditional mortgages is and the development of the market for reverse mortgages:

> The importance of trading down as a form of equity release depends heavily on financial and mortgage markets access and regulations, as well as on the availability of public housing and long term care accommodation. In most European countries, financial instruments that allow equity release are unavailable or relatively uncommon, and cheap public housing is a scarce resource (particularly

for the elderly), so trading down may well be the only way to generate a cash flow out of the available home equity.

(Angelini et al. 2011, p. 101)

Home ownership is sometimes referred to as the 'fourth pillar' of welfare provision and social insurance in later life (Bradbury 2013; Brownfield 2014; Kemeny 2001). It should come as no surprise, then, that reverse mortgages are also promoted as a source of funding for long-term care (Andrews and Oberoi 2015). This particular use of housing equity must be seen in the context of the ageing in place or 'ageing at home' concept—a key element of policy initiatives on behalf of older people in many countries. Ageing in place is 'the ability of older people to live in their own home and community safely, independently, and comfortably, regardless of age, income or level of intrinsic capacity' (WHO 2015, p. 36). Adaptations and modifications to the property may be needed in case of onset of disability or health problems, and new technologies (and markets) are being created to allow care and monitoring 'from a distance' thus enabling older people remain in their homes.

The use of home equity to pay for care has repercussions on the market for long-term care insurance: Davidoff (2010) showed that in the United States of America, home equity is a substitute for long-term care insurance because the main reason for home equity borrowing among older people would be as a buffer against the eventuality of a health shock or needing care. In contrast, Stucki (2006) suggested that for some older people, home equity might not cover much of the long-term care costs but could be used to pay for long-term care insurance thus turning both products complementary to each other.

One premise behind the adoption of the ageing-in-place concept as a central tenet of public policies for older people (particularly with regard to housing, but also social care, community development, and income) is that it reflects older people's preferences—and, thus, equity release instruments would contribute to the fulfilment of these preferences. Without delving into the reasons as to why the opinions and preferences of older people seem to be given such a peculiarly strong weight in this context as opposed to many other realms of policy, it is worth noting that it is advisable not to make broad generalisations about later life, and even more so not to base policy upon such broad generalisations. A burgeoning literature has nuanced and qualified the oft-repeated statement that ageing in place is 'what older people prefer'. Preferences for housing alternatives depend on cultural characteristics (Ær 2006; Andersen 2011; Jansen 2014) and change over the life course (Abramsson and Andersson 2016; Granbom et al. 2014) according to gender and marital status (Barry et al. 2018) and social support (Pastalan 1990; Tang and Lee

2011). They also vary, especially, along with changes in health status and in income and wealth. Concerning health, for example, Fernandez-Carro (2016) reported a low desirability to age in place among frail older people in Spain. Income and wealth are two strong determinants of the area of residence and its amenities, so Golant (2008) wondered about the suitability of the ageing-in-place concept in relation to older people on low incomes in the United States of America, many of whom dwell in non-decent or unfit housing stock. This is a good moment to emphasise, once again, that heterogeneity increases with chronological age; this is an empirical fact that also applies to the housing preferences of older people and which makes a one-size-fits-all approach to policy unlikely to be successful (Wiles et al. 2012).

In relation to long-term care, ageing in place is usually presented in binary 'either/or' terms, as the sole alternative to moving into sheltered accommodation or a nursing home. However, it is more appropriate to frame ageing in place—and the role of housing equity in this regard—as part of the 'continuum of care' (Weil and Smith 2016), see Volume II, Chap. 7. Besides, it is important to think of 'place' beyond a house (and beyond merely a container!) and incorporate considerations of ageing in the wider context of the community (Scharlach 2017) and the density of meaningful relationships supported in the community (Boyle et al. 2015). For example, Hillcoat-Nalletamby and Ogg (2014) reported that dissatisfaction with the immediate home environment (not necessarily with the wider neighbourhood) is a strong predictor of the desire to move among older people in Wales, and Smith et al. (2017) reported adverse mental and physical health effects on older people of ageing in place in newly gentrified neighbourhoods in the United States of America. In turn, Severinsen et al. (2016) studied why many older people in Australia seem to live in 'unsuitable' places (and are proud of their places of residence!). Seen from this wider geographical perspective, the 'romantic canonical narrative associated with the policy of "ageing in place"' (Vasara 2015) would give way to a more robust approach to the analysis of its strengths and limitations, and with this, the microeconomic decision to release housing equity as well as its consequences would be more comprehensively understood.

It is not coincidental that ageing in place has such a policy support given that it is seen as a less costly option for the provision of care services compared to institutionalised alternatives. However, this common knowledge has been contested by the only systematic review on the topic, on the grounds of the low methodological quality of the studies (Graybill et al. 2014).

7.3 Housing and Poverty in Later Life

Apart from a funding source for general consumption and caregiving services, housing equity is seen as a vehicle towards the alleviation or the reduction in the risk of poverty in later life. In contrast, mortgage indebtedness is positively associated with financial distress.

Most studies focused on the role of reverse mortgages with respect to poverty alleviation:

- Using data from 1991, Kutty (1998) estimated that 29 per cent of older people living in poverty in the United States of America could have been raised above the poverty line by means of a reverse mortgage—see also Mayer and Simons (1994).
- In Korea, with data from 2013, Heo et al. (2016) estimated that reverse mortgages would bring down poverty rates among older people from 46.7 per cent to 27.3 per cent.
- Moscarola et al. (2015) estimated the fall in poverty rates among people aged 65 years or over if they took out a reverse mortgage in nine European countries—for example, if 70 per cent of the average housing wealth were converted to cash, poverty rates in later life would descend by 25 per cent in Spain or 18 per cent in Belgium.

As in any other topic in economics, institutional and cultural factors are important—in this case, for the levels of home ownership and the effectiveness of reverse mortgages as an anti-poverty tool. We referred to the first point in the previous section. Regarding the latter, cultural differences have a mediating role:

- Comparing Spain with the United States of America, Costa-Font et al. (2010) noted that whereas roughly the same proportion of older people in both countries expressed a willingness to contract a reverse mortgage to finance basic needs in case infirmity or disability hit them, a much larger proportion of older people in the United States of America would be driven by a desire to finance higher consumption and quality of life compared to Spain—the authors surmised that the difference lies in the fact that leaving bequests is a deeply rooted cultural trait among older people in the latter country.
- In turn, Toussaint (2013) reported that in the Netherlands, contracting a reverse mortgage to finance long-term care is culturally more acceptable

than as a means of additional income in later life: housing equity would be perceived as a hedge against age-related shocks rather than a source of regular additional income.

Furthermore, two institutional settings influence how much drawing housing equity may reduce poverty in later life: taxation and the type of payment of the borrowed amounts:

- Compared to the optimistic projections by Kutty for the United States of America, Hancock (1998) estimated a more subdued impact on poverty reduction in the United Kingdom (the poverty rate among older people would have fallen from 67 per cent to 64 per cent), which according to Ong (2008) could be explained by different tax treatments in these countries.
- Venti and Wise (1991) considered that in the United States of America a lump-sum transfer would lift a significantly larger proportion of older people out of poverty compared to monthly payments.

Concerning mortgage debt, the most common indicator is the loan-to-value ratio: the ratio between the amount of the mortgage loan on the borrowing date and the purchase value of the property (Charupat et al. 2012; Williams 1987). Higher average loan-to-value ratios indicate that home buyers borrow a higher proportion of the purchase value of the property. There is a negative association between the proportion of homeowners aged 50 or over who report financial difficulties and typical loan-to-value ratios across developed countries (Angelini et al. 2011). Average loan-to-value ratios are rather crude indicators of the level of development of the mortgage markets in a country. Therefore, the inverse statistical relationship between loan-to-value ratios and financial hardship suggests that a higher proportion of older people are in financial distress in countries where the market for mortgages is not well developed compared to countries with well-developed market mortgages. Angelini et al. opine that downsizing and buying equity release instruments are related to the level of development of mortgage markets, and that this can explain the link between loan-to-value ratios and the proportion of older people with difficulty making ends meet.

The chronological age and life expectancy of the individual is also important, particularly when the focus is not on the 'income-poor, housing-rich' older people but on 'income-poor, housing-poor' agents—that is, those living below the poverty line and with little housing equity to rely on: the shorter their life expectancy, the more effective a reverse mortgage would be to lift them out of poverty for the rest of their lifetimes.

The expansion of home ownership mentioned in the previous section has reduced tenure inequality among older people, but not necessarily housing wealth inequality: especially, market-based regimes based on down-payments and mortgages would have contributed to an expansion in housing wealth inequality among people aged 50 years or over during the period. Using the classic classification of welfare states by Esping-Andersen (1990)—see Chap. 7, Volume II—Doling and Elsinga (2013) estimated that Mediterranean and Liberal countries have the highest levels of net housing equity as a percentage of GDP whilst Corporatist and Social Democratic countries (with the exceptions of Austria and France) have the lowest levels. Also, housing wealth represent the highest share of total household wealth for older people in Mediterranean countries compared to other welfare regimes.

Earlier on I referred to the 'life-course perspective'. One concern among many policy makers throughout the developed countries since the 2008 financial crisis is that younger birth cohorts, whom according to the life-cycle hypothesis should be buying their first houses, are either renting or staying in their parental homes. The main source of policy concern is the possibility that these individuals, as well as those on low income for much—if not all—of their lives, would not have access to housing equity they could draw upon in their later years, raising the spectre of poverty and deprivation.

7.3.1 Housing and Risk in Later Life

Home ownership reduces eviction risk and rent risk and avoids the 'fundamental rental externality' (Angelini et al. 2011). Henderson and Ioannides (1983) studied this externality, which stems from the fact that tenants do not have to face the social marginal cost of maintenance of their dwellings, which leads these agents to take less care of the houses compared to home owners who occupy their own properties. As a result, the operating costs to cover for the depreciation of rented dwellings are higher than those of owner-occupied properties. This externality would be one rationale for the public promotion of home ownership.

We mentioned above that home owners are not fully immunised against the risk of eviction, for example, by being in arrears on property taxes. However, in this subsection, I want to concentrate on the gravest of housing-related eventualities: homelessness in later life.

The definition of homelessness varies by country, so measures are not easy to compare internationally (OECD—Social Policy Division—Directorate of Employment, Labour and Social Affairs 2017). In some countries a broad definition of homelessness has been adopted, whilst in others (e.g. Austria or Canada) not even sleeping rough is included. Another indicator, living with friends on a temporary basis for lack of housing, is considered homelessness in the United Kingdom and Spain, but not in the United States of America or Sweden. The same applies to people who live in emergency sheltered accommodation for lack of housing: they are not homeless according to the definitions applied in Japan, Latvia, or Austria, for example.

In the United States of America, about 1.5 million people experienced sheltered homelessness (i.e. resided in an emergency shelter or in transitional accommodation due to lack of housing) at least once in their lifetimes, and around 550,000 are homeless on any single night. Of these, roughly 25 per cent are aged 50 or over (Solari et al. 2017). About half of these older people enduring homelessness became homeless in later life (Brown et al. 2016; Crane et al. 2005; Lee et al. 2016). The homeless population is ageing: in the United States of America, the median age of adult homeless people is 50. This is not a reflection of the ageing of the population: the trend is steeper, the problem of homelessness in later life is getting worse than demographic measures would predict (Culhane et al. 2010).

However, home ownership does not fully prevent older people from becoming homeless. The probability of facing homelessness is higher among renters, but Rota-Bartelink and Lipmann (2007) identified a significant proportion of homeless older people who had lived in their own homes before becoming homeless in Australia, and Laere et al. (2009) reported that evictions from their own homes was the main pathway into homelessness for older homeless people in the Netherlands. Regarding renters, eviction for rental arrears is one of the five main reasons for becoming homeless in later life in the United Kingdom (Crane et al. 2006) and Australia (Rota-Bartelink and Lipmann 2007).[6] For the United States of America, there is empirical evidence that eviction rates are not associated with the chronological age (Desmond and Gershenson 2017). However, Shinn et al. (2007) reported that half of seventy-nine people aged 55 or over in New York City lived 'conventional lives'—that is, long periods of employment and residential stability—before becoming homeless.

Older homeless people, like people experiencing deprivation and poverty, are usually hard to reach; homelessness in later life is also hard to 'see' institutionally (Loison-Leruste et al. 2015). A French study on homeless people aged 50–64, poignantly but accurately, defined this group as 'too young for the streets, too old for a retirement home'.[7] (Rouay-Lambert 2006).

7.4 Housing and Pensions

I mentioned above that housing is considered by some authors as the 'fourth pillar' of welfare systems, based on the possibility that older home owners may use housing equity as a pension. Housing equity can be turned into income in retirement by moving down to a cheaper property, by selling and moving into rented accommodation, and by buying an equity release product. This welfare role of housing income might reduce direct public spending on older people—this is the hypothesis firstly put forth by Kemeny (1981). Apart from this link between housing and pensions, there is also a possible flip side: people who would like to retire but cannot because they are still paying off a mortgage on their houses. This section discusses both economic situations.

7.4.1 Housing Income and Public Spending on Older People

The so-called asset-based welfare—which is 'mostly relevant to the elderly, as they tend to be income-poor but (housing) asset-rich' (Delfani et al. 2014, p. 658)—is a poverty eradication policy based on the premise that it would be better for the poor to accumulate assets than to receive income support:

> The idea of investment as superior to consumption is a principle long recognized in business and individual financial affairs, but strangely, it has not been recognized in welfare policy. Welfare policy for the poor has impeded rather than promoted investment. Income-based policy has sustained the economically weak, but it has not helped strengthen them. Therefore, a new welfare policy is required, a policy based on asset accumulation.
>
> (Sherraden 1991, p. 190)

One of the main types of assets whose accumulation is targeted by this policy approach is housing. This active policy purposefully seeks that people on low incomes increase their asset holdings—not only in the shape of housing, but also financial, health, and education. However, there are a number of issues of interest to us here:

- Given that the accumulation phase of the life cycle takes places at younger chronological ages, asset-based anti-poverty measures do not tackle current and even medium-term poverty in later life but aim at the longer term. Wel-

fare policies focused on older people are weakly based on asset building—as though older people would be 'past' asset accumulation—and more upon income support, and are age-regressive (i.e. with transfers diminishing with chronological age) (Huang 2015).
- When asset-based policies do aim at existing cohorts of older people in poverty, the implementation of policy instruments such as tax exemptions and benefits on home ownership and retirement individual accounts tend to be regressive (Hirayama 2010; Ronald et al. 2017)—an 'asset-building policy for the non-poor, but not for the poor' (Sherraden and Page-Adams 1995, p. 67).
- There may be also be a more tacit, less pro-active, *de facto* welfare policy approach, in which the governments simply recede and reduce welfare spending on older people in view of large home ownership rates, overlooking the regressive consequences of such a decision.

Either as a result of actively promoted policies or not, there might be a trade-off between home ownership and welfare spending on older people. The first studies that found a negative correlation between home ownership rates and public spending on older people—(Castles 1998; Doling and Horsewood 2011; Kemeny 2005)—resorted to cross-sectional observations in mainly developed countries. The conclusion was that housing income and, in particular, pension income would be substitutes.

However, even the strongest of correlations is silent about the direction of causation. In the case of an inverse relationship between home ownership and public spending on older people, is it that higher spending reduces the motive to buy a property earlier in the life cycle as a means of procuring income in retirement, or is it that high rates of home ownership reduce the demand for or the pressure on governments to spend on older people? Doling and Horsewood (2011) investigated the presence of causation beyond the statistical association between home ownership and state pensions using a longitudinal data set for seventeen developed countries between 1980 and 2003. These authors reported a strong correlation, and also that changes in house prices led to changes in home ownership, which eventually influenced the levels of public spending on state pensions: an increase in house prices above their long term was followed by a reduction in public spending between two to three years later. However, Delfani et al. (2014) qualified these results on focusing upon their institutional basis and opined that the link depends on the organisation of both the housing and pension systems in each country. In this respect, Delfani et al. (2015) found that in systems in which home ownership is the most popular tenure type, low-income households exhibit larger home ownership rates than similar households in other housing systems and that the ownership of a home offers the former some protection against deprivation in later life. As Lux and Sunega (2014, p. 51) concluded:

> Some practices effectively implemented in one environment (a social-democratic or a social-market housing system) cannot necessarily function effectively in another environment (a liberal housing system).

7.4.2 Housing and the Retirement Decision

Bloemen (2011) reported a significant association between having an outstanding mortgage and remaining in the labour market instead of retiring among a sample of men aged 48–64 years between 1995 and 2001 in the Netherlands. Mortgage debt is also a significant factor among retirees in the decision to return to work (Lahey et al. 2006). Furthermore, Mann (2011) looked at household debt in general, not only mortgage indebtedness, and found a significant relationship between debt accumulation and labour market participation in later life.

Given the expectation that housing wealth may become a source of income in later life, changes in housing prices could have a bearing on retirement decisions. This hypothesis was confirmed in studies on data from the United States of America by Farnham and Sevak (2015) looking into data for 1992–2004, Szinovacz et al. (2013) using data between 2006 and 2008 and Zhao (2018) with data for 2007–2009. Furthermore, some studies identified a gender-specific effect of changes in housing wealth on retirement: using 2011 data for China in 2011, Fu et al. (2016) reported that increases in house prices tended to reduce female labour force participation but had no impact on males. Also, Ondrich and Falevich (2016) studied data for the United States of America between 1992 and 2010 and found that declines in house prices as a result of the 2008 financial crisis reduced the probability of retiring only among men (by a sizeable proportion, between 14 and 17 per cent).

Nevertheless, a number of studies failed to find any significant or reported merely minor effects of house prices on retirement (Gorodnichenko et al. 2013; Skinner 1996).

Review and Reflect

1. The drop in consumption levels after retirement can be seen as part of a life-cycle consumption pattern or as a separate economic phenomenon. Discuss in this context the following assertion:

   > ...there is little reason to believe that the mechanisms that are most pertinent to resolving those puzzles are the mechanisms most pertinent to accounting for the consumption hump.
   >
   > *(Bullard and Feigenbaum 2007, p. 2308)*

2. Consider the following comments by Gotman in relation to barriers to purchase reverse mortgages:

   > "Deeply held beliefs" to combat include: housing wealth not regarded as a fungible asset; a desire to leave a bequest; and, saving home equity as insurance for emergencies. Those beliefs, supposedly rooted in a fear of impoverishment embodied in the Great Depression experience, and ascribed to what is called the "depression era mentality", are regarded as a generational phenomenon. In the United States as well as in other countries, obstacles to the extension of the reverse mortgage market are similarly ascribed to the current older generation's way of life and its anchored frugality. Consequently, educational efforts are pointed towards younger generations, among whom consumption credit is assumed to be already taken for granted, and who are ready for a "paper wealth" economy.
   >
   > *(Gotman 2011, p. 98)*

   The 'educational efforts' mentioned by Gotman can be construed as libertarian paternalism—see Chap. 9 in Volume IV. Provide examples of policy initiatives to increase the demand for reverse mortgages, and evaluate both their possible effectiveness and ethical stance.

   To what extent would the 2008 crisis act as another 'generational phenomenon' among younger cohorts?

3. Referring to the prospects in the United States of America, Angelini et al. (2011, p. 88) warned that 'In old age the location choices of some baby-boomers might become an issue if services are located in more densely located areas.'
   Comment.

4. Reflect on the following assertion:

   > ...politicians have been deeply aware of the electoral appeal of extending the reach of home ownership and know very well that much of that appeal lies in catering to the motive for the life cycle asset accumulation.
   >
   > *(Castles 1998, p. 7)*

(continued)

5. Table 7.1 lists two classes of housing income, in kind and in cash. Regarding this distinction, discuss the following quote:

> The identification of the significance of the incomes in kind and in cash that are derivable from a dwelling owned outright by the occupier is critical to the analysis ...it is not tenure itself that is important but rather the financial characteristics and performance of owner-occupied housing ...Imagine a society in which over an extended period, say two or three decades, house prices fell in real terms and generally performed notably less well than comparable forms of investment, such as shares. In such circumstances the income in kind which older people enjoyed by virtue of owning their home outright would be low, because the capital value and thus notional rental value would be low. For the same reason, the income in cash that could possibly be realised would also be low. In this respect, what is crucial therefore is not the home itself, nor the form of tenure, nor even its ability to distribute across the life cycle, but the expectation and the reality that the distribution has been effective in terms of the amount of income invested in early years in the life cycle and the rate of return by later years.
> (Doling and Horsewood 2011, pp. 169–170)

6. Consider the following quotation related to asset-based welfare policy:

> ...people think and behave differently when they are accumulating assets, and the world responds to them differently as well. More specifically, assets improve economic stability; connect people with a viable, hopeful future; stimulate development of human and other capital; enable people to focus and specialize; provide a foundation for risk taking; yield personal, social, and political dividends; and enhance the welfare of offspring.
> (Sherraden 1991, p. 148)

   Do you agree? Specifically, do you agree that *older* people think and behave differently when they are accumulating assets?

7. Provide reasons in support of the following statement:

> The housing system itself may...become an important barrier to the successful implementation of policy promoting ageing in place.
> (Lux and Sunega 2014, p. 51)

8. Comment on the following application of a critical approach to the 'consumer society':

> The same consumerist pressures that associate the idea of "care" with an inventory of consumer commodities such as "orange juice, milk, frozen

(continued)

> *pizza and microwave ovens" strip the families of their social-ethical skills and resources, and disarm them in their uphill struggle to cope with the new challenges; challenges aided and abetted by the legislators, who attempt to reduce state financial deficits through the expansion of the "care deficit" ("cutting funds for single mothers, the disabled, the mentally ill, and the elderly").*
>
> *(Bauman 2007, p. 140)*

9. Do you agree with this quote?

   > *That we are what we have ...is perhaps the most basic and powerful fact of consumer behavior.*
   >
   > *(Belk 1988, p. 139)*

   The assertion is included in one of the most widely cited papers in consumer research, which is hardly cited in economics. Why do you think this could be the case?
10. Comment:

    > *...retirement has become a time of growth when identity is broadened, expressed, and completed through consumption as individuals devote time and resources to contemplate and reorient their place in the world after rearing children and exiting primary jobs. These forays into self-development in retirement are facilitated by a cultural context in which identity experimentation is increasingly acceptable and common and in which there is a plentitude of attractive consumption options. In a postmodern world, everyone, including retirees, can make and remake their identities over the courses of their lifetimes and choose to what degree the new identities are consistent with the old.*
    >
    > *(Schau et al. 2009, p. 256)*

Notes

1. *Source*: own estimation using data from 'Household Income and Wealth, Australia, 2015–16', Australian Bureau of Statistics.
2. Re-estimated after omitting pension accounts.
3. Source: https://www.bls.gov/cpi/.
4. See also Feinstein and McFadden (1989) and Megbolugbe et al. (1997).
5. For a detailed description of alternative housing equity release products, see Reifner et al. (2009).

6. The other causes are bereavement, health problems, disputes with neighbours and co-tenants, and a relationship breakdown—and, in Australia, gambling-related debt.
7. 'Trop vieux pour la rue, trop jeunes pour la maison de retraite'.

References

Abramsson, Marianne and Eva Andersson (2016). "Changing preferences with ageing - Housing choices and housing plans of older people". In: *Housing, Theory and Society* 33.2, pages 217–241.

Ær, Thorkild (2006). "Residential choice from a lifestyle perspective". In: *Housing, Theory and Society* 23.2, pages 109–130.

Agarwal, Sumit et al. (2006). "An empirical analysis of home equity loan and line performance". In: *Journal of Financial Intermediation* 15.4, pages 444–469.

Ahrend, Rudiger, Boris Cournède, and Robert Price (2008). "Monetary Policy, Market Excesses and Financial Turmoil". In: *OECD Economic Department Working Papers* 597. URL: http://dx.doi.org/10.1787/244200148201.

Andersen, Hans Skifter (2011). "Motives for tenure choice during the life cycle: the importance of non-economic factors and other housing preferences". In: *Housing, Theory and Society* 28.2, pages 183–207.

Andersson, Eva K, Marianne Abramsson, and Bo Malmberg (2018). "Patterns of changing residential preferences during late adulthood". In: *Ageing & Society* pages 1–30.

Andrews, Doug and Jaideep Oberoi (2015). "Home equity release for long-term care financing: an improved market structure and pricing approach". In: *Annals of Actuarial Science* 9.1, pages 85–107.

Angelini, Viola, Agar Brugiavini, and Guglielmo Weber (2011). "Does downsizing of housing equity alleviate financial distress in old age?" In: *The Individual and the Welfare State* Edited by Axel Börsch-Supan et al. Heidelberg: Germany: Springer, pages 93–101.

Angelini, Viola, Anne Laferrère, and Guglielmo Weber (2011). "Homeownership in old age at the crossroad between personal and national histories". In: *The Individual and the Welfare State* Edited by Axel Börsch-Supan et al. Heidelberg: Germany: Springer, pages 81–92.

Attanasio, Orazio P et al. (2012). "Modelling the demand for housing over the life cycle". In: *Review of Economic Dynamics* 15.1, pages 1–18.

Bank, European Central (2018). *Households and non-financial corporations in the euro area: third quarter of 2017 Press Release 12 January 2018* Frankfurt am Main: Germany.

Barry Arro et al. (2018). "The meaning of home for ageing women living alone: An evolutionary concept analysis". In: *Health & g care in the Community* 26.3, e337–e344.

Bauman, Zygmunt (2007). *Consuming Life* Cambridge: United Kingdom: Polity Press.

Belk, Russell W (1988). "Possessions and the extended self". In: *Journal of Consumer Research* 15.2, pages 139–168.

Beltratti, Andrea and Claudio Morana (2010). "International house prices and macroeconomic fluctuations". In: *Journal of Banking & Finance* 34, pages 533–545.

Bloemen, Hans G (2011). "The effect of private wealth on the retirement rate: an empirical analysis". In: *Economica* 78.312, pages 637–655.

Boyle, Alexandra, Janine L Wiles, and Robin A Kearns (2015). "Rethinking ageing in place: the 'people' and 'place' nexus". In: *Progress in Geography* 34.12, pages 1495–1511.

Bradbury Bruce (2013). "Income Inequality: Economic Disparities and the Middle Class in Affluent Countries". In: edited by Janet Gornick and Markus Jäntti. Studies in Social Inequality Stanford, CA: United States of America: Stanford University Press, pages 334–361.

Brown, Rebecca T et al. (2016). "Pathways to homelessness among older homeless adults: results from the HOPE HOME study". In: *PloS one* 11.5. URL: https://doi.org/10.1371/journal.pone.0155065.

Brownfield, Christine (2014). "The fourth pillar the role of home equity release in retirement funding". In: *Actuaries Institute Financial Services Forum, Scoring Goals in a Changing World. 5–6 May 2014, Sydney: Australia*.

Bullard, James and James Feigenbaum (2007). "A leisurely reading of the life-cycle consumption data". In: *Journal of Monetary Economics* 54.8, pages 2305–2320.

Castles, Francis G (1998). "The really big trade-off: home ownership and the welfare state in the New World and the Old". In: *Acta Politica* 33.1, pages 5–19.

Charupat, Narat, Huaxiong Huang, and Moshe Milevsky (2012). *Strategic Financial Planning Over the Lifecycle: A Conceptual Approach to Personal Risk Management* New York, NY: United States of America: Cambridge University Press.

Chiang, Shu Ling and Ming Shann Tsai (2016). "Analyzing an elders desire for a reverse mortgage using an economic model that considers house bequest motivation, random death time and stochastic house price". In: *International Review of Economics & Finance* 42, pages 202–219.

Chou, Kee-Lee, Nelson WS Chow, and Iris Chi (2006). "Willingness to consider applying for reverse mortgage in Hong Kong Chinese middle-aged homeowners". In: *Habitat International* 30.3, pages 716–727.

Clark, William AV and Jun L Onaka (1983). "Life cycle and housing adjustment as explanations of residential mobility". In: *Urban studies* 20.1, pages 47–57.

Costa-Font, Joan, Joan Gil, and Oscar Mascarilla (2010). "Housing wealth and housing decisions in old age: sale and reversion". In: *Housing studies* 25.3, pages 375–395.

Crane, Maureen, Anthony M Warnes, and Ruby Fu (2006). "Developing homelessness prevention practice: combining research evidence and professional knowledge". In: *Health & social care in the community* 14.2, pages 156–166.

Crane, Maureen et al. (2005). "The causes of homelessness in later life: findings from a 3-nation study". In: *The Journals of Gerontology Series B: Psychological Sciences and Social Sciences* 60.3, S152–S159.

Culhane, Dennis P, Stephen Metraux, and Jay Bainbridge (2010). *The age structure of contemporary homelessness: Risk period or cohort effect?* Working Paper. Philadelphia, PA: United States of America, pages 1–28.

Davidoff, Thomas (2010). "Home equity commitment and long-term care insurance demand". In: *Journal of Public Economics* 94.1, pages 44–49.

Davidoff, Thomas, Patrick Gerhard, and Thomas Post (2017). "Reverse mortgages: What homeowners (don't) know and how it matters". In: *Journal of Economic Behavior & Organization* 133, pages 151–171.

De Deken, Johan, Neda Delfani, and Caroline Dewilde (2012). "Relation entre retraite et propriété immobilière depuis 1990" In: *Retraite et société* 1, pages 33–57.

Delfani, Neda, Johan De Deken, and Caroline Dewilde (2014). "Home-Ownership and Pensions: Negative Correlation, but no Trade-off". In: *Housing Studies* 29.5, pages 657–676.

—— (2015). "Poor because of low pensions or expensive housing? The combined impact of pension and housing systems on poverty among the elderly". In: *International Journal of Housing Policy* 15.3, pages 260–284.

Desmond, Matthew and Carl Gershenson (2017). "Who gets evicted? Assessing individual, neighborhood, and network factors". In: *Social science research* 62, pages 362–377.

Doling, John (1976). "The family life cycle and housing choice". In: *Urban Studies* 13.1, pages 55–58.

Doling, John and Marja Elsinga (2012). "Housing as income in old age". In: *International Journal of Housing Policy* 12.1, pages 13–26.

—— (2013). *Demographic Change and Housing Wealth:: Home-owners, Pensions and Asset-based Welfare in Europe* Dordrecht: The Netherlands: Springer.

Doling, John and Nick Horsewood (2011). "Home Ownership and Pensions: Causality and the Really Big Trade-off". In: *Housing, Theory and Society* 28.2, pages 166–182.

Duca, John V, John Muellbauer, and Anthony Murphy (2010). "Housing markets and the financial crisis of 2007–2009: lessons for the future". In: *Journal of financial stability* 6.4, pages 203–217.

Esping-Andersen, Gøsta (1990). *The Three Worlds of Welfare Capitalism* Cambridge: United Kingdom: Polity Press.

Farnham, Martin and Purvi Sevak (2015). "Housing Wealth and Retirement Timing". In: *CESifo Economic Studies* 62.1, pages 26–46.

Feinstein, Jonathan and Daniel McFadden (1989). "The dynamics of housing demand by the elderly: Wealth, cash flow, and demographic effects". In: *The Economics of Aging* Edited by David A Wise. Chicago, IL: United States of America: University of Chicago Press, pages 55–68.
Fernandez-Carro, Celia (2016). "Ageing at home, co-residence or institutionalisation? Preferred care and residential arrangements of older adults in Spain". In: *Ageing & Society* 36.3, pages 586–612.
Fokkema, Tineke, Jenny Gierveld, and Peter Nijkamp (1996). "Big cities, big problems: Reason for the elderly to move?" In: *Urban Studies* 33.2, pages 353–377.
Fornero, Elsa, Mariacristina Rossi, and Maria Cesira Urzí Brancati (2016). "Explaining why right or wrong: (Italian) Households do not like reverse mortgages". In: *Journal of Pension Economics & Finance* 15, pages 180–202.
Frank, Jacquelyn Beth (2002). *The Paradox of Aging in Place in Assisted Living*. Westport, CT: United States of America: Bergin & Garvey.
Fu, Shihe, Yu Liao, and Junfu Zhang (2016). "The effect of housing wealth on labor force participation: Evidence from China". In: *Journal of Housing Economics* 33, pages 59–69.
Golant, Stephen M (2008). "Commentary: Irrational exuberance for the aging in place of vulnerable low-income older homeowners". In: *Journal of Aging & Social Policy* 20.4, pages 379–397.
Gorodnichenko, Yuriy Jae Song, and Dmitriy Stolyarov (2013). *Macroeconomic determinants of retirement timing* NBER Working Paper 19638. Cambridge, MA: United States of America: National Bureau of Economic Research.
Gotman, Anne (2011). "Towards the end of bequest? The life cycle hypothesis sold to seniors". In: *Civitas-Revista de Ciências Sociais* 11.1, pages 93–114.
Granbom, Marianne et al. (2014). "Residential normalcy and environmental experiences of very old people: changes in residential reasoning over time". In: *Journal of Aging Studies* 29, pages 9–19.
Graybill, Erin M, Peter McMeekin, and John Wildman (2014). "Can aging in place be cost effective? A systematic review". In: *PloS One* 9.7. doi: doi.org/10.1371/journal.pone.0102705.
Green, R. and P.H Hendershott (1996). "Age, housing demand, and real house prices". In: *Regional Science and Urban Economics* 26, pages 465–480.
Green, Richard K and Hyojung Lee (2016). "Age, demographics, and the demand for housing, revisited". In: *Regional Science and Urban Economics* 61, pages 86–98.
Hancock, Ruth (1998). "Can housing wealth alleviate poverty among Britain's older population?" In: *Fiscal Studies* 19.3, pages 249–272.
Hanewald, Katja, Thomas Post, and Michael Sherris (2016). "Portfolio Choice in Retirement—What is The Optimal Home Equity Release Product?" In: *Journal of Risk and Insurance* 83.2, pages 421–446.
Hansen, Eigil Boll and Georg Gottschalk (2006). "What makes older people consider moving house and what makes them move?" In: *Housing, Theory and Society* 23.01, pages 34–54.

Hasu, Eija (2018). "Housing decision-making process explained by third agers, Finland: "we didn't want this, but we chose it" " In: *Housing Studies* pages 1–18.

Haurin, Donald and Stephanie Moulton (2017). "International perspectives on homeownership and home equity extraction by senior households". In: *Journal of European Real Estate Research* 10.3, pages 245–276.

Henderson, J Vernon and Yannis M Ioannides (1983). "A model of housing tenure choice". In: *The American Economic Review* 73.1, pages 98–113.

Heo, Yong-Chang, Seungjae An, and Baeg Eui Hong (2016). "Reverse Mortgage as an Income Stabilizer for the Elderly in Korea". In: *Asian Social Work and Policy Review* 10.1, pages 103–112.

Hillcoat-Nalletamby Sarah and Jim Ogg (2014). "Moving beyond 'ageing in place': older people's dislikes about their home and neighbourhood environments as a motive for wishing to move". In: *Ageing & Society* 34.10, pages 1771–1796.

Hiller, Norbert and Oliver W Lerbs (2016). "Aging and urban house prices". In: *Regional Science and Urban Economics* 60, pages 276–291.

Hirayama, Yosuke (2010). "The role of home ownership in Japan's aged society". In: *Journal of Housing and the Built Environment* 25.2, pages 175–191.

Huang Jin and Greenfield, Jennifer (2015). "Asset Development among Older Adults. A Capability Approach". In: *Financial capability and asset holding in later life: A life course perspective* Edited by Margaret S Sherraden and Nancy Morrow-Howell. New York, NY: United States of America: Oxford University Press, pages 139–160.

IBISWorld (2017). *Reverse Mortgage Providers in Australia Industry Market Research Report* Technical report. Melbourne: Australia.

Jäger, Philipp and Torsten Schmidt (2017). "Demographic change and house prices: Headwind or tailwind?" In: *Economics Letters* 160, pages 82–85.

Jansen, Sylvia JT (2014). "Different values, different housing? Can underlying value orientations predict residential preference and choice?" In: *Housing, Theory and Society* 31.3, pages 254–276.

Kaminsky Graciela L and Carmen M Reinhart (1999). "The twin crises: the causes of banking and balance-of payments problems". In: *American economic review* 89.3, pages 473–500.

Kemeny Jim (1981). *The Myth of Home Ownership* London: United Kingdom: Routledge and Kegan Paul.

—— (1992). *Housing and Social Theory* London: United Kingdom: Routledge.

—— (1995). *From Public Housing to the Social Market. Rental policy strategies in comparative perspective* London: United Kingdom: Routledge.

—— (2001). "Comparative housing and welfare: Theorising the relationship". In: *Journal of Housing and the Built Environment* 16.1, pages 53–70.

—— (2005). " "The Really Big Trade-Off". between Home Ownership and Welfare: Castles' Evaluation of the 1980 thesis, and a Reformulation 25 Years on". In: *Housing, Theory and Society* 22.2, pages 59–75.

Kutty Nandinee K (1998). "The scope for poverty alleviation among elderly homeowners in the United States through reverse mortgages". In: *Urban studies* 35.1, pages 113–129.

Laere, Igor R van, Matty A de Wit, and Niek S Klazinga (2009). "Pathways into homelessness: recently homeless adults problems and service use before and after becoming homeless in Amsterdam". In: *BMC Public Health* 9.3. URL: https://doi.org/10.1186/1471-2458-9-3.

Lahey Karen Eilers, Doseong Kim, and Melinda L Newman (2006). "Full retirement? An examination of factors that influence the decision to return to work". In: *Financial Services Review* 15.1, pages 1–19.

Leamer, Edward E (2007). *Housing is the business cycle* NBER Working Paper 13428. Cambridge, MA: United States of America.

Lee, Christopher Thomas et al. (2016). "Residential patterns in older homeless adults: Results of a cluster analysis". In: *Social Science & Medicine* 153, pages 131–140.

Leviton, Roberta (2002). "Reverse Mortgage Decision-Making". In: *Journal of Aging & Social Policy* 13.4, pages 1–16.

Li, Qiang and Satish Chand (2018). "Unlocking Housing Equity for Pensions in Urban China". In: *Journal of Housing For the Elderly* pages 1–15.

Lisack, Noëmie, Rana Sajedi, and Gregory Thwaites (2017). *Demographic trends and the real interest rate* Staff Working Paper 701. London: United Kingdom: Bank of England.

Loison-Leruste, Marie, Marion Arnaud, and Benoît Roullin (2015). *Les personnes de 50 ans ou plus utilisant des services d'hébergement et de distribution de repas pour sans-domicile. Etude pour l'Observatoire national de la pauvreté et de l'exclusion sociale* Technical report. Paris: France.

Lowe, Stuart G, Beverley A Searle, and Susan J Smith (2012). "From housing wealth to mortgage debt: the emergence of Britain's asset-shaped welfare state". In: *Social Policy and Society* 11.1, pages 105–116.

Lux, Martin and Petr Sunega (2014). "The impact of housing tenure in supporting ageing in place: exploring the links between housing systems and housing options for the elderly". In: *International Journal of Housing Policy* 14.1, pages 30–55.

Mankiw, N Gregory and David N Weil (1989). "The baby boom, the baby bust, and the housing market". In: *Regional Science and Urban Economics* 19, pages 235–258.

Mann, Allison (2011). "The effect of late-life debt use on retirement decisions". In: *Social Science Research* 40.6, pages 1623–1637.

Mayer, Christopher and Katerina Simons (1994). "Reverse mortgages and the liquidity of housing wealth". In: *Journal of the American Real Estate and Urban Economics Association* 22.2, pages 235–255.

Megbolugbe, Issac, Jarjisu Sa-Aadu, and James Shilling (1997). "Oh, yes, the elderly will reduce housing equity under the right circumstances". In: *Journal of Housing Research* 8.1, pages 53–74.

Merrill, Sally R, Meryl Finkel, and Nadine Kutty (1994). "Potential beneficiaries from reverse mortgage products for elderly homeowners: An analysis of AHS data". In:

Journal of the American Real Estate and Urban Economics Association 22.2, pages 257–299.

Milligan, Christine (2009). *There's No Place Like Home: Place and Care in an Ageing Society* Geographies of Health. Farnham: United Kingdom: Ashgate Publishing Limited.

Morris, Alan (2016). *The Australian Dream: Housing Experiences of Older Australians* Clayton: Australia: CSIRO Publishing.

Moscarola, Flavia Coda et al. (2015). "Reverse mortgage: a tool to reduce old age poverty without sacrificing social inclusion". In: *Ageing in Europe Supporting Policies for an Inclusive Society* Edited by Axel Börsch-Supan et al. Berlin: Germany: De Gruyter, pages 235–244.

Moulton, Stephanie, Donald R Haurin, and Wei Shi (2015). "An analysis of default risk in the Home Equity Conversion Mortgage (HECM) program". In: *Journal of Urban Economics* 90, pages 17–34.

Nakajima, Makoto and Irina A Telyukova (2017). "Reverse mortgage loans: A quantitative analysis". In: *The Journal of Finance* 72.2, pages 911–950.

National Statistics, Office for (2015). *Compendium Wealth in Great Britain Wave 4: 2012 to 2014. Main results from the fourth wave of the Wealth and Assets Survey covering the period July 2012 to June 2014* Newport: United Kingdom.

Nishimura, Kiyohiko G (2011). *This time may truly be different: Balance sheet adjustment under population ageing* Speech Prepared for the Panel "The Future of Monetary Policy". at the 2011 American Economic Association Annual Meeting in Denver. Tokyo: Japan.

OECD—Social Policy Division—Directorate of Employment, Labour and Social Affairs (2017). *HC3.1 Homeless Population* Technical report. Paris: France.

Ondrich, Jan and Alexander Falevich (2016). "The Great Recession, housing wealth, and the retirement decisions of older workers". In: *Public Finance Review* 44.1, pages 109–131.

Ong, Rachel (2008). "Unlocking housing equity through reverse mortgages: The case of elderly homeowners in Australia". In: *International Journal of Housing Policy* 8.1, pages 61–79.

Park, Soonyoun et al. (2017). "The dynamic effect of population ageing on house prices: evidence from Korea". In: *Pacific Rim Property Research Journal* 23.2, pages 195–212.

Pastalan, Leon A, editor (1990). *Aging in place: The role of housing and social supports.* New York, NY: United States of America: The Haworth Press.

Pitkin, J.R. and D Myers (1994). "The specification of demographic effects on housing demand: avoiding the age-cohort fallacy". In: *Journal of Housing Economics* 3, pages 240–250.

Reifner, Udo et al. (2009). *Study on Equity Release Schemes in the EU, Part 1: General Report* Technical report. Hamburg: Germany.

Reinhart, Carmen M and Kenneth S Rogoff (2008). "Is the 2007 US sub-prime financial crisis so different? An international historical comparison". In: *American Economic Review* 98.2, pages 339–44.

Ronald, Richard, Christian Lennartz, and Justin Kadi (2017). "What ever happened to asset-based welfare? Shifting approaches to housing wealth and welfare security". In: *Policy & Politics* 45.2, pages 173–193.

Rossi, Peter Henry (1955). *Why families move: a study in the social psychology of urban residential mobility* Glenceo, IL: United States of America: The Free Press.

Rota-Bartelink, Alice and Bryan Lipmann (2007). "Causes of homelessness among older people in Melbourne, Australia". In: *Australian and New Zealand Journal of Public Health* 31.3, pages 252–258.

Rouay-Lambert, Sophie (2006). "La retraite des anciens SDF. Trop vieux pour la rue, trop jeunes pour la maison de retraite". In: *Les Annales de la recherche urbaine*. Volume 100. 1, pages 136–143.

Roy Noémie et al. (2018). "Choosing between staying at home or moving: A systematic review of factors influencing housing decisions among frail older adults". In: *PloS one* 13.1. URL: https://doi.org/10.1371/journal.pone.0189266.

Scharlach, Andrew (2017). "Aging in Context: Individual and Environmental Pathways to Aging-Friendly Communities The 2015 Matthew A. Pollack Award Lecture". In: *The Gerontologist* 57.4, pages 606–618.

Schau, Hope Jensen, Mary C Gilly and Mary Wolfinbarger (2009). "Consumer identity renaissance: the resurgence of identity-inspired consumption in retirement". In: *Journal of Consumer Research* 36.2, pages 255–276.

Severinsen, Christina, Mary Breheny and Christine Stephens (2016). "Ageing in unsuitable places". In: *Housing Studies* 31.6, pages 714–728.

Sherraden, Michael (1991). *Assets and the Poor: A New American Welfare Policy* New York, NY: United States of America: Routledge.

Sherraden, Michael and Deborah Page-Adams (1995). "Asset-based Alternatives in Social Policy". In: *Increasing Understanding of Public Problems and Policies: Proceedings of the 1995 National Public Policy Education Conference* Oak Brook, IL: United States of America: Farm Foundation, pages 65–83.

Shinn, Marybeth et al. (2007). "Predictors of homelessness among older adults in New York City: Disability economic, human and social capital and stressful events". In: *Journal of Health Psychology* 12.5, pages 696–708.

Skinner, Jonathan (1996). "Is Housing Wealth a Sideshow?" In: *Advances in the Economics of Aging* Edited by David A Wise. Chicago, IL: United States of America: University of Chicago Press, pages 241–272.

Smith, Richard J, Amanda J Lehning, and Kyeongmo Kim (2017). "Aging in place in gentrifying neighborhoods: Implications for physical and mental health". In: *The Gerontologist* 58.1, pages 26–35.

Smith, Susan J and Beverly A Searle (2010). *The Blackwell Companion to the Economics of Housing: The Housing Wealth of Nations* Blackwell Companions to Contemporary Economics. Chichester: United Kingdom: Wiley-Blackwell.

Smith, Susan J, Beverly A Searle, and Gareth Powells (2010). "Introduction". In: *The Blackwell Companion to the Economics of Housing: The Housing Wealth of Nations* Edited by Susan J Smith and Beverly A Searle. Chichester: United Kingdom: John Wiley & Sons, pages 1–27.

Solari, Claudia D et al. (2017). *The 2016 Annual Homeless Assessment Report (AHAR) to Congress Part 2: Estimates of Homelessness in the United States* Technical report. Washington, DC: United States of America.

Spanko, Alex (2018). *Seniors' Home Equity Grows by $149B in Fourth Quarter March 27th, 2018.* URL: https://reversemortgagedaily.com/2018/03/27/seniors-home-equity-grows-by-149b-in-fourth-quarter/.

Speare, Alden Jr (1974). "Residential satisfaction as an intervening variable in residential mobility". In: *Demography* 11.2, pages 173–188.

Stucki, Barbara R (2006). "Using reverse mortgages to manage the financial risk of long-term care". In: *North American Actuarial Journal* 10.4, pages 90–102.

Szinovacz, Maximiliane E, Lauren Martin, and Adam Davey (2013). "Recession and expected retirement age: Another look at the evidence". In: *The Gerontologist* 54.2, pages 245–257.

Takáts, Eld (2012). "Aging and house prices". In: *Journal of Housing Economics* 21.2, pages 131–141.

Tang, Fengyan and Yeonjung Lee (2011). "Social support networks and expectations for aging in place and moving". In: *Research on Aging* 33.4, pages 444–464.

Taylor, John B (2009). "The financial crisis and the policy responses: An empirical analysis of what went wrong". In: *A Festschrift in Honour of David Dodge's Contributions to Canadian Public Policy: Proceedings of a Conference Held by the Bank of Canada, November 2008* Edited by Bank of Canada. Ottawa: Canada, pages 1–18.

Toussaint, Janneke (2013). "Mortgage-equity release: the potential of housing wealth for future Dutch retirees". In: *Journal of Housing and the Built Environment* 28.2, pages 205–220.

Vasara, Paula (2015). "Not ageing in place: Negotiating meanings of residency in age-related housing". In: *Journal of aging studies* 35, pages 55–64.

Venti, Steven F and David A Wise (1990a). "But they don't want to reduce housing equity". In: *Issues in the Economics of Aging* Edited by David A Wise. Chicago, IL: United States of America: University of Chicago Press, pages 13–29.

— (1991). "Aging and the income value of housing wealth". In: *Journal of Public Economics* 44.3, pages 371–397.

— (2004). "Aging and housing equity: Another look". In: *Perspectives on the Economics of Aging* Edited by David A Wise. Chicago, IL: United States of America: University of Chicago Press, pages 127–180.

Weil, Joyce and Elizabeth Smith (2016). "Revaluating aging in place: from traditional definitions to the continuum of care". In: *Working With Older People* 20.4, pages 223–230.

WHO (2015). *World Report on Ageing and Health* Geneva: Switzerland: World Health Organisation.

Wiles, Janine L et al. (2012). "The Meaning of "Aging in Place". to Older People". In: *The Gerontologist* 52.3, pages 357–366.

Williams, Alex O (1987). *Managing Risk in Mortgage Portfolios* Westport: CT: United States of America: Quorum Books.

Wind, Barend, Philipp Lersch, and Caroline Dewilde (2017). "The distribution of housing wealth in 16 European countries: Accounting for institutional differences". In: *Journal of Housing and the Built Environment* 32.4, pages 625–647.

Wolff, Edward (2017). *Deconstructing Household Wealth Trends in the United States, 1983 to 2016. For Presentation at the First WID World Conference* Paris: France.

Yang, Fang (2009). "Consumption over the life cycle: How different is housing?" In: *Review of Economic Dynamics* 12, pages 423–443.

Zhao, Bo (2018). "Too poor to retire? Housing prices and retirement". In: *Review of Economic Dynamics* 27, pages 27–47.

Part III

Behavioural Economics and Ageing

8

Behavioural Economics and Individual Ageing

> **Overview**
> This chapter acts as a primer on behavioural economics and ageing: it introduces the basic concepts studied in behavioural economics, including the anomalies and departures from the rational choice framework, along with a discussion of the main empirical findings.

Behavioural economics is a sub-discipline of economics that investigates the implications of the incorporation of advances into economic reasoning which clash with basic assumptions upon which mainstream economics models are built. Mainstream economics assumes that agents are rational (e.g. that they always maximise their benefit and minimise their costs); behavioural economics draws upon findings from, especially, cognitive and social psychology on how people make decisions—and it has mounted experimental evidence that shows that agents are less rational than they are assumed to be—for a comparison between behavioural and mainstream economics, see Tomer (2007). In a nutshell,

> …behavioral economists tend to emphasize experimental evidence, validation of modeling assumptions, synergies between psychology and economics, and skepticism regarding strong rationality assumptions.
>
> (Laibson 1998, p. 107).

One gateway into the vast modern[1] behavioural economics literature is to present a list of 'anomalies' or departures from the mainstream assumptions and discuss their implications for well-known models and frameworks and the policies based thereupon.[2] An anomaly, in this context, is an empirical finding that requires the adoption of 'implausible assumptions' to accommodate it or to provide an explanation (Thaler 1988). Most of the anomalies stem from cognitive heuristics or short-cuts people resort to when making decisions. (Formally, Lemaire (2016, p. 24) defined a heuristic as 'a non-systematic rule used to solve a problem, formulate a judgement, or make a decision.') Some of these heuristics are simply spontaneous, intuitive rules of thumbs—'intuitive' in the sense that they directly reflect impressions (Kahneman 2003). Both these simple heuristics as well as the more complex cognitive short-cuts inexorably lead to systematic departures from the predictions based on the assumption of rationality—these deviations from the canonical mainstream rational economic man model are considered 'errors' or 'biases'.

Munro (2009) distinguishes between three types of anomalies: intrinsic, extrinsic, and extreme event anomalies. Intrinsic anomalies are contradictions with tacitly adopted assumptions of the theory of rational choice. Extrinsic anomalies are contradictions with expressed assumptions and stated axioms of the theory of rational choice. Extreme event anomalies are contradictions with the theory of rational choice except in the cases where certain parameters adopt such extreme values that particular behaviour and observations can be accommodated within the rationality framework.

Within behavioural economics (and behavioural sciences in general), a burgeoning research sub-field is the investigation of the differences in those anomalies introduced by chronological age: once a particular departure from rationality is identified and validated in a number of experiments among, say, university students aged under 25 years, this line of research investigates whether the anomaly remains present with no change among older people or whether it changes with chronological age.

For example, in a study of 135 subjects aged between 12 and 90 years, Tymula et al. (2013) reported that chronological age was significantly associated with the probability of making inconsistent choices: the older the individual, the stronger the departure from making consistent (i.e. rational) choices. The study of whether behavioural biases persist unchanged into later life or if not, how they vary, has become a burgeoning field of academic endeavour.

In addition, Agarwal et al. (2009) found a U-shaped relationship between chronological age and errors in financial decisions with younger and older individuals more likely to, for example, incur costs for late payment of credit cards or overdraft fees or face higher interest rates on home equity loans than middle-aged agents. The authors opined that the higher probability of financial mistakes among older people reflects a reduction in analytical functioning due to the decline in cognitive ability with advancing chronological age.

Some older people may rely more on emotions to make decisions and inform their judgements concerning risk than younger adults. In part, this may reflect, according to Finucane (2008), some deterioration in cognitive abilities such as memory or mental processing speed, which advances with chronological age, given that these abilities are associated with deliberative 'System 2' (see below) processes.

However, the literature also reports that whilst older people are more likely to apply heuristic mechanisms to inform their decision-making, they are also able to resort to more complex decision-making strategies if they have enough time to process information and if they have the right motivation.

Besede et al. (2012) analysed the application of four heuristic devices among older people:

- payoff evaluation: the estimation of the probabilities of costs and benefits associated with each alternative
- tallying: a disregard of probabilities when basing decisions (which, for example, leads to placing excessive weight onto events with low probability of occurrence and under-weighing high-probability events)
- lexicographic ordering: choosing the most probable option and, in case of equi-probability, using the second most probable state, and so on
- elimination of dominated options: discarding the least desirable options in a piecemeal fashion

Besede et al. concluded that older people are more likely to use tallying, which is the heuristic mechanism most prone to manipulation of the four devices.

In this chapter, I follow the 'anomalies' expositional approach with particular emphasis on the economic implications brought about by the process of individual ageing. I am not going to review all the biases identified in the literature,[3] but I am only going to focus on the most salient ones in relation to individual ageing. We will encounter a number of additional applications of behavioural economics in the rest of the book.

8.1 Prospect Theory

In mainstream microeconomics, the preferences of the economic agents are axiomatically defined as complete and transitive—see, for example, Mas-Colell et al. (1995), Jehle and Reny (2011), and Varian (2014). Completeness in relation to preferences means that a choice between two alternatives, X and Y, can only result in either one being preferred over the other or in both being equal to each other. Transitivity means that when three alternatives are compared (imagine the third one, apart from X and Y is Z), if a person prefers X over Y and Y over Z, this implies that she prefers X over Z.[4]

Jehle and Reny (2011, p. 6) stated that transitivity is 'perhaps the most controversial of all the axioms describing consumer choices' and Mas-Colell et al. (1995, pp. 6–7) opined that 'substantial portions of economic theory would not survive if economic agents could not be assumed to have transitive preferences' given that the assumption of transitivity 'goes to the heart of the concept of rationality'. Continuing with the metaphor, prospect theory (Kahneman and Tversky 1979) is a dagger that goes through the assumption in the heart of the rationality axiom. Prospect theory, 'the first theory for decision under uncertainty that is both theoretically sound and empirically realistic' (Wakker 2010, p. 7), posits that the utility or value of alternatives is derived from perceived gains and losses instead of the final amounts or assets. The calculated probabilities of each option are replaced in the mental calculations by decision weights, which are based on choices between prospects and which are usually lower than the probabilities of occurrence except for events or choices with low probability that agents tend to overweight.

Imagine your wealth (W) amounts to $1000, and you are given the opportunity to take part in a game that includes two outcomes, x and y: either you have a 50 per cent chance (i.e. $p = 0.5$) of earning $600 (and a 50 per cent chance—$q = 0.5$—of not earning anything) or you get $300 with total certainty (i.e. $p = 1; q = 0$). The canonical expected utility approach declares that any economic agent would be indifferent between these two options. Both prospects can be estimated by means of this equation:

$$EU = p \cdot UW + x + q \cdot U(W + y) \tag{8.1}$$

Or, in general,

$$EU = \sum_{i=-m}^{i=m} p_i \cdot U(W + x_i) \tag{8.2}$$

In the first case, we have $EU = \frac{1}{2} \cdot U1000 + 600 + \frac{1}{2} \cdot U(1000 + 0)$ and in the second, $EU = U1000 + 300$. In both cases we get $EU = U1300$. Similarly if the game included losses as well as, or instead of, earnings. However, what Kahneman and Tversky found was that the prospect of losing a given amount is worth more than the prospect of winning the same amount—i.e. that the detriment in utility of a loss exceeds the increase in utility of gaining the same amount—which is known as loss aversion. Under loss aversion, agents seek risk when facing a loss prospect but exhibit risk aversion for gains. Goods are worth more to economic agents when they face the possibility of losing them or giving them up compared to the prospect of obtaining or gaining them. For example, 80 out of 95 people preferred the option of winning $3000 with certainty than that of winning $4000 with an 80 per cent chance (and a 20 per cent probability of not winning anything), but almost everyone preferred the option of losing $4000 with an 80 per cent probability (and a 20 per cent chance of not losing anything) than losing $3000 for sure. Therefore, these authors proposed a different equation to represent how people actually evaluate prospects under risk, where $v()$ denote the subjective probabilities or decision weights,

$$V = \pi(p) \cdot v(x) + \pi(q) \cdot v(y) \tag{8.3}$$

when either $p + q < 1$, or $x \leq 0 \leq y$ or $x \geq 0 \geq y$. Or, more generally,

$$V = v(y) + \pi(p) \cdot [v(x) - v(y)] \tag{8.4}$$

When $p + q = 1$ and either $x > y > 0$ or $x < y < 0$, Eq. (8.3) becomes:

$$V = \sum_{i=-m}^{i=m} p_i \cdot v(x_i) \tag{8.5}$$

In words,

the value of a strictly positive or strictly negative prospect equals the value of the riskless component plus the value-difference between the outcomes, multiplied by the weight associated with the more extreme outcome.

(Kahneman and Tversky 1979, p. 276)

The novel part in this approach is the assumption that the 'value-difference between the outcomes' is weighted when making the decision, which implies that the amount that is riskless carries no decision weight. In other words, the assumption is that 'values are attached to changes rather than to final states, and that decision weights do not coincide with stated probabilities' (Kahneman and Tversky 1979, p. 277). The importance of changes or differences in amounts rather than of absolute magnitudes is related to findings from the psychology of perception—Goldstein (2001, 2013) and Grondin (2016).

On accepting the Nobel prize for Economics, Kahneman (2003, p. 1455) wrote:

> From the vantage point of a student of perception, it is quite surprising that in standard economic analyses the utility of decision outcomes is assumed to be determined entirely by the final state of endowment, and is therefore reference-independent.

Kahneman found that assumption surprising because by then several experiments had confirmed that 'people often demand much more to give up an object than they would be willing to pay to acquire it' (Kahneman et al. 1991, p. 194). In other words, that economic agents are subject to the endowment effect. The endowment effect is an explanation for the difference between how much agents are willing to pay for goods they do not have and how much they are willing to accept as an offer for what they do have—that is, the 'willingness to pay - willingness to accept' gap (Horowitz and McConnell 2002),[5] although legal ownership is not a necessary condition for the effect to come about: psychological ownership (i.e. anticipating ownership or imagining touching a good, even touching an image of it) can also trigger endowment effects (Morewedge and Giblin 2015; Pierce et al. 2001; Shu and Peck 2011).

Furthermore, the perception and therefore valuation of the monetary changes depend not only on the amount of the variation itself (i.e. whether wealth, say, increases by $100 or $1000) but also of the reference point (i.e. whether the stock of wealth, level of consumption, etc., before the variation amounts to, say, $500 or $5,000,000)—which is known as reference dependence.

Prospect theory also posits that the difference between, say, winning $100 or $200 is perceived as greater than the difference between winning $5100 and $5200—which is known as diminishing sensitivity. Diminishing sensitivity means that above a reference point, the marginal value of increases in monetary values diminishes with their magnitude, and similarly for the marginal value of decreases below the same reference point—see also (Wong and Kwong 2005).

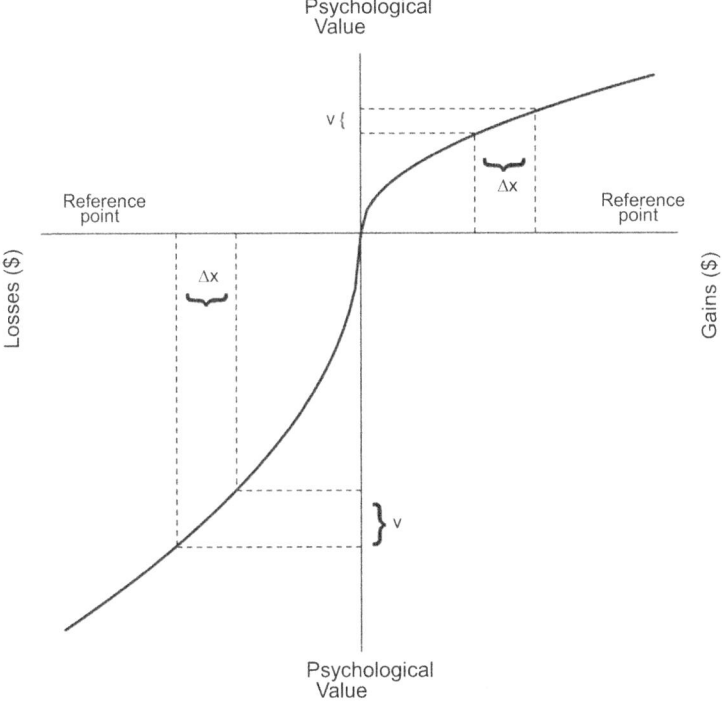

Fig. 8.1 Prospect theory value function. *Source: Figure is illustrative, prepared with mock data*

Figure 8.1 depicts the function of the value of gains above and losses below a reference point (represented by the horizontal axis) with the magnitude of those changes.

The value function has two parameters, α and λ, where α measures the degree of risk aversion, which affects the shape of the function all over its domain, and λ measures the degree of loss aversion and affects the shape in the loss region only. Its general form is:

$$V_x = \begin{cases} x^\alpha, & \text{if } x \leq 0 \\ -\lambda \cdot (-x)^\alpha, & \text{if } x < 0 \end{cases} \tag{8.6}$$

Figures 8.2, 8.3, and 8.4 show the effects of different levels in each parameter on the shape of the value function.

Figure 8.1 shows that there is a kink in the value function—that is, an utility function that changes with respect to a given point—at the reference point,

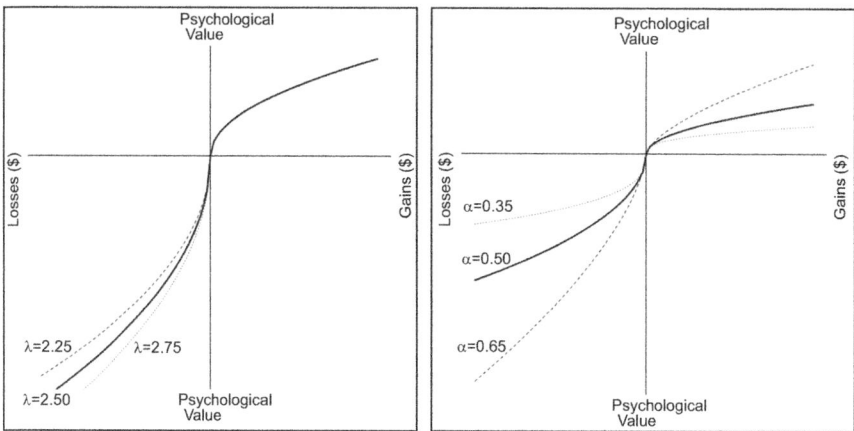

Fig. 8.2 Prospect theory value function with different loss aversion and risk aversion parameters

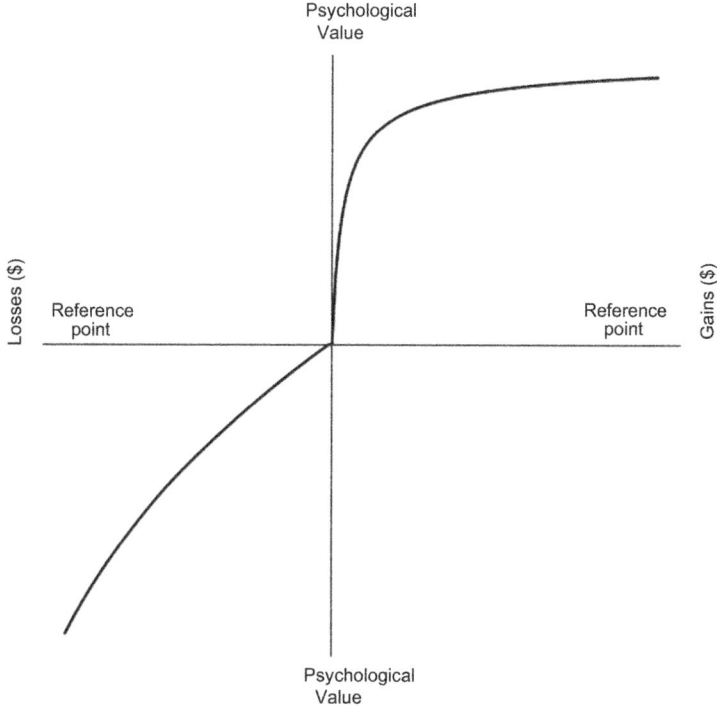

Fig. 8.3 Prospect theory value function for older people (Watanabe and Shibutani 2010, fig. 3)

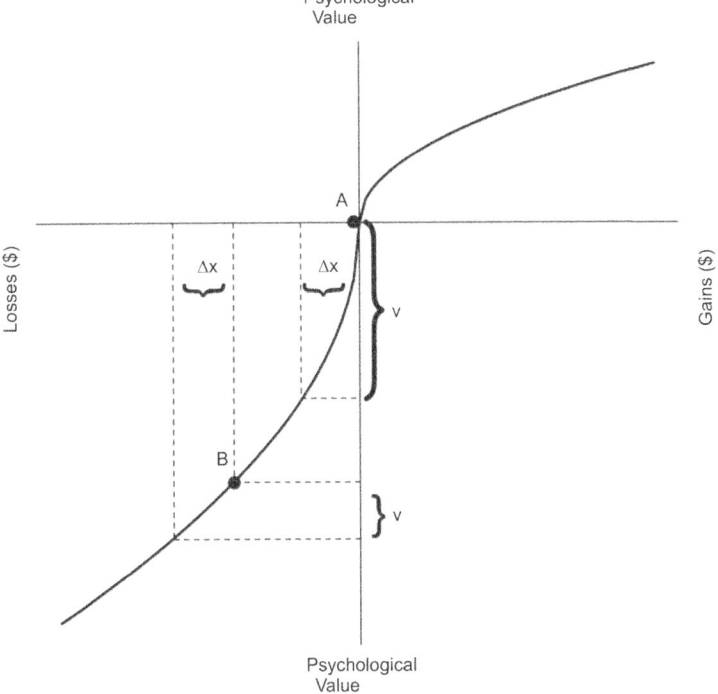

Fig. 8.4 Sunk cost effect

where agents change from risk seeking to loss aversion. Loss aversion is related to another anomaly, the status quo bias (Samuelson and Zeckhauser 1988), also known as inertia. Reference points usually correspond, at least psychologically, to the current situation—that is, the status quo. Hence, gains and losses are valued in relation to the status quo. Prospect theory predicts that any expected losses from the current situation would be valued higher than any expected gains: agents would be biased towards the status quo.

In a study on the value of life annuities, Brown et al. (2017) found that agents were willing to buy annuities at lower prices than they were willing to sell them and that the spread (i.e. the difference) between sale and purchase prices increased with lower cognitive ability and financial literacy. The authors failed to find any endowment or status quo biases (see Chap. 8 in this volume): they surmised, instead, that the anomalies responded to a reluctance to enter into a transaction when it is difficult to estimate its value. Kovalchik et al. (2005) also failed to find any significant endowment effects in an experiment studying the gap between willingness to pay and willingness to accept payment in a group of people aged 70–95 years compared to a group aged 18–26. However, the status quo bias has proved useful to increase participation rates' trials of interventions among patients with myocardial infarction: Mehta et al. (2016) found that an opt-out clause increased participation compared to an opt-in approach (see below), an effect that was independent of the chronological age of the participants.

8.2 Framing Effects

In addition to the reference point, behavioural economics proposes that decisions depend on how the choices are framed or presented—that is, on their 'architecture'. Because economic agents evaluate options by making comparisons, how alternatives are presented affects the yardsticks the decisions are based upon and, therefore, even if the expected gains and losses remain unchanged, the choice may differ according to how the options are framed. This is known as the framing effect (DellaVigna 2009).

Remember that gains do not carry the same weight as losses of exactly the same amount, so framing an outcome or the probability of its occurrence positively would not have the same effect as framing it negatively. Furthermore, framing a risk in relative terms is more efficacious with regard to influencing choice than framing it in absolute terms (Malenka et al. 1993). Most studies of framing effects in risk communication have been carried out in connection with health messages. This is very important for the public communication of risk, as it can be either involuntarily misused or deliberately manipulated: for example, using absolute risk to communicate losses and relative risk to communicate gains misinforms the public, affecting decision-making (Spiegelhalter and Pearson 2010).

Levin et al. (1998) distinguished three types of framing effects:

- Risky choice framing. Faced with the option between a sure alternative and a risky alternative with exactly the same expected outcome or value, if the choice is framed positively (e.g. lives saved, money earned, additional expected years of life ahead), the sure option is more likely to be chosen. Instead, if the choice is negatively framed (e.g. lives, money, or potential number of years lost), the risky option gets chosen more often. There are two categories of risky choice framing: unidirectional, which can be risk-aversion augmenting or risk-seeking augmenting, and bi-directional (Wang 1996). Unidirectional risky choice framing is present when a given preference (e.g. the sure option against the risky one) is manifest under

positive or negative framing situations. Bi-directional risky choice framing is present when there is a reversal in preference depending on whether the situation is framed positively or negatively.
- Attribute framing. This framing effect applies to evaluations of one single attribute or a 'yes/no' decision or opinion. Think of the half-full versus half-empty glass. Evaluations or opinions about, say, a result or performance vary according to whether the information is positively framed (e.g. in terms of rate of success or percentage of hits) or negatively: medical procedures are more likely to be approved and treatments more likely to be observed if the possible outcomes are framed in terms of survival rates rather than mortality rates.
- Goal framing. Framing a choice as part of a goal towards achieving a positive outcome does not have the same impact on decision-making as framing it as part of a goal towards avoiding a negative result. Women tend to engage more in breast self-examination routines if they are exposed to positively worded information or messages (e.g. self-examination makes detecting a tumour early more likely) than a negatively framed information or messages (e.g. not self-examining makes it less likely to detect a tumour early).

In an experimental study, Kim et al. (2005) found that older people were more susceptible to both negative and positive framing effects than younger participants. The authors surmised that this may be a consequence of a loss in cognitive resources (the authors used a combined measure of working memory and speed), which makes older people to rely more on heuristic information processing devices and rules of thumb to decide on cognitively low demanding matters to save mental energy for more important tasks; see also Hess et al. (2001).

Mikels and Reed (2009) reported that younger and older participants in an experiment about decision-making in conditions of uncertainty showed risk aversion regarding gains but that only the younger group exhibited risk-seeking attitudes in the negative or loss domain: older people did not demonstrate a framing decision-making bias.

Similarly, Watanabe and Shibutani (2010) failed to find any risky-choice framing effects among people aged 65 or over in a study in Japan compared to participants aged between 20 and 64. Risk aversion was found in both chronological age groups for positively framed options, whilst the older group also exhibited risk aversion in the negatively framed condition. In other words, the older participants exhibited risk aversion regardless of how the options were framed. This suggests that the value function of older individuals would not accord to the general proposition of prospect theory. The authors surmised that this could be of a greater importance placed on emotional needs, intuition, and feeling among people in later life, rather than on calculation and deliberation. They proposed the value function for older people in Fig. 8.3.

Pachur et al. (2017) compared the decision quality under risk[6] and the degree of risk aversion in a group of sixty individuals aged 18–30 years with another group of sixty people aged 63–88 years. The older group exhibited less decision quality, possibly denoting reduced cognitive ability. In addition, the older adults showed more risk-seeking behaviour in the positive, gain domains than the younger group. This finding was not related to cognitive ability but to differences in affect. Older people showed a stronger motivational reorientation towards a 'positivity bias' than younger people—enhancing positive and diminishing negative information (Mather and Carstensen 2005).

In a meta-analysis of studies of risky choice framing, Mata et al. (2011) found that the reported age differences in various studies depended on the content or task involved in the experiment. When it came to decisions from experience (i.e. when the experimental design and conditions allowed for learning about the relationship between the outcomes and their probabilities of occurrence), older individuals tend to be more risk-seeking than younger adults—see also Wiesiolek et al. (2014) for the relationship between learning, decisions under risk, and ageing. Rolison et al. (2013) also reported varying domain-specific attitudes to risk with chronological age and gender: risk-taking in financial matters falls with chronological age among men. Women become more risk-taking until middle age in the social domain (which includes decisions such as disagreeing with an authority figures, speaking one's mind about an unpopular subject, or moving to a city far away from the extended family, etc.), to increasingly avoid taking risks the older they get—see Pu et al. (2017), who also reported no framing effects among older people in a life-saving scenario but found similar framing effects as younger people in a financial scenario.

However, another meta-analysis (Best and Charness 2015), which included not only different scenarios or domains but also size effects (e.g. whether the experiments asked about deaths or financial losses in their hundreds or tens of thousands), concluded that younger adults are more likely to choose the risky options when positively framed than older adults. However no age effects were detected for negatively framed scenarios, except if high amounts were involved in which case younger adults also would be more likely to choose the risky alternatives.

A study on the effects of positive and negative messages in health information brochures found that older adults found positively framed brochures more informative than younger adults. However, older people were more likely to misremember negatively framed[7] messages as positive, concluding:

> Perhaps the most effective way to present health information with a longer lasting impact for older adults is through positive messages emphasizing the benefits gained by certain health behaviors
>
> (Shamaskin et al. 2010, p. 750)

In the context of buying an annuity, Hu and Scott (2007) reported a framing effect that made annuities more desirable when framed as an insurance product against longevity risk than as a financial product (see also Bockweg et al. 2017 and Brown 2007, 2009).

8.3 Anchoring Effect

When given options that have values attached, people tend to choose the first of those values—this is known as the anchoring effect, for the value acts as an 'anchor' for the choice; see Furnham and Boo (2011) for a literature review. The first value acts as a reference point, priming the decision, although it may not have anything to do with the options in question: the cue may be completely arbitrary (Ariely 2009; Kahneman et al. 1999). In one experiment with almost 500 university students in the United States of America, Simonson and Drolet (2004) asked one randomly chosen group to indicate the highest price they would be willing to pay for items such as a toaster or a backpack. The other group was given the same task but before that they were asked to write down the last two digits of their social security number. Needless to say, how much anyone is willing to pay for a toaster has absolutely nothing to do with the last two digits of her social security number. However, participants with security numbers with the last two digits ending below fifty expressed a willingness to pay prices significantly lower for the items than those with security numbers with the last two digits ending above fifty.

Wilson et al. (1996) pointed out that the anchoring heuristic can explain anything from why people tend to believe that past events were inevitable to assessments of how useful a good or pleasurable an experience has been, from predictions of future performance or likelihood of future events to responses to general knowledge questions and concluded: 'In this age of specialization and fragmentation of psychology, it is rare to find a single, relatively simple process that explains such diverse phenomena' [p. 387].

An experimental study on decision-making under risk used three alternative priming scenarios—for example, a context favourable to risk seeking as individuals engaging in such behaviour were rewarded—but found no priming effects among older participants (Wood et al. 2016).

8.4 Priming

Priming is a heuristic bias similar to the anchoring effect in that it involves some sort of cue but with priming, the anchor could be an image, a colour, a word—as Kahneman (2011) vividly explains, the word 'eat' primes 'soup' whilst the word 'wash' primes 'soap' or, say, the first item that catches a consumer's sight on entering a shop: first impressions do count! Furthermore, a recent experience can act as a priming cue if it activates some mental association or representation in a person's memory that influences her behaviour.

Priming not only affects decision-making but also other types of behaviour and actions; for example, performance. In a celebrated study, Bargh et al. (1996) university students were divided into two groups, and each group was given the task of constructing grammatically correct sentences out of lists of scrambled words. The words given to one of the groups were related to ageist attitudes and stereotypes, except slowness, whereas the other group worked with a neutral set of words. After completing this task, the students were dismissed but the experiment did not finish there: the students were not aware that they were being recorded while walking down the corridor. Those students who were exposed (i.e. 'primed') to ageism-related words walked significantly more slowly than the students in the other group.

Although presented as a study on framing, Thomas and Millar (2011) should be understood as a study on priming effect. In this experiment, the authors asked participants to carry out a decision task (a choice between a risky and a sure bet) and a processing task (a mathematical calculation) and, in between the tasks, they primed the participants by encouraging them either to 'think like a gambler/scientist' or to use their 'initial reactions and gut feelings'. Not only did Thomas and Millar find that both the younger and older groups were affected by priming equally but also a reduction in the susceptibility to priming effects in both groups by means of the same de-priming techniques.

8.5 Sunk Cost Effect

Costs incurred in the past and that cannot be recovered are known as sunk or retrospective costs. In contrast, prospective costs are costs that may be incurred or not in the future. Mainstream microeconomic theory states that only prospective costs are relevant for economic decisions. However, behavioural economists found that when it comes to considering costs for decision-making, economic agents do not let bygones be bygones: sunk costs, although already buried in the past, influence decisions. The sunk cost effect, also known as 'escalation of commitment', is defined as 'paying for the right to use a good or service will increase the rate at which the good will be utilized, ceteris paribus' (Thaler 1980, p. 47). It manifests in 'a greater tendency to continue an endeavor once an investment in money, effort, or time has been made' (Arkes and Blumer 1985, p. 124).

Teger (1980) provides a familiar example: imagine you have spent a considerable amount of money in relation to the market value of your car to get an engine overhaul. A week later the transmission starts causing trouble and you need to spend more money on the car to fix it. Alternatively, you can sell it. But is the latter an alternative indeed after spending so much on the engine? If you sold the car, would it not feel as though you wasted all the money it cost you to overhaul the engine? Teger refers to this phenomenon as having 'too much invested to quit'.

The sunk cost anomaly can be explained in relation with reference dependence. Remember that according to prospect theory, decisions depend on the reference point. After incurring in a sunk cost, the reference point changes along the value function into the losses domain (from point A to point B in Fig. 8.4); further losses from the new reference point do not generate as large negative effects in value as if they were measured from the original reference point:

There seems to be a neurological basis for the susceptibility to the sunk cost anomaly as studies in neuroscience have detected activity in certain areas in the brain associated with sunk cost effects (Haller and Schwabe 2014; Zeng et al. 2013).

We mentioned above that older people exhibit a higher 'positivity bias' in information processing than younger adults. Because the sunk cost effect is related to loss avoidance, having a lower negativity bias would make older people less susceptible to the sunk cost effect. This is precisely what the empirical evidence shows (Bruine de Bruin et al. 2014; Roth et al. 2015; Strough et al. 2008, 2011). Furthermore, investment-related information tends to have a lower perceived salience in the goals and decisions of older people compared to younger adults, which can also explain the association between chronological age and sunk cost effects.

8.6 Mental Accounting

You have finished your meal at a restaurant, ask for the bill, produce your wallet, and look for a banknote. You see you have two notes of the same denomination, each for more than the total bill. You take out one and give it to the waiter. Which banknote? Oh, it doesn't matter, does it? They are of the same denomination and, as far as you know, neither of them are counterfeit. One note in your wallet is equal to any other as a medium of exchange: in economics, this is known as the 'fungibility' of money, which indicates that its units are interchangeable. As money is fungible, we can add together money earned from different sources, by different people, from different periods of time, in different currencies, and of different forms. One corollary of the fungibility assumption is that the marginal propensity to consume (or to save) is the same regardless of the type of wealth or source of income or stage in the life course when the money was earned or obtained.

But, is it? US-born writer Gertrude may have insisted that 'Rose is a rose is a rose is a rose',[8] but behavioural economics has rejected the assumption that money is money is money: money is not fungible (Abeler and Marklein 2008; Barberis et al. 2001; Shefrin and Thaler 1988; Thaler 1999). The marginal propensity to consume may vary according to the source of income or time when it was obtained; something ruled out by mainstream models. As Browning and Crossley (2001, p. 4) put it: 'What the life-cycle framework does rule out is "rule of thumb" behavior, in which households simply spend a fixed fraction of their income. It also rules out many psychological or behavioral explanations …'.

Imagine you go to a casino with 10 dollars and you win another 10. The 10 dollars you took to the casino with you are not the same as the 10 dollars you won (Belsky and Gilovich 2000). This bias is known as mental accounting, 'one of the most common and costly money mistakes -the tendency to value some dollars less than other and thus to waste them' (Belsky and Gilovich 2000, p. 33), which has been defined as 'the set of cognitive operations used by individuals and households to organize, evaluate, and keep track of financial activities' (Thaler 1999, p. 183)—see also Thaler (1990).

The life-cycle (LC) hypothesis treats all the different types of wealth alike: wealth is fungible. Behavioural economics, as we have seen, has provided evidence against this assumption: economic agents treat income and wealth differently according to their source, timing, and so on. Informed by these findings, an alternative to the LC hypothesis has been proposed: the behavioural life cycle.

Wealth in the LC hypothesis is part of the budget restriction agents are subject to. In the behavioural LC model, the budget restriction does not enter as one variable or amount but in as many different asset types as mental accounting processes may distinguish. The consumption function now depends on different types of assets, and for each one the agent will have a different marginal propensity to consume in contrast to the LC hypothesis whose assumption is that an agent will show the same marginal propensity to consume from any type of asset. Furthermore, the budget restriction also includes income set apart from wealth or accumulated savings. According to the LC hypothesis, income and wealth are indistinct: both variables make up the budget. The behavioural LC model predicts that the marginal propensity to consume will be higher for income than wealth: consumption is more sensitive to changes in income than in wealth.

In an example (Thaler 1990), two agents are born on the same day, exhibiting identical lifetime earning-consumption profiles, and retire on the same day. They should have, according to the LC hypothesis, the same accumulated wealth. If one of these agents has $100,000 in pension wealth, this means that the other agent should have $100,000 in other assets: a perfect offset not only in the obvious algebraic sense but in an economic sense as well. According to the LC hypothesis, the effect of pension saving or wealth on other types of saving assets should be equal to minus one. Extensive evidence shows that this is not the case: pension wealth is not a perfect, not even a close, substitute for other types of wealth (Case et al. 2005; Schooley and Worden 2008; Venti and Wise 1990b). For example, analysing data between 1969 and 1971 of over 11,000 household heads living in the United States of America who were aged between 57 and 62 years in 1969 Levin (1998) found compelling evidence in favour of the behavioural LC model, and in concluding threw down the following gauntlet:

> Any defense of the conventional life-cycle model will have to explain the anomalous results obtained here. Although that may prove possible, the behavioural life-cycle model which incorporates problems of self-control into a standard life-cycle framework is a relatively simple explanation of the patterns found in this data.
>
> (Levin 1998, p. 82)

8.7 Myopia

Myopia refers to time inconsistent preferences, which translate into neglecting the future and giving undue importance to short-term gains and pleasures at the expense of receiving larger benefits in the long run (Strotz 1955). For example, a study found that the biggest barrier to saving among people aged 22–29 years is their desire of 'living well now' without concern about the future (Bryan and Lloyd 2014; Bryan et al. 2011; Foster 2017; Pettigrew et al. 2007). The framework of the multiple selves (briefly mentioned in Volume I Chap. 9) proposes that myopia is caused by agents' seeing their future selves as 'strangers' (Ersner-Hershfield et al. 2008).

In an experiment involving 265 individuals aged between 23 and 88 years, Bechara et al. (2002) reported that the older the subjects, the more responsive or sensitive they were to the prospect of receiving a future reward regardless of the fact that the decision would carry an immediate punishment.[9] The authors failed to find that myopia would increase with chronological age; rather, they found '…a sort of "hypersensitivity to reward" whereby participants are drawn more strongly to the higher reward', which is explained by a tendency among older people to prefer positive-valenced information and to maximise positive emotion in the present. Using the same instrument, Fein et al. (2007) studied 164 people aged from 18 to 85 and reported that older participants took less advantageous decisions in that they tended to favour options with larger immediate rewards despite long-term adverse consequences, compared to younger participants.

In turn, Bauer et al. (2013) suggested that decisions in later life usually associated with 'myopic' behaviour may be the result of an 'age-related increase in hypersensitivity to reward' rather than a lack of sensitivity to future disadvantage.

8.8 Lack of Willpower

A phenomenon also leading to departures from rational behaviour is lack of willpower, also referred to as 'bounded self-control' or 'limited willpower'. Ainslie and Haendel (1983) provided this famous example: Imagine no uncertainty. Faced with a choice between a prize of £50 today versus £100 in 6 months, most people choose the first option; faced with a choice between £50 in 12 months versus £100 in 18 months, most people choose the second option. In both cases, the time gap or lapse is the same (6 months) and the amounts are also the same, so a rational agent should show consistency: having chosen the first option in the first case, she is expected to choose the first option in the second case as well. What's going on? Due to weakness of will or lack of willpower, she succumbs to the tempting power of 'now'.

Hyperbolic discounting, 'present bias', and delayed gratification were introduced in Volume I, Chap. 8, where we also encountered famous prayer by Augustine of Hippo (i.e. St Augustine): 'Grant me chastity and continence, but not yet' (Hippo 1998, p. 145). Strotz (1955) developed a theoretical framework in which an agent exhibiting time inconsistent preferences either does not realise of the inter-temporal reversal and consequently behaves in a profligate or miserly way, or does realise and resorts to a strategy of either pre-commitment or consistent planning. The lack of willpower or weakness of will can be overcome by either of these strategies.

The most famous example comes from ancient Greek literature: in Homer's poem *The Odyssey*, Odysseus (also known as Ulysses) made a pact with his sailors as their ship approached the sirens who, with their beguiling songs, used to lure any passing sailors to their deaths. Knowing about this, Ulysses orders the ears of the sailors to be anointed with wax and, more to the point, knowing about his own lack of willpower that would make him succumb to the sirens' chants, he orders the sailors to bind him to a mast (Homer 1990, Bk 12: Ln 47–54). Strotz presents the case of an economic agent in a given period, say $t = 0$, maximising her utility over the rest of her lifetime subject to a budget constraint; that is, the same problem introduced in the life-cycle hypothesis. In the LC framework (and in mainstream economics in general), it is assumed that an economic agent will follow the consumption profile determined by the maximisation exercise: once the amounts that maximise her utility are determined, she will stick to that temporal plan, for any deviations above or below the optimal consumption levels in each period would reduce her lifetime utility. However, the key issue that Strotz points out is that this is the plan she will *start* on, the plan at time $t = 0$; she may change her mind—and there is nothing 'irrational' in reconsidering, re-evaluating, or having second thoughts. In particular, the time discount factor may change with the passing of time: 'To-day it will be rational for a man to jettison his "optimal" plan of yesterday …because to-day he is a different person with a new discount function' (Strotz 1955, p. 173). Precisely, only if the discount rate is constant over time, the original plan will be followed throughout; this is the assumption that behavioural economists have rejected. In general, economic agents do not exhibit temporally fixed discount functions. Odysseus survives thanks to an indirectly rational means, pre-commitment: 'binding oneself is a privileged way of resolving the problem of weakness of will, the main technique for achieving rationality by indirect means' (Elster 1979, p. 37)—see also Elster (2000). Strotz opined that planning ahead is an effective strategy against inconsistent time preferences. In this regard, one of the desired outcomes of the process of socialisation of children in Western cultures is self-control and delay of gratification (though this value has a class gradient, with lower income families placing less importance on it than middle-class families) (Singh 2015). In contrast, Elster doubted the effectiveness of planning ahead, arguing the lack of willpower would prevent the agent from complying with her plans; instead, pre-commitment, which he defines as a strategy against inconsistency rather than irrationality, is a more plausible option to 'impose' a current decision on our future selves.

Studies on lack of willpower and ageing tend to focus on healthy eating, saving for retirement, and physical activity.[10] The evidence suggests that lack of willpower is a potent barrier for older people. For example, Appleton et al. (2010) and Kearney and McElhone (1999) reported that in Northern Ireland and the United Kingdom in general, a lack of willpower was the most important self-perceived barrier to healthy eating among individuals of retirement age (see also Lara et al. 2014). It is worth noting the heightened importance of the lack of willpower as a barrier to saving for retirement in the current political context of an increasing shift onto individuals of the risk of inadequate retirement income (see García 2006).[11] No studies have looked at the role of weakness of the will on making inconsistent decisions over time—whether the grip of bounded self-control on decision-making loosens with chronological age.

8.9 Complexity

I mentioned above that some biases arise from the application of mental heuristics. Why would economic agents resort to these cognitive short-cuts? One influential answer was proposed by the US 'all-rounder' social scientist[12] Herbert Simon: the sheer complexity of the information and calculus needed to make optimal decisions exceeds human mental processing capacity; as a consequence economic agents do not engage in 'maximising' behaviour but in 'satisficing' behaviour. That is, they seek and settle for satisfactory solutions that are not necessarily optimal; they do not exhibit rationality but 'bounded' rationality (Simon 1947, 1955, 1959). It is worth bearing in mind that 'maximising' (or more accurately, 'optimising') is less likely the older the economic agent and that it is not only a more taxing process than seeking a 'good enough' solution, but is potentially more regrettable (Bruine et al. 2016).

Maximisation involves an extensive search of and a thorough comparison between alternatives to choose the best option, which is time consuming and cognitively demanding. Ill-defined goals, time limitations, or a huge number of alternatives may even make maximising counter-productive. Older people prefer to consider fewer options and less information compared to younger adults (Besede et al. 2012; Reed et al. 2008).

Apart from bounded rationality, economic agents exhibit bounded willpower and bounded self-interest (Jolls et al. 1998). Bounded willpower is the manifestation of weakness of will, as in the case of Ulysses and the sailors. Bounded self-interest expresses the finding that agents do care about others, including strangers, in ways that go beyond the conceptualisation of altruism: for example, an economic agent would wish to treat others fairly if they treat her fairly:

> …the agents in a behavioral economic model are both nicer and (when they are not treated fairly) more spiteful than the agents postulated by neoclassical theory
> (Jolls et al. 1998, p. 1479).

Whilst rational choice theory assumes that the larger the number of available options the better for consumers, behavioural economists beg to differ informed by studies from cognitive psychology on 'choice overload' (Iyengar and Lepper 2000) or the 'tyranny of choice' (Schwartz 2000) which report that the proliferation of alternatives (of either products or actions) to choose from decreases the utility or satisfaction agents experience or derive from the chosen option and may de-motivate agents to engage in the mental experience of choosing altogether.

The literature suggests that choice overload is independent of chronological age: it affects younger and older individuals alike. This is the conclusion from a study by Tanius et al. (2009) who looked into the effect of the size of the choice set of private medical insurance and found that age was not significant. In a similar vein, Frey et al. (2015) presented three experiments with younger and older people with different levels of information load and demand on cognitive resources. With a few options, there were no discernible age effects but as the number of payoff distributions increased, older individuals underperformed: the result, therefore, was less a consequence of choice alternatives but of the complexity and cognitive demands the options imposed.

Complex choice architecture is very consequential in many areas related to ageing and economics, from annuities to retirement plans to healthcare insurance. Regarding private healthcare or medication plans, Hanoch et al. (2011) reported that older individuals whose main expressed interest was financial (i.e. keeping costs as low as possible), when facing more options to choose from were more likely not to choose the lowest cost plan. In turn, Iyengar and Kamenica (2010) and Sethi-Iyengar et al. (2004) reported that being offered a higher number of private retirement savings plan alternatives was associated with a lower participation rate: the likelihood of contributing to any plan was inversely correlated with the number of available plans.

8.10 Same Findings, Other Approaches

Several authors have found similar biases and errors in judgement among older people though not all of the studies resort to explanations from behavioural economics. For example, neuroeconomics draws upon advances in neuroscience to understand how age-related changes in the brain may modify economic decision-making and behaviour. To illustrate,

- Harris et al. (2016) and Josef et al. (2016) suggest that personality traits are less stable in childhood and older age than in middle age, with stability therefore following a U-shaped curve over time, possibly as a consequence of the effects of life transitions and reduced cognitive as well as physical ability in later life.
- Grubb et al. (2016) reported that age-related changes in the brain, specifically in the right posterior parietal cortex, was associated with risk aversion or intolerance, rather than chronological age alone.
- Rutledge et al. (2016), by contrast, suggested that the increase in risk aversion in later life may be due to decreases in the density of dopamine, which is a neurotransmitter that is related with the reward and pleasure centres in the brain and which is also associated with normal ageing.
- Seaman et al. (2017) reported that, compared to younger adults, older people are more prone to accepting positively skewed gambles (i.e. 25 per cent chance of a large gain) relative to symmetric (i.e. 50/50 chance) gambles, which makes older people more susceptible of falling prey of financial scams and fraud schemes. The authors found that the differences can be explained by variations in the activity in a brain region called the nucleus accumbens; see also Han et al. (2016).
- There is evidence that myopic preferences are related to brain activity: studies have identified a region in the brain—the ventromedial prefrontal cortex—which is less activated in people who show disregard for the future when thinking about the future than the level of activity shown in those

who do not have the same level of disregard (Clark et al. 2008b; Mitchell et al. 2011).
- Research has shown that prospective thoughts about future events (see Chap. 1 in this volume) activate different regions in the brain depending on whether they are pleasant (say, winning the lottery) or unpleasant (incurring in a financial loss) (Gilbert and Wilson 2007).

Though related to this line of research, behavioural economics is less interested in the neurological bases of decision-making and primarily focuses, instead, on behavioural responses to external stimuli and internal emotions—on more malleable influences of economic decisions. Behavioural economists are more interested in whether and how the behaviour, decisions, and choices of older people can be influenced nonetheless.

Notes

1. Modern behavioural economics is considered to have started with the papers by Tversky and Kahneman (1974) and Kahneman and Tversky (1979). For a historical view of behavioural economics, see Heukelom (2014). Belsky and Gilovich (2000) is a non-technical introduction.
2. Although see Rachlin (1995) for a criticism to this approach.
3. For book-length treatments of behavioural economics, see Altman (2012), Wilkinson and Klaes (2012), and Cartwright (2014).
4. I am not going to delve into the weak and strong axioms of revealed preferences and other theoretical points, which I leave to any interested readers to consult in any intermediate or advanced textbook in microeconomics, such as Mas-Colell et al. (1995), Jehle and Reny (2011), and Varian (2014). For thorough, though advanced, book-length discussions, see Wakker (2010) and Chambers and Echenique (2016).
5. See Plott and Zeiler (2005) for a criticism.
6. Decision quality is 'the frequency with which the decision-maker chooses the option with the higher expected value ...associated with each possible outcome of an option' (Pachur et al. 2017, p. 504).
7. Or 'valenced', which in psychology refers to the 'intrinsic attractiveness or aversiveness' of events, objects, and situations (Frijda 1986, p. 207).
8. In the poem 'Sacred Emily' (Stein 2012).
9. The experiment applied a computer-based psychological task known as the Iowa gambling task; rewards and punishment were administered in the shape

of game money. See Bechara et al. (1994, 2005) and Damasio (1994) to learn more about this task.
10. On lack of willpower and retirement saving, see Ellen et al. (2012), Millar and Devonish (2009), Raaij (2016), Visco (2009), and the sub-section on the Save More Tomorrow programme below.
11. For lack of willpower as a barrier to physical activity in later life, see Dye and Wilcox (2006), Kim (2015), and Schreier et al. (2016).
12. In the words of the Committee who awarded Herbert Alexander Simon the Sveriges Riksbank Prize in Economic Sciences in Memory of Alfred Nobel in 1978:

Simon's scientific output goes far beyond the disciplines in which he has held professorships - political science, administration, psychology and information sciences. He has made contributions in, among other fields, science theory, applied mathematical statistics, operations analysis, economics, and business administration. In all areas in which he has conducted research, Simon has had something of importance to say...

(*The Prize in Economics* 1978 2014)

References

Abeler, Johannes and Felix Marklein (2008). *Fungibility, labels, and consumption.* IZA Discussion Paper 3500. Bonn: Germany: Institute for the Study of Labor (IZA).
Agarwal, Sumit et al. (2009). "The Age of Reason: Financial Decisions over the Life Cycle and Implications for Regulation". In: *Brookings Papers on Economic Activity*, pages 51–101.
Ainslie, George and Varda Haendel (1983). "The motives of the will". In: *Etiologic aspects of alcohol and drug abuse*. Edited by Keith A Gottheil Edward and Druley, Thomas Skoloda, and Howard Waxman. Springfield, IL: United States of America: Charles C Thomas, pages 119–140.
Altman, Morris (2012). *Behavioral Economics For Dummies*. Mississauga, ON: Canada: Wiley.
Appleton, Katherine M et al. (2010). "Barriers to increasing fruit and vegetable intakes in the older population of Northern Ireland: low levels of liking and low awareness of current recommendations". In: *Public health nutrition* 13.4, pages 514–521.
Ariely, Dan (2009). *Predictably Irrational: The Hidden Forces that Shape Our Decisions*. London: United Kingdom: HarperCollins.
Arkes, Hal R and Catherine Blumer (1985). "The psychology of sunk cost". In: *Organizational behavior and human decision processes* 35.1, pages 124–140.
Barberis, Nicholas, Ming Huang, and Tano Santos (2001). "Prospect theory and asset prices". In: *The Quarterly Journal of Economics* 116.1, pages 1–53.

Bargh, John A, Mark Chen, and Lara Burrows (1996). "Automaticity of social behavior: Direct effects of trait construct and stereotype activation on action". In: *Journal of personality and social psychology* 71.2, pages 230–244.
Bauer, AS et al. (2013). "Myopia for the future or hypersensitivity to reward? Age-related changes in decision making on the Iowa Gambling Task". In: *Emotion* 13.1, pages 19–24.
Bechara, Antoine, Sara Dolan, and Andrea Hindes (2002). "Decision-making and addiction (part II): myopia for the future or hypersensitivity to reward?" In: *Neuropsychologia* 40.10, pages 1690–1705.
Bechara, Antoine et al. (1994). "Insensitivity to future consequences following damage to human prefrontal cortex". In: *Cognition* 50.1-3, pages 7–15.
Bechara, Antoine et al. (2005). "The Iowa Gambling Task and the somatic marker hypothesis: some questions and answers". In: *Trends in cognitive sciences* 9.4, pages 159–162.
Belsky, Gary and Thomas Gilovich (2000). *Why Smart People Make Big Money Mistakes-and how to Correct Them: Lessons from the New Science of Behavioral Economics*. New York, NY: United States of America: Simon & Schuster.
Besede, Tibor et al. (2012). "Age effects and heuristics in decision making". In: *Review of Economics and Statistics* 94.2, pages 580–595.
Best, Ryan and Neil Charness (2015). "Age differences in the effect of framing on risky choice: A meta-analysis". In: *Psychology and aging* 30.3, pages 688–698.
Bockweg, Christian et al. (2017). "Framing and the annuitization decision. Experimental evidence from a Dutch pension fund". In: *Journal of Pension Economics & Finance*, pages 1–33. DOI: https://doi.org/10.1017/S147474721700018X.
Brown, Jeffrey R (2007). *Rational and Behavioral Perspectives on the Role of Annuities in Retirement Planning*. NBER Working Paper 13537. Cambridge, MA: United States of America: National Bureau of Economic Research.
—— (2009). "Understanding the role of annuities in retirement planning". In: *Overcoming the saving slump: how to increase the effectiveness of financial education and saving programs*. Edited by Annamaria Lusardi. Chicago, IL: United States of America: University of Chicago Press, pages 178–208.
Brown, Jeffrey R et al. (2017). "Cognitive constraints on valuing annuities". In: *Journal of the European Economic Association* 15.2, pages 429–462.
Browning, Martin and Thomas F. Crossley (2001). "The life-cycle model of consumption and saving". In: *Journal of Economic Perspectives* 15.3, pages 3–22.
Bruine de Bruin, Wändi, Andrew M Parker, and JoNell Strough (2016). "Choosing to be happy? Age differences in "maximizing" decision strategies and experienced emotional well-being". In: *Psychology and aging* 31.3, pages 295–300.
Bruine de Bruin, Wändi, JoNell Strough, and Andrew M Parker (2014). "Getting older isn't all that bad: Better decisions and coping when facing "sunk costs"". In: *Psychology and aging* 29.3, pages 642–647.
Bryan, Mark and James Lloyd (2014). *Who Saves for Retirement? 2: Eligible non-savers*. Technical report. London: United Kingdom.

Bryan, Mark et al. (2011). *Who saves for retirement*. Technical report. London: United Kingdom.

Cartwright, Edward (2014). *Behavioral Economics*. Routledge Advanced Texts in Economics and Finance. Abingdon: United Kingdom: Taylor & Francis.

Case, Karl E, John M Quigley, and Robert J Shiller (2005). "Comparing wealth effects: the stock market versus the housing market". In: *Advances in macroeconomics* 5.1. DOI: https://doi.org/10.2202/1534-6013.1235.

Chambers, Christopher and Federico Echenique (2016). *Revealed Preference Theory*. Econometric Society Monographs. Cambridge: United Kingdom: Cambridge University Press.

Clark, L et al. (2008b). "Differential effects of insular and ventromedial prefrontal cortex lesions on risky decision-making". In: *Brain* 131.5, pages 1311–1322.

Damasio, Antonio R (1994). *Descartes' Error: Emotion, Reason and the Human Brain*. New York, NY: United States of America: G. P. Putnam's Sons.

DellaVigna, Stefano (2009). "Psychology and economics: Evidence from the field". In: *Journal of Economic Literature* 47.2, pages 315–72.

Dye, Cheryl J and Sara Wilcox (2006). "Beliefs of low-income and rural older women regarding physical activity: You have to want to make your life better". In: *Women & Health* 43.1, pages 115–134.

Ellen, Pam Scholder, Joshua L Wiener, and M Paula Fitzgerald (2012). "Encouraging people to save for their future: Augmenting current efforts with positive visions of the future". In: *Journal of Public Policy & Marketing* 31.1, pages 58–72.

Elster, Jon (1979). *Ulysses and the Sirens: Studies in Rationality and Irrationality*. Cambridge: United Kingdom: Cambridge University Press.

Elster, Jon (2000). *Ulysses Unbound: Studies in Rationality, Precommitment, and Constraints*. Cambridge: United Kingdom: Cambridge University Press.

Ersner-Hershfield, Hal, G Elliott Wimmer, and Brian Knutson (2008). "Saving for the future self: Neural measures of future self-continuity predict temporal discounting". In: *Social cognitive and affective neuroscience* 4.1, pages 85–92.

Fein, George, Shannon McGillivray, and Peter Finn (2007). "Older adults make less advantageous decisions than younger adults: Cognitive and psychological correlates". In: *Journal of the International Neuropsychological Society* 13.3, pages 480–489.

Finucane, Melissa L (2008). "Emotion, affect, and risk communication with older adults: challenges and opportunities". In: *Journal of Risk Research* 11.8, pages 983–997.

Foster, Liam (2017). "Young people and attitudes towards pension planning". In: *Social Policy and Society* 16.1, pages 65–80.

Frey, Renato, Rui Mata, and Ralph Hertwig (2015). "The role of cognitive abilities in decisions from experience: Age differences emerge as a function of choice set size". In: *Cognition* 142, pages 60–80.

Frijda, Nico H (1986). *The Emotions*. Studies in Emotion and Social Interaction. Cambridge: United Kingdom: Cambridge University Press.

Furnham, Adrian and Hua Chu Boo (2011). "A literature review of the anchoring effect". In: *The Journal of Socio-Economics* 40.1, pages 35–42.

García, María Teresa M (2006). "Individual responsibility for the Adequacy of Retirement Income". In: *Pensions: An International Journal* 11.3, pages 192–199.

Gilbert, Daniel T and Timothy D Wilson (2007). "Prospection: Experiencing the future". In: *Science* 317.5843, pages 1351–1354.

Goldstein, E. Bruce, editor (2001). *Blackwell Handbook of Perception*. Blackwell Handbooks of Experimental. Psychology Oxford: United Kingdom: Wiley.

—— (2013). *Sensation and Perception*. Belmont, CA: United States of America: Cengage Learning.

Grondin, Simon (2016). *Psychology of Perception*. Cham: Switzerland: Springer.

Grubb, Michael A et al. (2016). "Neuroanatomy accounts for age-related changes in risk preferences". In: *Nature communications* 7. DOI: https://doi.org/10.1038/ncomms13822.

Haller, Ariane and Lars Schwabe (2014). "Sunk costs in the human brain". In: *Neuroimage* 97, pages 127–133.

Han, S Duke et al. (2016). "Grey matter correlates of susceptibility to scams in community-dwelling older adults". In: *Brain imaging and behavior* 10.2, pages 524–532.

Hanoch, Yaniv et al. (2011). "Choosing the right medicare prescription drug plan: the effect of age, strategy selection, and choice set size". In: *Health Psychology* 30.6, pages 719–727.

Harris, Mathew A et al. (2016). "Personality stability from age 14 to age 77 years". In: *Psychology and aging* 31.8, pages 862–874.

Hess, Thomas M, Daniel C Rosenberg, and Sandra J Waters (2001). "Motivation and representational processes in adulthood: the effects of social accountability and information relevance". In: *Psychology and Aging* 16.4, pages 629–642.

Heukelom, Floris (2014). *Behavioral Economics: A History*. Historical Perspectives on Modern Economics. Cambridge: United Kingdom: Cambridge University Press.

Hippo, Augustine of (1998). *The Confessions*. Oxford World's Classics. Oxford: United Kingdom: Oxford University Press.

Homer (1990). *The Odyssey [c. 8C B.C.](translation by Richmond Lattimore)*. Great Books of the Western World. Chicago, IL: United States of America: Encyclopaedia Britannica Inc.

Horowitz, John K and Kenneth E McConnell (2002). "A Review of WTA/WTP Studies". In: *Journal of Environmental Economics and Management* 44, pages 426–447.

Hu, Wei-Yin and Jason S Scott (2007). "Behavioral obstacles in the annuity market". In: *Financial Analysts Journal* 63.6, pages 71–82.

Iyengar, Sheena S and Emir Kamenica (2010). "Choice proliferation, simplicity seeking, and asset allocation". In: *Journal of Public Economics* 94.7, pages 530–539.

Iyengar, Sheena S and Mark R Lepper (2000). "When choice is demotivating: Can one desire too much of a good thing?" In: *Journal of personality and social psychology* 79.6, pages 995–1006.

Jehle, Geoffrey Alexander and Philip J. Reny (2011). *Advanced Microeconomic Theory*. Harlow, Essex: United Kingdom: Pearson Education Limited.

Jolls, Christine, Cass R Sunstein, and Richard Thaler (1998). "A Behavioral Approach to Law and Economics". In: *Stanford Law Review* 50, pages 1471–1550.

Josef, Anika K et al. (2016). "Stability and change in risk-taking propensity across the adult life span". In: *Journal of personality and social psychology* 111.3, pages 430–450.

Kahneman, Daniel (2003). "Maps of Bounded Rationality: Psychology for Behavioral Economics". In: *The American Economic Review* 93.5, pages 1449–1475.

—— (2011). *Thinking, Fast and Slow*. London: United Kingdom: Penguin Books.

Kahneman, Daniel, Jack L Knetsch, and Richard H Thaler (1991). "Anomalies: The endowment effect, loss aversion, and status quo bias". In: *The Journal of Economic Perspectives* 5.1, pages 193–206.

Kahneman, Daniel, Ilana Ritov, and David Schkade (1999). "Economic Preferences or Attitude Expressions?: An Analysis of Dollar Responses to Public Issues". In: *Journal of Risk and Uncertainty* 19.1-3, pages 203–235.

Kahneman, Daniel and Amos Tversky (1979). "Prospect Theory: An Analysis of Decision under Risk". In: *Econometrica* 47.2, pages 263–292.

Kearney, JM and S McElhone (1999). "Perceived barriers in trying to eat healthier. Results of a pan-EU consumer attitudinal survey". In: *British Journal of Nutrition* 81.S1, S133–S137.

Kim, Sunghan et al. (2005). "Framing effects in younger and older adults". In: *The Journals of Gerontology Series B: Psychological Sciences and Social Sciences* 60.4, P215–P218.

Kim, YoungHee (2015). *An Analysis of Barrier Factors of Physical Activity among Old Adults*.

Kovalchik, Stephanie et al. (2005). "Aging and decision making: A comparison between neurologically healthy elderly and young individuals". In: *Journal of Economic Behavior & Organization* 58.1, pages 79–94.

Laibson, David (1998). "Comment on 'Personal Retirement Saving Programs and Asset Accumulation. Reconciling the Evidence' by James M Poterba, Steven F Venti, and David A Wise". In: *Frontiers in the Economics of Aging*. Edited by David A Wise. Chicago, IL: United States of America: The University of Chicago Press, pages 106–124.

Lara, Jose, Leigh-Ann McCrum, and John C Mathers (2014). "Association of Mediterranean diet and other health behaviours with barriers to healthy eating and perceived health among British adults of retirement age". In: *Maturitas* 79.3, pages 292–298.

Lemaire, Patrick (2016). *Cognitive Aging: The Role of Strategies*. Abingdon: United Kingdom: Routledge.

Levin, Irwin P, Sandra L Schneider, and Gary J Gaeth (1998). "All frames are not created equal: A typology and critical analysis of framing effects". In: *Organizational behavior and human decision processes* 76.2, pages 149–188.

Levin, Laurence (1998). "Are assets fungible?: Testing the behavioral theory of life-cycle savings". In: *Journal of Economic Behavior & Organization* 36.1, pages 59–83.

Malenka, David J et al. (1993). "The framing effect of relative and absolute risk". In: *Journal of general internal medicine* 8.10, pages 543–548.

Mas-Colell, Andreu, Michael D Whinston, and Jerry R Green (1995). *Microeconomic Theory*. Oxford: United Kingdom: Oxford University Press.

Mata, Rui et al. (2011). "Age differences in risky choice: A meta-analysis". In: *Annals of the New York Academy of Sciences* 1235.1, pages 18–29.

Mather, Mara and Laura L Carstensen (2005). "Aging and motivated cognition: The positivity effect in attention and memory". In: *Trends in cognitive sciences* 9.10, pages 496–502.

Mehta, Shivan J et al. (2016). "Participation Rates With Opt-out Enrollment in a Remote Monitoring Intervention for Patients With Myocardial Infarction". In: *JAMA Cardiology* 1.7, pages 847–848.

Mikels, Joseph A and Andrew E Reed (2009). "Monetary losses do not loom large in later life: Age differences in the framing effect". In: *Journals of Gerontology Series B: Psychological Sciences and Social Sciences* 64.4, pages 457–460.

Millar, Michael and Dwayne Devonish (2009). "Attitudes, savings choices, level of knowledge and investment preferences of employees toward pensions and retirement planning: Survey evidence from Barbados". In: *Pensions: An International Journal* 14.4, pages 299–317.

Mitchell, Jason P et al. (2011). "Medial prefrontal cortex predicts intertemporal choice". In: *Journal of cognitive neuroscience* 23.4, pages 857–866.

Morewedge, Carey K and Colleen E Giblin (2015). "Explanations of the endowment effect: an integrative review". In: *Trends in cognitive sciences* 19.6, pages 339–348.

Munro, Alistair (2009). *Bounded Rationality and Public Policy: A Perspective from Behavioural Economics*. The Economics of Non-Market Goods and Resources. Dordrecht: The Netherlands: Springer Netherlands.

Pachur, Thorsten, Rui Mata, and Ralph Hertwig (2017). "Who dares, who errs? Disentangling cognitive and motivational roots of age differences in decisions under risk". In: *Psychological science* 28.4, pages 504–518.

Pettigrew, Nick et al. (2007). *Live now, save later? Young people, saving and pensions*. Research Report. London: United Kingdom.

Pierce, Jon L, Tatiana Kostova, and Kurt T Dirks (2001). "Toward a theory of psychological ownership in organizations". In: *Academy of Management Review* 26.2, pages 298–310.

Plott, Charles R and Kathryn Zeiler (2005). "The Willingness to Pay-Willingness to Accept Gap, the "Endowment Effect", Subject Misconceptions, and Experimental Procedures for Eliciting Valuations". In: *The American Economic Review* 95.3, pages 530–545.

Pu, Bingyan, Huamao Peng, and Shiyong Xia (2017). "Role of Emotion and Cognition on Age Differences in the Framing Effect". In: *The International Journal of Aging and Human Development*. DOI: https://doi.org/10.1177/0091415017691284.

Raaij, W Fred van (2016). "Saving Behavior". In: *Understanding Consumer Financial Behavior. Money Management in an Age of Financial Illiteracy*. New York, NY: United States of America: Springer, pages 33–44.

Rachlin, Howard (1995). "Behavioral economics without anomalies". In: *Journal of the experimental analysis of behavior* 64.3, pages 397–404.

Reed, Andrew E, Joseph A Mikels, and Kosali I Simon (2008). "Older adults prefer less choice than young adults". In: *Psychology and aging* 23.3, pages 671–675.

Rolison, Jonathan J et al. (2013). "Risk-taking differences across the adult life span: a question of age and domain". In: *Journals of Gerontology Series B: Psychological Sciences and Social Sciences* 69.6, pages 870–880.

Roth, Stefan, Thomas Robbert, and Lennart Straus (2015). "On the sunk-cost effect in economic decision-making: a meta-analytic review". In: *Business research* 8.1, pages 99–138.

Rutledge, Robb B et al. (2016). "Risk taking for potential reward decreases across the lifespan". In: *Current Biology* 26.12, pages 1634–1639.

Samuelson, William and Richard Zeckhauser (1988). "Status quo bias in decision making". In: *Journal of risk and uncertainty* 1.1, pages 7–59.

Schooley, Diane K and Debra Drecnik Worden (2008). "A behavioral life-cycle approach to understanding the wealth effect". In: *Business Economics* 43.2, pages 7–15.

Schreier, Maria Magdalena et al. (2016). "Fitness training for the old and frail". In: *Zeitschrift für Gerontologie und Geriatrie* 49.2, pages 107–114.

Schwartz, Barry (2000). "Self-determination: The tyranny of freedom". In: *American psychologist* 55.1, pages 79–88.

Seaman, Kendra L et al. (2017). "Individual differences in skewed financial risk-taking across the adult life span". In: *Cognitive, Affective, & Behavioral Neuroscience* 17.6, pages 1232–1241.

Sethi-Iyengar, Sheena, Gur Huberman, and Wei Jiang (2004). "How much choice is too much? Contributions to 401 (k) retirement plans". In: *Pension design and structure: New lessons from behavioral finance* 83, pages 84–87.

Shamaskin, Andrea M, Joseph A Mikels, and Andrew E Reed (2010). "Getting the message across: age differences in the positive and negative framing of health care messages". In: *Psychology and Aging* 25.3, pages 746–751.

Shefrin, Hersh M and Richard H Thaler (1988). "The behavioral life-cycle hypothesis". In: *Economic inquiry* 26.4, pages 609–643.

Shu, Suzanne B and Joann Peck (2011). "Psychological ownership and affective reaction: Emotional attachment process variables and the endowment effect". In: *Journal of Consumer Psychology* 21.4, pages 439–452.

Simon, Herbert A (1947). *Administrative behavior. A study of decision-making processes in administrative organization*. New York, NY: United States of America. Macmillan Co.

—— (1955). "A behavioral model of rational choice". In: *The Quarterly Journal of Economics* 69.1, pages 99–118.
—— (1959). "Theories of decision-making in economics and behavioral science". In: *The American Economic Review* 49.3, pages 253–283.
Simonson, Itamar and Aimee Drolet (2004). "Anchoring effects on consumers' willingness-to-pay and willingness-to-accept". In: *Journal of consumer research* 31.3, pages 681–690.
Singh, Arun Kumar (2015). *Social Psychology*. Delhi: India: PHI Learning.
Spiegelhalter, DJ and M Pearson (2010). *2845 ways to spin the risk*. URL: http://understandinguncertainty.%20org/node/233.
Stein, Gertrude (2012). *Geography and Plays*. Madison, WI: United States of America: University of Wisconsin Press.
Strotz, Robert Henry (1955). "Myopia and inconsistency in dynamic utility maximization". In: *The Review of Economic Studies* 23.3, pages 165–180.
Strough, JoNell, Leo Schlosnagle, and Lisa DiDonato (2011). "Understanding decisions about sunk costs from older and younger adults' perspectives". In: *Journals of Gerontology Series B: Psychological Sciences and Social Sciences* 66.6, pages 681–686.
Strough, JoNell et al. (2008). "Are older adults less subject to the sunk-cost fallacy than younger adults?" In: *Psychological Science* 19.7, pages 650–652.
Tanius, Betty E et al. (2009). "Aging and choice: applications to Medicare Part D". In: *Judgment and Decision Making* 4.1, pages 92–101.
Teger, Allan I (1980). *Too Much Invested to Quit*. Pergamon general psychology series. Elmsford, NY: United States of America: Pergamon Press Inc.
Thaler, Richard (1980). "Toward a positive theory of consumer choice". In: *Journal of Economic Behavior & Organization* 1.1, pages 39–60.
Thaler, Richard H (1988). "Anomalies: The winner's curse". In: *The Journal of Economic Perspectives* 2.1, pages 191–202.
—— (1990). "Anomalies: Saving, fungibility, and mental accounts". In: *The Journal of Economic Perspectives* 4.1, pages 193–205.
—— (1999). "Mental accounting matters". In: *Journal of Behavioral decision making* 12.3, pages 183–206.
The Prize in Economics 1978 (2014). Press Release. URL: http://www.nobelprize.org/nobel_prizes/economic-sciences/laureates/1978/press.html.
Thomas, Ayanna K and Peter R Millar (2011). "Reducing the framing effect in older and younger adults by encouraging analytic processing". In: *Journals of Gerontology Series B: Psychological Sciences and Social Sciences* 67.2, pages 139–149.
Tomer, John F (2007). "What is behavioral economics?" In: *The Journal of Socio-Economics* 36.3, pages 463–479.
Tversky Amos and Daniel Kahneman (1974). "Judgment under uncertainty: Heuristics and biases". In: *Science* 185.4157, pages 1124–1131.
Tymula, Agnieszka et al. (2013). "Like cognitive function, decision making across the life span shows profound age-related changes". In: *Proceedings of the National Academy of Sciences* 110.42, pages 17143–17148.

Varian, Hal R (2014). *Intermediate Microeconomics: A Modern Approach*. New York, NY: United States of America: W. W. Norton.

Venti, Steven F and David A Wise (1990b). "Have IRAs increased US saving?: Evidence from consumer expenditure surveys". In: *The Quarterly Journal of Economics* 105.3, pages 661–698.

Visco, Ignazio (2009). "Retirement saving and the payout phase". In: *OECD Journal: Financial Market Trends* 2009.1, pages 143–162.

Wakker, Peter P (2010). *Prospect Theory: For Risk and Ambiguity*. Cambridge: United Kingdom: Cambridge University Press.

Wang, Xiao Tian (1996). "Framing effects: Dynamics and task domains". In: *Organizational behavior and human decision processes* 68.2, pages 145–157.

Watanabe, Satoshi and Hirohide Shibutani (2010). "Aging and decision making: Differences in susceptibility to the risky-choice framing effect between older and younger adults in Japan". In: *Japanese Psychological Research* 52.3, pages 163–174.

Wiesiolek, Carine Carolina, Maria Paula Foss, and Paula Rejane Beserra Diniz (2014). "Normal aging and decision-making: a systematic review of the literature of the last 10 years". In: *Jornal Brasileiro de Psiquiatria* 63.3, pages 255–259.

Wilkinson, Nick and Matthias Klaes (2012). *An Introduction to Behavioral Economics*. Basingstoke: United Kingdom: Palgrave Macmillan.

Wilson, Timothy D et al. (1996). "A new look at anchoring effects: basic anchoring and its antecedents". In: *Journal of Experimental Psychology: General* 125.4, pages 387–402.

Wong, Kin Fai Ellick and Jessica Kwong (2005). "Comparing two tiny giants or two huge dwarfs? Preference reversals owing to number size framing". In: *Organizational Behavior and Human Decision Processes* 98.1, pages 54–65.

Wood, Meagan, Sheila Black, and Ansley Gilpin (2016). "The effects of age, priming, and working memory on decision-making". In: *International journal of environmental research and public health* 13.1. DOI: https://doi.org/10.3390/ijerph13010119.

Zeng, Jianmin et al. (2013). "An fMRI study on sunk cost effect". In: *Brain research* 1519, pages 63–70.

9

Behavioural Economics and Policy

> **Policy**
> This chapter presents the public policy relevance of the findings from behavioural economics. Topics include forms of paternalism (libertarian, constitutionally constrained, autonomy-enhancing, and asymmetric). It closes with a discussion of the Save More Tomorrow programme.

Imagine you are a policy maker who has bought into the persuasive evidence brought together by behavioural economists. Would you try and prevent people from making biased decisions? Would you try and change the 'architecture' of the choices they face so that they do not fall prey to potential cognitive anomalies? The proponents of behavioural economics respond in the affirmative. They promote an ethical position called 'libertarian paternalism'.

Paternalism is the inducing to behave or choose according to what a third person thinks is good for an agent. It may involve the restriction of the freedom of an individual against her will by means of coercion to pursue her own good. However, it may use other non-coercive means, such as persuasion, financial incentives, or design changes. Note that the induction of the behaviour or choice has to be:

- against the will of the agent (i.e. a choice they would not have made otherwise). If the choice that is affected is not substantially voluntary or knowledgeable—where the meaning of 'substantially enough' depends on the potential harm the choice or action in question may cause to the agent—then this influence on freedom is known as 'soft' as opposed to 'hard' paternalism. The latter is the doctrine that justifies interventions on behaviours or choices that are substantially knowledgeable or voluntary (Arneson 2005).

- for her own good. We can see that, for example, preventing an individual from causing harm to others is not paternalistic.

9.1 Libertarian Paternalism

Libertarian paternalism is a variant of soft paternalism that favours intervening in how alternatives are presented without—in theory—restricting the choice set and the individuals' freedom to choose (Sunstein and Thaler 2003; Thaler and Sunstein 2003). For example, given the status quo bias, if the legislation imposes the expression of voluntary consent for donations of organs when an individual dies, this could be instrumented by either an opt-out option where, unless a person refuses, she is classified as a donor or an opt-in option where only people who give explicit consent are classified as donors. The existence of the status quo bias suggests that an opt-out option would significantly increase organ donations compared to an opt-in alternative. Nothing prevents, in the former case, individuals not willing to donate their organs from opting out—no coercion is involved, the freedom to choose is not being curtailed—except the status quo bias, and so few will. Differences in how alternatives are presented have been found to have powerful effects on the rates of organ donation consent across countries (Johnson and Goldstein 2003).

I said that 'in theory' the number of alternatives is not restricted, because paternalist libertarians do not object, in principle, to removing harmful options or reducing the choice set.

Going back to the example of consenting for organ donation, notice that it is not more costly to, say, tick a box to opt in or to opt out: the transaction costs are the same in either case. However, sometimes the changes in choice architecture do modify the transaction costs compared to other choice design alternatives: a libertarian paternalistic policy may steer people towards a more costly though more environmentally friendly transport or energy option, for example. In these cases, the 'libertarian' element of the proposal or intervention may be compromised.

Dunn asked, in the context of older people with cognitive impairment where the disquisition rests upon a third-party definition of what is best for an older individual on the one hand and respecting her autonomy on the other:

Can paternalistic intervention ever be justified in overriding the wishes of the elderly in order to promote their well-being or prevent harm to them?

(Dunn 1999, p. 60)

And she responded affirmatively. Libertarian paternalism is not concerned with people with an impaired ability to make decisions—for a paternalistic intervention in such cases would not be morally objectionable[1]—but focuses on competent agents who, when it comes to making decisions, do not choose options according to the rational choice theory but behave, instead, as behavioural psychology and economics purports they would do: incurring in all sorts of biases and recurring to different heuristics. The call to a libertarian paternalist action is 'the fact that in many domains, people lack clear, stable or well-ordered preferences' (Sunstein and Thaler 2003, p. 1161).

9.1.1 Nudge

On introducing the concept of 'nudge', Thaler and Sunstein defined it as:

...any aspect of the choice architecture that alters people's behavior in a predictable way without forbidding any options or significantly changing their economic incentives.

(Thaler and Sunstein 2008, p. 6)

and went on to explain (op. cit., p. 6) that 'To count as a mere nudge, the intervention must be easy and cheap to avoid. Nudges are not mandates. Putting fruit at eye level counts as a nudge. Banning junk food does not.'

Nudges do not have to be paternalistic (see Raihani (2013) for some examples), but if their motive is paternalistic, then they are libertarian paternalistic interventions (Hansen 2016; Sunstein 2014).

One objection to libertarian paternalist policies is that policy makers try to impose their views of what is best for individual agents or for society disregarding the individuals' opinions and social preferences, which are swiftly dismissed as 'irrational'. Here is an example of such a concern:

Not only do libertarian paternalism and nudges manipulate our choices, but more importantly, they claim to do so in our interests while furthering others. ...this disregard for people's true interests is a natural legacy of the way that both mainstream and behavioral economists think about decision-making: a deliberative process, however complex, guided by an overly simplistic goal. These simplistic goals allow economists to build complicated models of decision-making, but economists neglect to question whether the goals and interests assumed in their models correspond to what real people value. They focus on the process more than the goal, and they end up missing the forest for the trees. In the end, they presume to know what people's interests are and to act to promote those interests—which is the most distressing problem with libertarian paternalism and nudges

(White 2013, p. 127)

A related objection is that '…typically work better in the dark. …If we explain the endowment effect to employees, they may be less inclined to Save More Tomorrow' (Bovens 2009, p. 217). In this sense, nudging is akin to institutional deceiving. If this is the case, libertarian paternalist policies and interventions based on behavioural economics would not be morally permissible. Bear in mind, notwithstanding, that there is some evidence that the behavioural effects of default options persist even after individuals are informed about their existence: in an online experiment, Loewenstein et al. (2015) invited individuals to complete hypothetical advance directives, that is, documents in which they expressed their preferences for end-of-life care measures (e.g. feeding tube insertion). Participants were randomly assigned into two groups, one of which was given an advance directive with a default option favouring prolonging life and the other group was presented with a document that included the opposite default option. Both groups were then randomly assigned again into two groups: one group was informed about the existence of the respective default option before completing the form, and the other group after completing it. No adverse responses to the disclosure of a default option was observed: neither did participants who were informed prior to the completion of the form try to resist the influence of the default nor those who were informed after completing the form tried to undo the effects of the default option—a phenomenon known as 'psychological reactance', that is, behaviour towards the restoration of autonomy, sense of control, or a freedom if they are lost or perceived to be threatened (Brehm 1966; Brehm and Brehm 1981). Incidentally, the evidence from social psychology, albeit scant, suggests that psychological reactance diminishes with increasing chronological age—see Añaños (2015) and Hong et al. (1994).

Nevertheless, other forms of paternalist policies, also informed by behavioural choice theory and behavioural economics, have been proposed to account for these potential pitfalls: constitutionally constrained paternalism, autonomy-enhancing paternalism, and asymmetric paternalism.

9.2 Constitutionally Constrained Paternalism

Proposed by Schubert (2014), this form of libertarian paternalism would reflect or track the constitutional preferences of the electorate, which would translate into a collective agreement that the government implements nudge and counter-nudge interventions to steer individuals away from their potentially deleterious first preferences—see also Heap (2017) for a similar argument.

Schnellenbach (2016) objected to the implicit premise that the outcome is all that matters when choosing so that nudging people towards utility-enhancing outcomes they would not choose without paternalistic libertarian interventions is always morally acceptable. The objection relies on the alternative premise, which has some backing in empirical evidence—see Lusk et al. (2013) and Marette et al. (2016)—that the procedural element of choice is also important: people value their autonomy to make decisions and suffer from a loss in welfare if their choice sets are somehow tampered with. As a result, Schnellenbach (2016) only found non-manipulative libertarian paternalist policies morally acceptable.

9.3 Autonomy-Enhancing Paternalism

Binder and Lades (2013) listed three problems with 'libertarian paternalism':

- how to identify what is best for individuals—that is, the ability of policy makers to judge if an agent's choice (i.e. her revealed preference) is 'informed' enough
- the emphasis on outcomes at the expense of autonomy
- the neglect of the effects of the interventions over time—in particular, the agents' ability to learn

At a different level, the authors also objected that libertarian paternalism equates the preservation of the freedom of choice with liberty and argued that another element of liberty is the ability to make informed decisions.

As a result, Binder and Lades proposed a remedial form of soft paternalism, autonomy-enhancing paternalism, which promotes 'behavioral interventions that help individuals to become better decision-makers and thus make better informed, less biased, and more autonomous choices over time that may better reflect their true preferences' (Binder and Lades 2015, p. 6). Libertarian paternalism attempts to influence preferences or behaviours by altering choice contexts without trying to reduce the cognitive biases people are prone to (not only the status quo bias as in the organ donation example, but framing, anchoring, etc.), but by making the most of them on behalf of the agents themselves (hence, its 'paternalistic' content). This is the objection raised by autonomy-enhancing paternalists.

According to the dual process cognitive theories (Stanovich 1999; Stanovich and West 2000), the human brain forms thoughts in two different ways or systems, labelled 'System 1', which corresponds to automatic, impulsive, fast, emotional, and frequent decisions, and 'System 2', which corresponds to reflective, conscious, slow, logical, and studious decisions—see also Evans (2003) and Kahneman (2011). Autonomy-enhancing paternalism criticises libertarian paternalism in that the latter seems to promote or at least make use of biases in System 1 decisions without enhancing System 2 mechanisms: '…behavioral interventions should foster critical thinking by strengthening System 2, weakening System 1, or encouraging that decisions are made in System 2 without affecting the strengths of the systems' (Binder and Lades 2015, p. 5).

Simplification and reductions in choice sets would be morally acceptable for autonomy-enhancing paternalists, whilst recurring to default rules to make use of inertia or status quo biases or relying on the effects of anchors would not. Finally, choice designs that do not take into account that agents can learn over time and thus miss the opportunity to reduce an existing cognitive bias would not be acceptable either.

9.4 Asymmetric Paternalism

Also known as 'cautious paternalism', this doctrine consists in the proposal to design interventions and regulations that offer or impose substantial gains and benefits to agents who incur in cognitive errors, without affecting the pay-offs of agents who choose or behave as the rational theory of choice model would predict. This doctrine is 'paternalistic' because the objective is to help agents to achieve their goals, and it is 'asymmetric' because it only influences those agents who would make irrational decisions, without impinging on the decisions of the other agents (Loewenstein et al. 2007). As Camerer et al. (2003, p. 1212) explained, asymmetric paternalism 'creates large benefits for those who make errors, while imposing little or no harm on those who are fully rational'. Default clauses for retirement saving or organ donations are two examples of asymmetric paternalism: these clauses impose no costs on those individuals who would save or donate regardless but does influence the outcomes for those agents who would not save or donate otherwise.

9.5 The Save More Tomorrow™ Programme

By the end of the 1990s, private firms in the United States of America were switching from schemes in which employees are promised a pre-specified retirement payment based on their earnings (so-called defined-benefit plans) to schemes in which retirement income is based on the contributions made by both employees and employers (i.e. 'defined-contribution' plans); see Volume III, Part II. In this way, the responsibility for saving towards retirement was being shifted onto the shoulders of the employees.

According to behavioural economics, economic agents are prone to procrastination, status quo bias, loss aversion, choice overload, and money illusion. Procrastination manifests in hyperbolic discounting (i.e. saving for retirement looks more attractive in the future than in the present), the status quo bias manifests in inertia, choice overload calls for simplicity and few options, and loss aversion manifests in weighing losses more highly than gains of the same magnitude. Money illusion, in turn, manifests when agents base their decisions on nominal values instead of real values—that is, on values not adjusted for inflation rather than on values after considering inflation. It is a 'bias in the assessment of the real value of economic transactions, induced by a nominal evaluation …due to the ease, universality, and salience of the nominal representation' (Shafir et al. 1997, p. 348). Money illusion is another anomalous behaviour that departs from the assumptions in mainstream economic models.

With such adverse influences on retirement saving, economists Thaler and Benartzi designed a programme to help employees increase their contributions into defined-contribution retirement plans. The Save More Tomorrow™ programme—or SMart—(Benartzi 2012; Thaler and Benartzi 2004) was first introduced in 1998 in a manufacturing firm with 315 members of staff. By 2012, more than 60 per cent of US companies were using the scheme.[2] Its 'basic idea is to give workers the option of committing themselves now to increasing their savings rate later, each time they get a raise' (Thaler and Benartzi 2004, p. S166).

Let's dissect this explanation. Workers needed to increase their contributions (i.e. savings rate) but that would translate into an immediate reduction in disposable income in exchange of a higher retirement income in the future, which was not a very enticing prospect according to the theories of behavioural economics. Furthermore, inertia was also acting against increasing contributions, for these would imply a change from whatever little they were already saving into their retirement. The employees would be less opposed to committing to higher pension contributions in the future than in the present due to the hyperbolic discounting and present bias. However, come that day in the future they would be likely to procrastinate again until 'mañana, mañana' because of inertia and the dent the additional savings would make in their disposable income. This demonstrates the nub of the time inconsistency of their temporal decision-making. However, Thaler and Benartzi used money illusion: what if instead of a fixed date in the future, the commitment were tied to the period when they got a pay rise? They could increase their savings without considering the money set apart for saving as a loss. Furthermore, as Hertwig and Grüne-Yanoff (2017) point out, the programme did not offer the workers a choice between consumption now and consumption later (as behavioural economists, the designers knew that workers would have chosen consumption now hands down) but between consumption now and consumption when salary increased in the future. Moreover, the choice was simple, no complex design scaffolding was erected.

The programme included three elements:

- The importance of increasing contribution rates in advance of the pay increase was explained to the employees. The timing had to be right, neither too close to the pay rise nor well in advance, due to hyperbolic discounting; as Rudzinska-Wojciechowska (2017, p. 5) explained,

  …the fact that the consequences of the decision were more temporally distant probably led people to give more weight to the benefits of saving and less to the costs of this decision, causing participation to increase

- If they signed up, the contribution rates went up with the first payment after the salary increase, to minimise loss aversion.
- The contribution rates would keep increasing with each pay rise until a fixed maximum, thus harnessing inertia (i.e. the employees would not be likely to opt out, despite increasing portions of their additional income going to a pension put and despite having the opt-out option open).

Of 315 employees, 162 joined the SMart programme. Their contributions went up from 3.5 per cent on average before the implementation to 13.6 per cent after the fourth pay rise. In contrast, forty-five employees declined signing up; their average pre-scheme saving rate was 6.1 per cent and fell to 5.9 per cent after the fourth pay increase. (The rest of the staff either did not contact a financial advisor or followed the advisor's recommendations outside the SMart scheme. Their contribution rates after the fourth salary increase were much lower than those who joined the programme.) In a nutshell, it worked.

However, it only worked for those employees who enrolled in the scheme, about 50 per cent of all workers. One problem with the scheme was that it presented a 'one-size-fit-all' design, when '…people's motivations to save for retirement undergo changes across the adult lifespan' (Rolison et al. 2017, p. 53); in particular, '…younger adults prioritize knowledge seeking in preparation for future possibilities whereas people approaching retirement prioritize realizing goals (e.g. making savings contributions)' (Rolison et al. 2017, p. 54).

Review and Reflect

1. In a revised version of his 2002 Nobel lecture, Daniel Kahneman stated:

   Economists often criticize psychological research for its propensity to generate lists of errors and biases, and for its failure to offer a coherent alternative to the rational-agent model. This complaint is only partly justified: psychological theories of intuitive thinking cannot match the elegance and precision of formal normative models of belief and choice, but this is just another way of saying that rational models are psychologically unrealistic. Furthermore, the alternative to simple and precise models is not chaos. Psychology offers integrative concepts and mid-level generalizations, which gain credibility from their ability to explain ostensibly different phenomena in diverse domains

   (Kahneman 2003, p. 1449)

2. On reviewing a celebrated paper by Arrow (1963) on health economics, Mark Pauly remarked that the paper

   …discussed concepts that made (and make) economists attentive but uncomfortable, like trust and morals. (These concepts do not tend to disquiet lawyers or policy makers.)

   (Pauly 2001, p. 830)

   I do not want to surmise whether lawyers or policy makers (I was going to add 'of all people' but refrained from doing so) get uneasy when trust and morals are discussed, but the question I do want readers to ponder about is why economist would. I am not talking of whether economists are being more untrustworthy or morally dubious than other professions, but from a theoretical point of view: why economists would not feel at ease with incorporating concepts such as trust and morals in their models

   (continued)

and theoretical lucubrations. Does anything in this chapter suggest otherwise?
3. Comment on the following assertion regarding the role of emotions on the time horizons of economic models and policies.

> In [the mainstream expected utility approach] only long-term consequences matter. Prospect theory, in contrast, is concerned with short-term outcomes, and the value function presumably reflects an anticipation of the valence and intensity of the emotions that will be experienced at moments of transition from one state to another...Which of these concepts of utility is more useful? The cultural norm of reasonable decision-making favors the long-term view over a concern with transient emotions. Indeed, the adoption of a broad perspective and a long-term view is an aspect of the meaning of rationality in everyday language...On the other hand, an exclusive concern with the long term may be prescriptively sterile, because the long term is not where life is lived. Utility cannot be divorced from emotion, and emotions are triggered by changes. A theory of choice that completely ignores feelings such as the pain of losses and the regret of mistakes is not only descriptively unrealistic, it also leads to prescriptions that do not maximize the utility of outcomes as they are actually experienced...
>
> (Kahneman 2003, p. 1457)

4. Research in behavioural economics resorts to experiments. How much would you say the following criticism applies to this sub-discipline of economics?

> Experimental social science can too easily become a caricature of real science, attempting to simplify or generalise the complexity, diversity and unpredictability of human beings with free choice. It simply cannot be studied without melding the theoretical with the empirical.
>
> (Weeks 1994, p. 345)

5. Discuss the following.
   The British economist John Maynard Keynes is apparently wrongly attributed the famous quip about changing his mind when facts change, implying other people did not proceed accordingly. As we saw in this chapter, among the basic assumptions of mainstream economic models are that economic agents are averse to risk and ambiguity, and that they discount the future at a rate that remains constant over time. What if, instead, it is demonstrated that they are risk-seeking, ready to embrace ambiguity, and that their preferences towards the future change as time goes by? Surely, if this is the case, any economist worth her salt should change her mind regarding the underpinnings of almost, if not all, her models, just as Keynes is attributed to have said he would do. Nevertheless, the ideas by behavioural economists seem, at best, to be trickling into the mainstream. This state of affairs reminds me of the following explanation about the reluctance by many economists

(continued)

to let go the assumption of rationality embedded in most macroeconomic models:

> The canonical paradigm that is taught to new Ph.D. students all over the world today is that people are like computers in the processing of information: They analyze all exogenous processes that impinge on the economy, break down the processes into statistical noise that is propagated through known impulse response functions, and calculate and implement their optimal response using dynamic programming. As a result, speculative markets are perfectly efficient and prices change through time only as a result of objective new information.
>
> It is of course absurd to imagine that people literally do such elaborate calculations, but the academic leap of faith has been that somehow people behave as if they do such calculations. Unfortunately, the daring of this leap of faith has largely been forgotten by economic theorists, who, after decades of such thinking in the literature, have the complacency to assume that these models are safely established and decorous.
>
> *(Akerlof and Shiller 2009)*

6. In a review of the first thirty years since the development of prospect theory in economics, Barberis (2013, p. 174) concluded: 'While prospect theory contains many remarkable insights, it is not ready-made for economic applications.'
   Why do you think, before reading Barberis's paper, that this could be the case?
7. Consider the following:

   > The strain of amalgamating different types of information into an overall decision may often force an individual to resort to judgmental strategies that do an injustice to his underlying system of values.
   >
   > *(Lichtenstein and Slovic 1973, p. 20)*

   Would such strain be bigger or smaller, generally speaking, the older the person who has to make a decision based on that information?
   Imagine the decision regarding buying an annuity, which types of biases might be triggered? And with which consequences?
8. In connection with myopia and time inconsistency, reflect upon the following thought:

   > The possible future selves have no power, save as frightening spectres to keep the current self in line.
   >
   > *(Posner 1997, p. 26)*

(continued)

9. We encounter the Norwegian philosopher Elster in this chapter. Consider this comparison between economists, on the one hand, and historians and sociologists on the other:

   In general I believe that the economists...tend to assume that preferences are mostly similar across time and space, whereas historians and sociologists are governed by their fear of committing anachronisms and imputing to premodern societies the preference patterns characteristic of their own.

   (Elster 1979, p. 115)

   To what extent do you think that economic theory would be affected if preferences happen to differ across time and space? Can economic theory accommodate 'middle range' theories Merton (1957)—see Volume I, Chap. 4?
10. This quotation also comes from Elster (1979):

    To postulate costs of information, costs of transaction, psychic costs or different time perspectives just to make the behaviour fit the theory is an unacceptable way out. Some independent evidence for these additional variables should always be given; the ad hoc hypotheses should have some independent predictive power.

    (Elster 1979, p. 156)

    Does behavioural economics provide 'independent evidence' for its additional explanatory variables and hypotheses? Should the same principle apply to any 'axiomatic' assumptions?
11. In several countries, housing wealth is considered an alternative to pension saving. However, the behavioural LC model would treat these two types of assets differently. The following argument adds a sociological interpretation for the lack of fungibility of housing and pension wealth:

    How people understand housing wealth (and debt) in relation to other forms of saving and insurance is furthermore complicated by the cultural and social constructions of 'home' vis-a-vis the experience of 'being in the market' ...and the reconfiguration of risk, choice and responsibility in homeownership and retirement savings ...

    (Clark et al. 2012, p. 1253)

    Do you think that apart from liquidity there are other relevant differences between housing and pension wealth to grant the concern expressed by the authors?
12. Consider the following statement (by two economists):

    Economists will and should be ignored if we continue to insist that it is axiomatic that constantly trading stocks or accumulating consumer debt

(continued)

> *or becoming a heroin addict must be optimal for the people doing these things merely because they have chosen to do it.*
>
> *(O'Donoghue and Rabin 2003, p. 186)*

Heroin addiction and increasing indebtedness are demerit goods—that is, 'bads'—that reduce utility or well-being. Would this objection also apply to 'normal' goods, and hence to mainstream economics in general?
Do you think that behavioural economics avoids this criticism? Why?

13. Consider the following proviso about the Save More Tomorrow™ initiative:

    > *a nudge like Thaler's Save More Tomorrow (Thaler and Benartzi 2004) will be effective if most people's discounting function is hyperbolic in shape, and doesn't change form under the intervention. If however people have widely differing discount functions, or their discount functions are not stable under intervention, then such a nudge is unlikely to be effective*
    >
    > *(Grüne-Yanoff 2017, p. 72)*

    Do discount functions change with chronological age? If so, how and what implications this may have for retirement saving schemes as such SMart?

14. Reflect on the following criticism raised at the SMart scheme:

    > *...the SMT nudge does not aim to foster people's competences. Instead, it skillfully designs an external choice architecture—involving automatic enrollment, projection of the choice to give up consumption into the near future, and dynamic adjustment of savings rates—that harnesses cognitive and motivational deficiencies to prompt behavior change ...the nudge approach steers behavior without taking the detour of honing new competences...*
    >
    > *(Hertwig and Grüne-Yanoff 2017, p. 978)*

    Would this criticism be applicable to libertarian paternalism in general? Do you agree that this is a source of concern? And, if so, how strongly do you object with this 'steering' without 'honing new competences'?

15. Hagman et al. (2015) distinguish between 'pro-self' and 'pro-social' nudges. The first type corresponds to nudges that steer economic agents away from irrational behaviour that goes against their own long-term interests. For example, the SMart retirement saving scheme is a pro-self nudge, because it is in the agent's interest to increase her pension contributions. Pro-social nudges, instead, steer economic agents away from behaviour that harms or reduces social, aggregate welfare but that increases the welfare of the individual. An example of a pro-social nudge is to enrol people by default, with an opt-out clause they are not expected to choose, in a pension scheme.

16. Here is a concern voiced by a group of economists regarding libertarian paternalist policies:

(continued)

> Unfortunately, benevolent institutional nudges, whether by the government or by other agents, will probably provide little protection for older adults. These benevolent nudges will often be outweighed by malevolent ones emanating from marketers and unscrupulous relatives.... Older adults with low financial literacy or significant cognitive impairment may be no match for highly incentivized parties with malevolent interests and ample opportunities to nudge in the wrong direction.
>
> *(Agarwal et al. 2009, p. 83)*
>
> Do you think that this concern is warranted? Provide examples of 'malevolent' nudges. Do you agree that they often outweigh 'benevolent' nudges? Why? If so, would this concern be applicable to other forms of paternalist interventions?

Notes

1. Outside certain forms of libertarianism, at least; see, for example, Szasz (2011).
2. Source: ' "Save more tomorrow' gives pensions a boost" ', Sophia Grene, January 8, 2012, Financial Times.

References

Agarwal, Sumit et al. (2009). "The Age of Reason: Financial Decisions over the Life Cycle and Implications for Regulation". In: *Brookings Papers on Economic Activity*, pages 51–101.

Akerlof, George A and Robert J Shiller (2009). "Disputations: Our New Theory Of Macroeconomics". In: *New Republic, May 8* URL: https://newrepublic.com/article/64866/disputations-our-new-theory-macroeconomics.

Añaños, Elena (2015). "La tecnologa del ńEyeTrackerż en adultos mayores: cómo se atienden y procesan los contenidos integrados de televisión". In: *Comunicar* 23.45, pages 75–83.

Arneson, Richard J (2005). "Joel Feinberg and the justification of hard paternalism". In: *Legal Theory* 11.3, pages 259–284.

Arrow, Kenneth J. (1963). "Uncertainty and the welfare economics of medical care". In: *American Economic Review* 13.5, pages 941–973.

Barberis, Nicholas C (2013). "Thirty years of prospect theory in economics: A review and assessment". In: *The Journal of Economic Perspectives* 27.1, pages 173–195.

Benartzi, Shlomo (2012). *Save More Tomorrow: Practical Behavioral Finance Solutions to Improve 401(k) Plans*. New York, NY: United States of America: Penguin.

Binder, Martin and Leonhard K Lades (2013). *Autonomy-Enhancing Paternalism*. Working paper 1304. Jena: Germany: Max Planck Institute of Economics.

—— (2015). "Autonomy-Enhancing Paternalism". In: *Kyklos* 68.1, pages 3–27.

Bovens, Luc (2009). "The Ethics of Nudge". In: *Preference Change: Approaches from Philosophy, Economics and Psychology*. Edited by Till Grüne-Yanoff and Sven Ove Hansson. Dordrecht: The Netherlands: Springer Netherlands, pages 207–219.

Brehm, Jack W (1966). *A Theory of Psychological Reactance*. Social psychology. New York, NY: United States of America: Academic Press.

Brehm, Sharon S. and Jack W Brehm (1981). *Psychological Reactance: A Theory of Freedom and Control*. New York, NY: United States of America: Academic Press.

Camerer, Colin et al. (2003). "Regulation for Conservatives: Behavioral Economics and the Case for "Asymmetric Paternalism"". In: *University of Pennsylvania law review* 151.3, pages 1211–1254.

Clark, Gordon L, Stephen Almond, and Kendra Strauss (2012). "The home, pension savings and risk aversion: intentions of the defined contribution pension plan participants of a London-based investment bank at the peak of the bubble". In: *Urban Studies* 49.6, pages 1251–1273.

Dunn, Caroline (1999). "The effect of ageing on autonomy". In: *Justice for Older People*. Edited by Harry Lesser. Amsterdam: The Netherlands: Editions Rodopi, pages 51–64.

Elster, Jon (1979). *Ulysses and the Sirens: Studies in Rationality and Irrationality*. Cambridge: United Kingdom: Cambridge University Press.

Evans, Jonathan St BT (2003). "In two minds: dual-process accounts of reasoning". In: *Trends in cognitive sciences* 7.10, pages 454–459.

Grüne-Yanoff, Till (2017). "Reflections on the 2017 Nobel Memorial Prize Awarded to Richard Thaler". In: *Erasmus Journal for Philosophy and Economics* 10.2, pages 61–75.

Hagman, William et al. (2015). "Public views on policies involving nudges". In: *Review of Philosophy and Psychology* 6.3, pages 439–453.

Hansen, Pelle Guldborg (2016). "The Definition of Nudge and Libertarian Paternalism: Does the Hand Fit the Glove?" In: *European Journal of Risk Regulation* 7.1, pages 155–174.

Heap, Shaun P Hargreaves (2017). "Behavioural public policy: the constitutional approach". In: *Behavioural Public Policy* 1.2, pages 252–265.

Hertwig, Ralph and Till Grüne-Yanoff (2017). "Nudging and boosting: Steering or empowering good decisions". In: *Perspectives on Psychological Science* 12.6, pages 973–986.

Hong, Sung-Mook et al. (1994). "Psychological reactance: Effects of age and gender". In: *The Journal of Social Psychology* 134.2, pages 223–228.

Johnson, Eric J and Daniel Goldstein (2003). "Do defaults save lives?". In: *Science* 302, pages 1338–1339.

Kahneman, Daniel (2003). "Maps of Bounded Rationality: Psychology for Behavioral Economics". In: *The American Economic Review* 93.5, pages 1449–1475.

—— (2011). *Thinking, Fast and Slow*. London: United Kingdom: Penguin Books.

Lichtenstein, Sarah and Paul Slovic (1973). "Response-induced reversals of preference in gambling: An extended replication in Las Vegas". In: *Journal of Experimental Psychology* 101.1, pages 16–20.

Loewenstein, George, Troyen Brennan, and Kevin G Volpp (2007). "Asymmetric paternalism to improve health behaviors". In: *JAMA* 298.20, pages 2415–2417.

Loewenstein, George et al. (2015). "Warning: You are about to be nudged". In: *Behavioral Science & Policy* 1.1, pages 35–42.

Lusk, Jayson L, Stephan Marette, and F Bailey Norwood (2013). "The paternalist meets his match". In: *Applied Economic Perspectives and Policy* 36.1, pages 61–108.

Marette, Stephan, Jayson L Lusk, and F Bailey Norwood (2016). "Choosing for others". In: *Applied Economics* 48.22, pages 2093–2111.

Merton, Robert K. (1957). *Social Theory and Social Structure*. New York, NY: United States of America: Free Press.

O'Donoghue, Ted and Matthew Rabin (2003). "Studying optimal paternalism, illustrated by a model of sin taxes". In: *The American Economic Review* 93.2, pages 186–191.

Pauly, Mark (2001). "Foreword". In: *Journal of Health Politics, Policy and Law [Special Issue: Kenneth Arrow and the Changing Economics of Health Care]* 26.5, pages 829–834.

Posner, Richard A (1997). "Are we one self or multiple selves? Implications for Law and Public Policy". In: *Legal Theory* 3, pages 23–35.

Raihani, Nichola J (2013). "Nudge politics: efficacy and ethics". In: *Frontiers in psychology* 4. https://doi.org/10.3389/fpsyg.2013.00972.

Rolison, Jonathan J, Yaniv Hanoch, and Stacey Wood (2017). "Saving for the future: Dynamic effects of time horizon". In: *Journal of Behavioral and Experimental Economics* 70, pages 47–54.

Rudzinska-Wojciechowska, Joanna (2017). "If you want to save, focus on the forest rather than on trees. The effects of shifts in levels of construal on saving decisions". In: *PloS one* 12.5. DOI: https://doi.org/10.1371/journal.pone.0178283.

Schnellenbach, Jan (2016). "A constitutional economics perspective on soft paternalism". In: *Kyklos* 69.1, pages 135–156.

Schubert, Christian (2014). "Evolutionary economics and the case for a constitutional libertarian paternalism -a comment on Martin Binder, "should evolutionary economists embrace libertarian paternalism?"". In: *Journal of Evolutionary Economics* 24.5, pages 1107–1113.

Shafir, Eldar, Peter Diamond, and Amos Tversky (1997). "Money illusion". In: *The Quarterly Journal of Economics* 112.2, pages 341–374.

Stanovich, Keith E (1999). *Who is rational?: Studies of individual differences in reasoning*. Mahwah, NJ: United States of America: Lawrence Erlbaum Inc. Publishers.

Stanovich, Keith E and Richard F West (2000). "Individual differences in reasoning: Implications for the rationality debate?" In: *Behavioral and brain sciences* 23.5, pages 645–665.

Sunstein, Cass R (2014). *Why Nudge?: The Politics of Libertarian Paternalism*. Storrs lectures on jurisprudence. New Haven, CT: United States of America: Yale University Press.

Sunstein, Cass R and Richard H Thaler (2003). "Libertarian paternalism is not an oxymoron". In: *The University of Chicago Law Review* 70.4, pages 1159–1202.

Szasz, Thomas (2011). *Faith in Freedom: Libertarian Principles and Psychiatric Practices*. New Brunswick, NJ: Unites States of America: Transaction Publishers.

Thaler, Richard H and Shlomo Benartzi (2004). "Save more tomorrow^{TM}: Using behavioral economics to increase employee saving". In: *Journal of Political Economy* 112.S1, S164–S187.

Thaler, Richard H and Cass R Sunstein (2003). "Libertarian paternalism". In: *The American Economic Review* 93.2, pages 175–179.

—— (2008). *Nudge: Improving Decisions about Health, Wealth, and Happiness*. New Haven, CT: United States of America: Yale University Press.

Weeks, David J (1994). "A review of loneliness concepts, with particular reference to old age". In: *International Journal of Geriatric Psychiatry* 9.5, pages 345–355.

White, Mark D (2013). *The Manipulation of Choice: Ethics and Libertarian Paternalism*. New York, NY: United States of America: Palgrave Macmillan.

Part IV

Political Economy

10

Economics and the Political Economy of Ageing

> **Overview**
>
> This chapter presents the understanding within economics of political economy and its relationship with economics and ageing. Topics include the median voter model, the fiscal leakage and elderly (or grey) power, and the role of interest groups around ageing issues.

10.1 Introduction

Many years ago, on reading an academic paper, I came upon a reference to a statistical technique I was unfamiliar with, hierarchical linear modelling. After a close look at the equations, I could recognise a random effects model—a technique I had been trained in as an economics undergraduate. When I looked up 'hierarchical models' in a library (this story predates the ubiquity of Internet searches), I learned that what, by and large, economists call random effects models is known, rather confusingly, as hierarchical, multilevel, or nested models ...depending on the discipline. In a sense, this part deals with the opposite problem: that different disciplines use the same term, 'political economy', but they refer to different concepts—in fact, to completely different lines of inquiry.

'Today there is little left in political economy into which the ageing problem does not intrude', wrote Mullan (2002, p. 3). This dictum could be turned around and would still ring true, because there is little left in ageing studies and policies into which political economy does not intrude. Both statements, however, beg one obvious question: what is political economy?

For the classical economists, such as Steuart, Senior, Verri, Smith, Ricardo, Say, or Stuart Mill, as well as for Marx and Pareto, political economy was what is now known, purely and simply,[1] as economics. So if economics is what economists do, as Jacob Viner allegedly remarked (Backhouse and Medema 2009), back then it was political economy what economists did. Sometime around the second part of the nineteenth century, 'political economy' became 'economics' as the epithet 'political' did not sound too 'scientific' (even though it referred to the polity, the body politic, the $\pi o \lambda \iota \tau \epsilon \iota \alpha$ of classical philosophy). Notwithstanding, the expression 'political economy' (or 'new political economy') later re-entered the lexicon of economics to denote a branch interested in the study of non-market, political, institutional, and collective behaviour by individuals and organisations[2]

Political economy applies

> ...an economic approach, constrained maximizing and strategic behavior by self-interested agents, to explain the origins and maintenance of political processes and institutions and the formulation and implementation of public policies. At the same time, by focusing on how political and economic institutions constrain, direct, and reflect individual behavior, it stresses the political context in which market phenomena take place and attempts to explain collective outcomes like production, resource allocation, and public policy in a unified fashion.
>
> (Alt and Alesina 1996, p. 645)

There is a different use of the term 'political economy' in economics: the opposite to orthodox (i.e. neoclassical) economics. Political economy or 'radical' political economy, in this view, comprises all forms of heterodox economics, including institutional, Marxist, post-Keynesian, Austrian, feminist, and ecological economics. Some authors expand the reach of political economics to include economic history, history of economics, and development economics to the heterodox approaches (Thornton 2016, Table 4.4). As Rothschild (1989, p. 3) opined—referring to this meaning of the term—political economy is '...characterised by a conscious opposition to the ruling neoclassic paradigm and various attempts to develop alternative approaches, methods, and theories' and went on to assert that political economists, though they disagree on many fronts, share a 'critical attitude towards 'pure" and neoclassical economics' [p. 11].

It is this assimilation of political economy with radical views what a group of gerontologists seized upon when they termed a critical approach to study later in life 'the political economy of ageing'. In the words of one of its main authors, the political economy of old age focuses on:

> …the social creation of dependent status, the structural relationship between the elderly and younger adults and between different groups of the elderly, and the socially constructed relationship between age, the division of labour and the labour-market.
>
> (Walker 1981, p. 75)

Two other prominent thinkers within this critical gerontological tradition[3] defined political economy of ageing as the line of inquiry that

> …takes as problematic the effects of social history, the world economy, capitalism and social class on the ageing process and the aged and the policy interventions designed for them.
>
> (Estes et al. 1982, p. 151)

Therefore, 'political economy' of ageing may refer to either the study of rational decision-making by and about older people and its influence on public policies, or the study of historical and socially constructed structural relationships of dependency and poverty in later life.

Gerontologists have also developed another sub-discipline, 'political gerontology', which looks into the political behaviour of older people not necessarily through a critical lens.

This chapter tackles the (orthodox) economists' ideas of ageing from a political economy perspective; the next chapter describes some of the social gerontologists' views on the subject.

10.2 Political Economy of Ageing: The Orthodox Economics View

Political economy, as developed by orthodox economists, addresses voting mechanisms and rules for arriving at collective decisions. For example, the implications for electoral results, decision-making, and policy of whether choices are made by majority rule or unanimity, whether any agent has veto power, whether there is a 'first past the post' rule, the possibility of holding run-off elections if no majority is obtained, the implications of franchise rules—that is, who can vote—and so forth. This branch of economics is also interested in the interplay between institutional and economic factors, including dynamic efficiency, time horizon, altruism, redistribution, crowding-out effects, and interest groups (Galasso 2008a; Galasso and Profeta 2002).

Political economy models are part of choice-based social theories of political attitudes and behaviour. There is a different family of theories that also attempt to explain attitudes and behaviour in the political realm: the structural-based family of theories. These theories are more prevalent in the 'political gerontology' sub-discipline of gerontology, which will be addressed in the following chapter.

In relation to economics and ageing, political economy models have been applied to consider, among others, the following topics:

- the extent to which the growing number of older people in ageing societies may be influencing the political process, and how, including:
- the design and adoption of age-related public policies (intergenerational redistribution, social security, health, education, childcare, etc.)
- the patterns of political participation by age

One key question is whether a person's voting decision is a function of her chronological age (Townley 1981). The main assumption in the literature is that it is; otherwise, I would have hardly written this chapter! However, as Davidson (2016, p. 57) remarked,

> The relationship between birth certificates and ballot papers can be a surprisingly complex variable for understanding political behaviour.

Political economists attempt to simplify this 'surprising' complexity by means of two groups of theoretical models: the majority voting and the interest group models. Galasso and Profeta (2002) explained[4] that the main difference between these families of models is that the former require that a majority of the electorate support the policy, whilst the latter contemplate the possibility that a minority group may enter in a coalition with other minority groups and carry through a given policy.

Breyer (1994, Table 1) presented the following classification of the main assumptions on which these models are based[5]:

- Decision rule. Direct democracy (one person, one vote), unanimity, interest group democracy, or indirect democracy (decision-makers optimise *weighted* utility, welfare, etc., functions)

- Voters' motivation. Either selfish or altruistic towards the older generation, their own family members, or all future generations
- Characteristics of the voters in a cohort. Either homogeneous or heterogeneous in income, wealth, or number of descendants
- Voters' perception of the validity of the decisions. Either indefinite or for one period (e.g. the 'honeymoon period', the 'first 100 days', etc.)
- Assumptions about the time structure. Either a static model or one with an infinite horizon of two or more overlapping generations

I am not going to cover all the different models stemming from the combinations of these assumptions, as it would demand a book-length treatment itself[6] even if it focused exclusively on older people. Rather, the following two sections present the backbone of the basic models from which most of the literature emerges.

10.2.1 Population Ageing and the Median Voter Model

In various parts along this textbook, we have come across a 'benevolent' politician, a decision-maker willing to maximise social welfare, happiness, quality of life, aggregate utility, and so on. In fact, we have assumed her existence more or less throughout the book, each time a 'social welfare function' or similar has been invoked and optimised. This politician has not only been benevolent, solely seeking that the aggregate of individuals the models called 'society' reached their maximum utility levels, but also alone, as she has always decided on the levels of the different variables without consulting, negotiating, debating, compromising, or, it goes with saying, without any setbacks and never conceding defeat. She has faced no conflict of interests, no individuals with their own personal agendas trying to influence hers, no other decision-making mechanisms than the optimisation of a function subject to one or more constraints. Politics was merely a set of technical procedures to turn her decisions into fruition. What? That this is yet another example of economists adopting wacky assumptions of the 'can opener' sort?[7] Well, perhaps, but it was also economists who, after the World War II set about to understand—using the tools of rational choice theory—how members of a committee vote, how collectives of people adopt decisions, and what the implications of various institutional mechanisms are for collective choice. Furthermore, these scholars became interested in whether incumbents manipulate short-term economic conditions before elections to favour their electoral chances; the role of pressure and interest groups; the trading of favours by legislators; the formation of coalitions; why, in countries where voting is not compulsory, people bother to vote at all; and many other topics. Crucially, they were not thinking of benevolent, altruistic decision-makers, but of selfish agents, with their own sets of preferences, influenced by interest groups with their own agendas: a 'new' political economy was born.[8,9]

Pre-eminent among these theoretical developments is the median voter model, which Congleton (2004, p. 383) went as far as to call 'a fundamental property of democracy'. This author also opined:

> There is no more transparent nor easily communicated explanation of political outcomes in a democracy than that all political outcomes reflect median voter preferences.
>
> (Congleton 2004, p. 383)

Sorry for the spoiler, but that 'all political outcomes reflect median voter preferences' is the main conclusion of the median voter theorem—perhaps, the most famous choice-based theory of political attitudes and behaviour.

Sometimes, in indirect democratic political systems, the electorate is asked to vote between two alternatives, as in a referendum. For example, on 27 and 28 November 1994, the Norwegian electorate, on 22 January 2012, the Croatian electorate voted on whether to join or not the European Union, and, on 23 June 2016, the British electorate voted on whether to withdraw from or to remain in the European Union. However, it is more common to decide between candidates from different political parties, or between candidates vying for representing a party in primary elections. Here, the electorate does not face a binary option but an array of possibilities: each candidate or party can be thought to represent a position along the multidimensional space of political options and variables (say, each candidate or party is more likely to adopt a particular position regarding income tax, publicly owned enterprises, migration, foreign policy, etc.). The assumption is that people vote for the candidate or party more aligned to their preferences.

To simplify, let's imagine that there is one single issue to consider rather than a multidimensional space of topics. We can think of the topic with highest salience—that is, the hottest topic of political discussion—during the pre-electoral period, say, public spending on a particular policy or project. No one talks of anything else; the election will be decided on this issue. Still, the electorate faces several alternatives regarding the amount of public monies to spend on that policy or project. Imagine there are three alternatives: $A1$, $A2$, and $A3$. We assume that each voter has one (and for the moment, only one) amount of choice in mind: voter A would rather the government spent

A1 to A2. This means that we are assuming that there are no two options about which this voter is indifferent. And the same with each voter: each person prefers one and only one option. Even without indifference about any options, there are many ways to order the preferences of a voter; to illustrate, the following schema shows the possible preferences (denoted by the symbol $\succ$) between three alternatives:

- $A1 \succ A2 \succ A3$
- $A1 \succ A3 \succ A2$
- $A2 \succ A1 \succ A3$
- $A2 \succ A3 \succ A1$
- $A3 \succ A1 \succ A2$
- $A3 \succ A2 \succ A1$

The possibilities quickly increase with the number of alternatives.[10] If we included indifference, the possibilities would grow even more (there would be thirteen possible preferences instead of six, given the three alternatives).[11]

Actually, again to simplify the presentation, let's turn those 'several' alternatives into continuous amounts along a line from 0 to 1 sorted in increasing order of spending. We assume, then, that everyone has made up their mind and we also assume that they will vote accordingly. That is, they will not vote strategically, but will vote for whatever option they prefer; this is known as the 'sincerity' assumption.

Each of the j voters prefers an amount v^j between 0 and 1; in symbols, $0 \leq v^j \geq 1$. Let's assume that the amount x renders voter A the highest utility, v_x^A; hence, she prefers this amount to any other. But the electoral outcome may not coincide with her wishes. We can assume that the further away the electoral result is from a voter's preference, the lower will be the utility she will experience (hence the median voter model is also known as the 'spatial voting model' for the centrality of the notion of preference distance).

So, we have that each voter prefers one and only one amount, that any other option reduces her utility, and that this reduction is a function of the distance between the different options and her preferred option. This is known as a single-peaked or unimodal preference; 'single-peaked' because starting from the option of choice, any other alternative makes the road all downhill in terms of utility.

Let's assume something else: individual preferences are symmetric, so that preferences decrease by the same extent as the options get farther away from

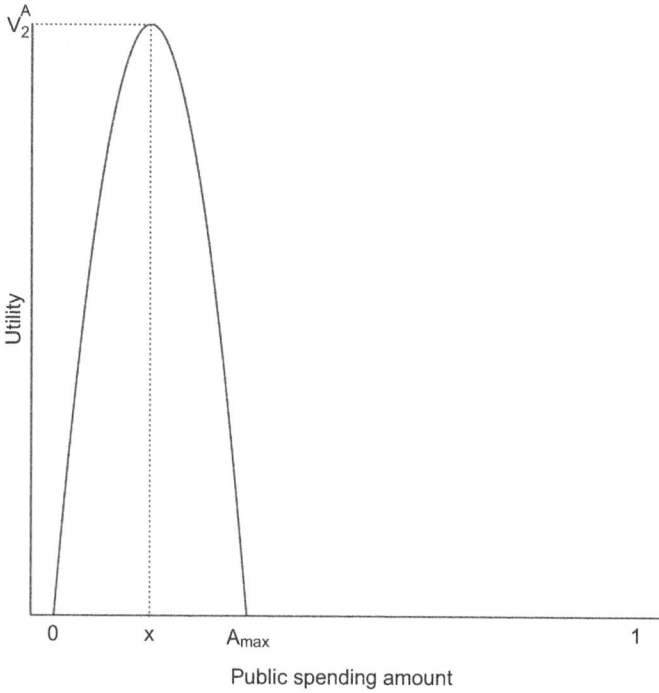

Fig. 10.1 Single-peaked preference of voter A. *Source: Figure is illustrative, prepared with mock data*

the preferred choice each way (i.e. spending more or spending less, in our example) (Figs. 10.1 and 10.2).

An individual's utility can be depicted as a bell-shaped curve with maximum at her preferred choice. For example, voter A prefers the amount A_x, and anything beyond A_{max} would not give her any utility at all:

Other voters will have different preferred amounts, but we assume they all have the same utility functions. For example, voter E prefers the government spent much more than voter A; anything below E_{min} would render no utility to her:

Figure 10.3 presents the utility functions for five voters.

Single-peaked preferences can be formalised thus. Let's assume voter A derives the maximum utility from option x, which we denote by $U^A(x)$. Now let's consider two other options, y and z. The voter's utility function is single-peaked if $U^A(y) \leq U^A(z)$ for $x \leq z \leq y$ or $y \leq z \leq x$:

Now a key assumption: there is a decision-maker (say, an autocratic ruler in an otherwise formal democracy), whose sole objective is to remain in power,

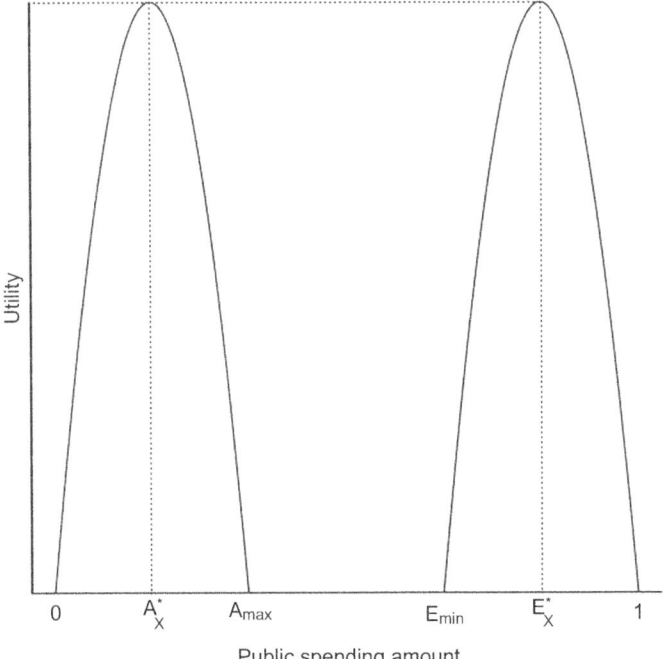

Fig. 10.2 Single-peaked preferences of two voters. *Source: Figure is illustrative, prepared with mock data*

for which he needs to win the election (i.e. what is sometimes known as an 'office-seeker' agent). Another assumption to throw in: the absence of intergenerational altruism, which means that voters do not care for their offspring (i.e. the utility of their offspring is not an element of their own utility function) (Fig. 10.4).

There's more assuming to do: this is the election to end all elections, a once-and-for-all decision; whatever option gets voted, it will remain a policy *for ever*. No U turns, no further electoral calls. (This assumption is called 'commitment'.)

Given all these assumptions, which amount of public spending will this electoral hopeful decide on?

We can obtain the number of voters that would vote for each amount—that is, the frequency of voters per alternative. In other words, we can get the distribution of public spending preferences in the electorate, a ranking of voters along their respective preferred choice.

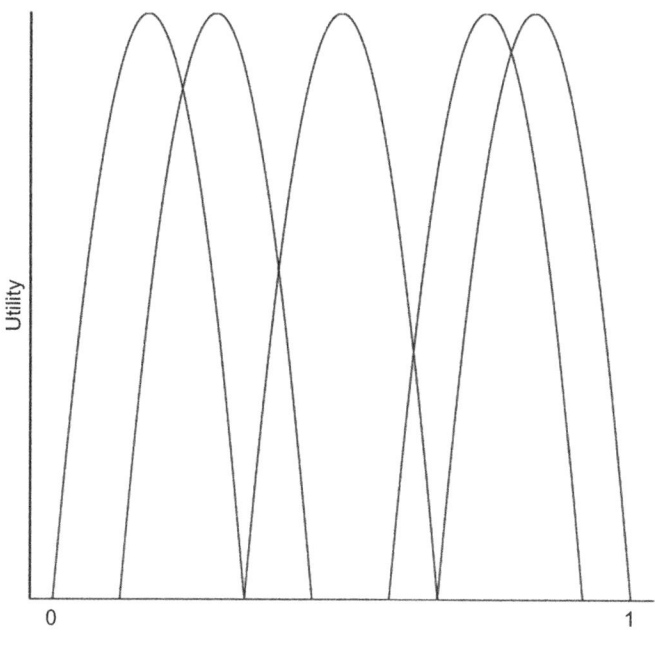

Fig. 10.3 Single-peaked preferences of five voters. *Source: Figure is illustrative, prepared with mock data*

Under all the assumptions listed above, the key voter is the *median* voter—that is, the voter whose preference corresponds to the median preference of the electorate (in this case, the median of the individual choices of public spending amount on the policy or project). With these assumptions, we can forget about optimising a long-run social welfare function; politicians just try and please the largest number of voters in their birth cohort for which, *in extremis*, they are only interested in pleasing one person, the median voter.[12]

Why? Imagine a candidate announcing that, if elected, she will spend the amount A_x. That would maximise A's utility but would alienate lots of voters who would rather the government spent more. The opposite would be true if the candidate announced she would spend E_x: most voters would not vote for such a large amount. If a candidate announced she would spend med_x, she would get the highest number of votes: 'the median voter's preference will emerge as the collective preference in a majority rule election' (Holcombe 1989, p. 117). The candidate (or the policy maker seeking to remain in power) will

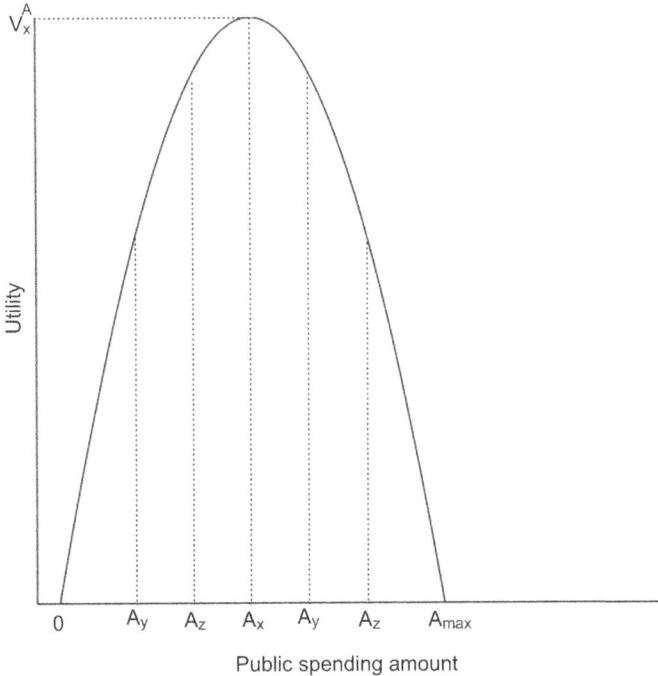

Fig. 10.4 Single-peaked preferences of five voters. *Source: Figure is illustrative, prepared with mock data*

choose the level of spending that maximises the utility of the median voter. This result, expressed formally, is known as the 'median voter theorem':

> If x is a single-dimensional issue, and all voters have single-peaked preferences defined over x, then xm, the median position, cannot lose under majority rule.
> (Mueller 2003, p. 86)

Imagine again that there are two options, x and y. Voter A prefers x, because $V_x^A > V_y^A$—that is, option x renders her more utility than option y. If voter A happens to be the median voter, option x is chosen by the majority of the electorate, by definition of the median. Now imagine that the median voter is voter M, who prefers y over x, as $V_y^M > V_x^M$. Now, it is option y which is chosen by the majority, again by definition of the median. Therefore, candidates will propose the option that coincides with the median voter's preference, because that is the winning option under majority voting.

This result should give pause to anyone interested in ageing. If a population is ageing, the median age is rising, which means that the chronological age of the median voter is rising. The median voter theorem predicts that policy choices preferred by older people will trample over the preferences of any other age group; older people will get whatever they may want because politicians seeking to get elected or re-elected will lean towards the median voter irrespective of the consequences for social welfare, happiness, national income, and so on. Moreover, once the preferences of the median voter are known, policy proposals (and decisions) by the different candidates, incumbents, and political parties will fully converge to these preferences, regardless of partisan affiliation, ideology, or any other consideration. With ageing median voters, democracies will become gerontocracies:

> In 2018, the median age will exceed the indifference age by one year and the proportion of those who are against the reform and who, instead, favour an extension of the pay-as-you-go system, will be 53.0%. Germans will then be trapped in their pension system, unable to realize further reforms in the direction of funding. Germany will be a gerontocracy: The old can, through their majority, exploit the young by increasing pension burdens.
>
> (Sinn and Uebelmesser 2003, p. 157)

That population ageing may increase aggregate public spending on age-related topics such as pensions is hardly an earth-shattering prediction, given that with a higher number and proportion of older people needs would also increase. However, the theories of gerontocracy based on the median voter theorem not only predict that a higher proportion of older people increases the size of public policy programmes aimed at older people, but that it may also increase the generosity of these programmes—in other words, that spending per older person increases along with the proportion of older voters. And this begs a relevant research question: does it?

Empirical Relevance

Every theoretical model hinges on a number of assumptions. Basic assumptions usually require that additional assumptions be introduced in the models, compounding the care with which theoretical economics output must be treated. The strong dependency of conclusions upon underlying assumptions is a characteristic of all models in theoretical economics. Once one or more assumptions are relaxed, the fundamental insights and conclusions of a model may change. For example, we concluded above that the median voter model predicts, under a number of simplifying assumptions, that political economic games in ageing societies are a prelude to gerontocracies. The gerontocracy thesis, according to Davidson (2016, p. 53),

> …resonates with, and feeds into, a developing political narrative that assumes electoral futures will be determined by wholly selfish rational choice older voters. This is constructed as a future threat that requires swift political action to reduce generosity and range of welfare spending for people in later life.

In other parts of this textbook, I mentioned authors that saw in population ageing a source of fiscal imbalance of calamitous proportions, a powerful spanner in the wheel of economic growth, and several other negative implications. Endorsing the gerontocracy thesis, Berry (2012) went even further in his foreboding: population ageing would 'threaten the practice of representative democracy in a profound sense' [p. 709], mostly by 'unravelling democratic legitimacy' [p. 721]. No less.

One of the various things the median voter model assumed is the absence of intergenerational altruism—the assumption of perfect intergenerational selfishness. Consider Gonzalez-Eiras and Niepelt (2008); this paper presents a model of social security populated with younger and older individuals in which total lack of altruism towards future generations was also assumed. Older people were supportive of the introduction of social security benefits and younger agents opposed the policy. But social security systems came into existence in this model because of an additional assumption, a theoretical 'twist': younger individuals were assumed not to be against these benefits as much as older people were in favour of them. The rationale behind this additional assumption was that younger people, though bearing the cost of the contributions levied on their income, realised they would benefit from some indirect gains. In particular, the contributions would reduce their savings, which would in turn reduce the demand for capital, thus increasing the price of capital goods. Capital goods will be in the hands of the younger generation when they become older in the next period, so this price increase would offset the impact of the social security contributions. Moreover, this capitalisation and monopolistic position in the supply of capital would allow younger individuals to call the shots when the time would come for them to trade with their own children yet to be born (see also Kotlikoff and Rosenthal 1993). As we can see, one additional assumption changed the conclusions of the model compared to the canonical version of the median voter theorem.

The median voter is a model of *demand*. Demand for regulation, policies, spending, transfers, and so on. Now, the median voter can be defined along any characteristic, not only chronological age as we assumed above. In fact, there are other variables more influential on political participation and voting behaviour and on the demand for public goods, among older people than chronological age. Health, for example, has been reported to be of higher concern than pensions among older people. Other variables, such as social class, gender, ethnicity, religion, or generation (in the Mannheim's sense; see Volume I, Chap. 2), cut across several domains. Therefore, even if we accept the theorem, it does not necessarily follow that the preferences of the electorate are to be distributed by chronological age. The median voter is not necessarily the voter of median age; chronological age is not the sole—not even the main—dividing line among interest groups and coalitions (Schulz and Binstock 2006; Street and Cossman 2006).

In principle, then, we would be treading on safer ground if we restricted the analysis to age-related areas of policy. This line of reasoning would suggest that the median voter model based on chronological age would not be applicable to policy topics in which age is not relevant, such as public spending on nuclear deterrence, for example, but to other issues such as pensions, healthcare, primary and secondary education, and so on. It should be no surprise, then, that it is on these areas where most of the theoretical and empirical work about ageing and the median voter has focused. Of course, even around these topics, people hold different views for other characteristics than their age, but at least the theorem provides testable propositions regarding the association between chronological age, political behaviour, and outcomes, once other relevant variables are controlled for.

The empirical evidence is mixed (Busemeyer et al. 2009). Many papers suggest that older people support, on average, policies that favour other age groups and which would not, directly, favour them, such as childcare services (Fullerton and Dixon 2010; Goerres and Tepe 2010; Prinzen 2016). Some authors have surmised that these preferences are not necessarily altruistic, but could be explained by selfish motives (e.g. older people would support higher spending on education as it would reduce crime and fear of crime, or because better schools in their locality would increase property prices, etc.).

When preferences for elderly spending was reported among older people, the effects tended to be small (Sørensen 2013). However, De Mello et al. (2017) reported a strong positive relationship between population ageing and support for public spending on pension and an equally strong negative association between ageing and support for public expenditure on education across thirty-four countries (the 'grey peril' hypothesis).

Boersch-Supan et al. (2011) investigated the relationship between the old-age dependency ratio (see Volume I, Chap. 5) in all regions in Europe and indicators of intergenerational cohesion, including financial transfers from parents to children, financial transfers from children and grandchildren to their parents and grandparents, and the absence or presence of conflict as seen from either parents and grandparents or children and grandchildren (the data are from 2004). The authors found no evidence that population ageing would strain intergenerational relations, and also that the higher the old-age dependency ratio, the more people would prefer higher taxes and spending more on social services, including among respondents younger than 30 years old (see also Börsch-Supan 2013). They concluded:

> …we cannot find even in the oldest regions of Europe the picture that some writers have painted for the United States: exploitation of the younger generation by the older, due to their economic and voting power. If Old Europe holds a lesson for the United States, it is that is no signs of gerontocracy exist even in regions as old today as the United States will be in 2030.
> (Boersch-Supan et al. 2011, p. 20)

If we assume pure selfishness within each generation, this result is counter-intuitive. Montén and Thum (2010) provided a theoretical explanation based on the notion of fiscal competition at local level. With ageing populations, the fiscal space of local governments would erode (see Volume II, Chap. 2), so they need to attract agents of working age (i.e. 'young' families) in order to maintain and increase their tax base. Therefore, they need to provide public goods attractive to, and design public policies aimed at, younger people. In theory, the intensification of population ageing would lead to higher levels of provision of public goods for younger people than under a scenario of no demographic change. Furthermore, these authors predict that the more unequally distributed is the older population across municipalities, the higher the level of pro-young public spending by gerontocracies will be.

A different explanation is given by Kemmerling and Neugart (2009). Most political economy models focus on one interest group—in our case, older people—and one median voter, here defined by chronological age. These authors studied the lobbying by financial services companies across eleven developed countries and looked into two particular reforms of the pension systems: the 1987 reform in the United Kingdom and the 2001 reform in Germany. In both cases, despite the ageing of the population and the increasing median chronological age of the electorates, compulsory pension contributions were lowered, affecting the interests of future older people. The authors proposed that this was the result of effective financial market lobbying in order to increase households' demand for private old-age savings products.

Another pitfall of median voter models is that they assume away institutional inertia—what is also known as 'path dependence' (Tepe and Vanhuysse 2009). Institutional inertia/path dependency can impede or delay that the preferences of the median voter or that older people's influence may translate into policies. Consider the models we described above. Their main policy choice variable, the thrust of the political game, was the level of social security contributions. Voters were divided on this topic and the government or the candidates were supposed to be free to set the level to that which maximised the utility of the median voter. However, governments inherit complex systems, legislation and institutional structures, which they cannot dismantle on an impulse. Moreover, workers who have made contributions to a system have gained vested pension rights.

The basic median voter model assumes that elections take place only once and that the promises of politicians are binding, indefinitely, over time: whatever they promise during the electoral race, they will deliver. These are, to say the least, extremely simplifying assumptions. A revised model specification may contemplate elections as a game played regularly in each period. Would this alternative assumption change the main prediction of the canonical model? To begin with, future elections introduce a degree of uncertainty in the model around their outcomes. 'What if the currently young generations change the rules of the game and affect 'our' pension income?', wonder the agents of working age who are making social security contributions. Hu (1982) introduced the assumption that current workers' expectation about the future level of social security contributions depends on the current level of their contributions and a stochastic term—that is, $\Gamma(\tau, \epsilon)$, where τ represents the contributions and ϵ the stochastic term. The assumption is that $\frac{\partial \Gamma}{\partial \tau} > 0$, that is, workers expect higher pension income as a result of their higher contributions to the social security system (and therefore to current retirees).[13] There is a risk, measured by ϵ, that younger workers repudiate the implicit intergenerational compact, manipulate their voting patterns, and increase their lifetime income and consumption. Driven by this income effect, they may proceed accordingly. However, increasing uncertainty about future social security contributions has the additional effect of raising precautionary saving towards funding for consumption in later life. Will, in the aggregate, the income effect or the precautionary saving effect prevail? Although ultimately an empirical matter, Hu assumed that the income effect is not so big that consumption increases. Therefore, the main prediction of the model is that this uncertainty reduces current consumption, lifetime consumption, and the contribution rates compared to what the median (middle-aged) voter desires. This author considered another source of uncertainty: demographic change. What if the future number of contributors to a PAYG pension scheme is not enough to provide contributions at a level current workers are funding? In this model, demographic uncertainty also reduces the median voter's chosen level of social security.

Boadway and Wildasin (1989a,b) extended this model by introducing a capital market but one in which agents face borrowing constraints and long voting cycles (i.e. multi-period life cycles). A currently median voter (of working age) would find that the longer the wait for a next election (or vote on social security contributions), the closer she will be to retirement, or the further into retirement. Therefore, longer voting cycles will translate into higher contribution rates.

There is a plethora of median voter models. The models that assume sequential games—that is, a dynamic environment in which actors play the same game in each period—show that one key parameter is whether voters expect their individual contribution influences the contributions that, on their behalf, may be made by future generations. If current voters do not expect to influence the behaviour of the next generation, they would not contribute to current retirees. Instead, if they expect that they may influence the next generation, they will act accordingly. One such influence could come in the shape of a sinking feeling: voters can feel, like the sword of Damocles hanging over their heads, that the next generation may threat onto withhold their contributions in retaliation for their lack of support towards the current generation of retirees. If this is what they expect, current voters will make positive transfers to current retirees (see Conesa and Krueger 1999; Cooley and Soares 1999; Cooley et al. 1999). This is understood as an intergenerational social contract (Boldrin and Rustichini 2000). Whether this expectation is warranted or not, studies have predicted increasing pension spending as a result of repeated games. For example, using simulations based on demographic projections, Galasso (2008b) predicted that social security contribution rates in Spain would increase from 21.3 per cent in 2000 to 45.5 per cent in 2050.

10.2.2 The Age of Policy Makers

A different approach is to focus on the chronological age of policy makers rather than voters. Would it be that older politicians take decisions that reflect age biases? This would support a charge against 'gerontocracies', irrespective of how they may be formed. This is what Atella and Carbonari (2017) set about to find out. These authors proposed a model in which the more impatient the ruling classes or political elites are in a country, the more the prospects of economic growth are harmed. The main transmission mechanism is that older decision-makers would favour short-term rewards compared to younger politicians, who would adopt a longer view on the consequences of policies. As short-termism among decision-makers would reduce support for, and spending on, human capital accumulation and information and communication technology, a majority of older decision-makers would have adverse consequences on productivity growth. The authors based their model on the premise that older people would tend to have higher subjective temporal discount rates (i.e. they would be more impatient, putting less value on future outcomes than younger people), from where they conjectured that gerontocracies would be harmful for economic growth. Atella and Carbonari reported significant and negative associations between the average chronological age of members of the national legislative houses in Denmark, Finland, France, Italy, Germany, Netherlands, and the United Kingdom between 1983 and 2004 and total factor productivity growth. They replicated this finding in a modified specification of the model that uses median age of the population instead of the average age of parliamentarians. Needless to say, terming higher average chronological ages of lawmakers' (whose average chronological age in the oldest of the three samples used in the paper was 48.74 years with a standard deviation of 1.99 anyway) 'gerontocracy' verges on the abuse of the language—although the term has been used, confusingly, to refer to a form of age-related political behaviour and to a form of political organisation (Eisele 1979). Nonetheless, this is initially an intriguing result that merits deeper analysis of the actual decision-making of legislators and its processes within each nation. Although behavioural economics literature has reported mixed results regarding myopic behaviour in later life, neuroeconomics and neuroimaging studies have found adverse ageing effects on the ventromedial prefrontal cortex (vmPFC), the region in the brain which is associated with tasks involving future discounting (see Chap. 8 in this volume). The implication that older parliamentarians would show a statistically significant deterioration of the vmPFC compared to younger parliamentarians, thus affecting their future orientation and biasing their decision-making regarding the longer term, is a fascinating hypothesis (which obviously cannot be granted from this study and which would be wholly inappropriate even to suggest at this stage…).

One crucial point that somehow gets lost, especially in the theoretical models of political participation and behaviour, and which, though dotted all over this textbook, is worth repeating (and repeating), is that older people exhibit great heterogeneity. Regarding older people and politics, Walker put it succinctly thus:

> [O]lder people do not necessarily share a common interest by virtue of their age alone which transcends all other interests. Thus, it is mistaken to regard senior citizens as a homogeneous group which might coalesce around or be attracted by one-dimensional politics of old age.
>
> (Walker 1999, p. 16)

Voting Cycles

So far we assumed that the transitive rule holds. The table above presented the preferences of a voter between three alternatives. Imagine that instead of one voter, we show the preferences of several voters between these three alternatives. The individual preferences comply with the transitive rule, but the collective or aggregate preference may not: we could get that some prefers option 1 to 2, others option 2 to 3, and yet other voters prefer option 3 to 1. If each group is of the same size, no option is favoured by a majority of voters. This situation is known as a voting cycle (also as the 'paradox of voting' or 'Condorcet paradox').

In theory, this paradox is not too unlikely. The probability of occurrence increases with the number of voters and alternatives, as Table 10.1 illustrates.

Furthermore, bundles of policies may generate the paradox:

Table 10.1 Probability of paradox of voting

		Number of voters				
		3	5	7	9	Limit
Number of alternatives	3	0.056	0.069	0.075	0.078	0.088
	4	0.111	0.139	0.15	0.156	0.176
	5	0.160	0.20	0.215		0.251
	Limit	1	1	1	1	1

cycling is always more likely when voters are permitted to vote on related aspects of public policy—for example, on the volume of expenditure of some publicly-provided service and on the fiscal institutions for funding such a service—than when they are restricted to voting on each dimension separately.

(Rowley 1984, p. 119)

In practice, there have been examples. See Bochsler (2010) for a real-life case of a voting cycle in the Swiss canton of Bern in 2004 and Kurrild-Klitgaard (2017) for cases in the 2016 Republican presidential primaries in the United States of America. Van Deemen (2014) assessed the empirical relevance of the voting paradox in general.

There are various technical conditions that can avoid the paradox[14] as well as a number of institutional arrangements to solve it, including:

- Issue-by-issue voting. In many legislative houses, committees discuss single topics and agree on bills that are expected to carry support from the majority in the house. This way, each committee addresses and votes on a unidimensional domain.
- Apart from issue-by-issue voting, another mechanism is to set the order of voting and sequential voting (Dekel and Piccione 2000).
- Another solution is agenda setting (Cox and McCubbins 2005; Jenkins and Monroe 2016; Romer and Rosenthal 1978). Setting the agenda includes the power to initiate policy (e.g. the European Commission has the monopoly to initiate policy in the European Parliament: no national government of a constituent country can initiate a proposal to modify existing legislation) and the power to prevent changes from being made (i.e. gatekeeping power or negative agenda control).
- Veto power. If the paradox arises, one agent (e.g. a president) has the power to tip the balance and break the cycle (McCarty and Poole 1995).
- Legislative bargaining and lobbying (Helpman and Persson 2001).
- Probabilistic voting (Banks and Duggan 2005; Coughlin 1992): the probability that a voter actually casts her vote may depend on how distant the policy alternatives are from her own position.

10.2.3 Elderly Power and Fiscal Leakage

In the context of the political economy of population ageing, there are two families of median voter models: the 'elderly power' and the 'fiscal leakage' hypotheses (Tepe and Vanhuysse 2009). The elderly (or 'grey' or 'senior') power hypothesis is based on the fact that for many age-related policies, older people benefit from the transfers or services but do not fully internalise the costs of these policies. The same is assumed about people who are close to older age (e.g. retirement), who may feel they are to obtain net benefits from such policies in the not-too-distant future. The fiscal leakage hypothesis, in turn, proposes that because population ageing implies that, per person, benefits and spending on services for older people will decrease and so will their profitability, people of working age will vote for reduced age-related spending unless older people form a majority. Therefore, different median voter models predict opposing views, depending on additional assumptions.

On reviewing the historical evidence in developed countries since the late 1980s, Kohli and Arza (2011, p. 260) concluded that the grey power hypothesis had been rejected 'partly due to its erroneously mechaniňcal model of voting preferences', as older people 'do not only vote in their own narrow self-interest' but 'are also interested in the well-being of their descendants and are net contributors in the intergenerational exchange with them'. (See also Fullerton and Dixon (2010), Hamil-Luker (2001) and Street (1997).)

A Model of Elderly Power

Let's consider the median voter along income or wealth dimensions. As discussed in Part II in this volume, the distribution of income or wealth is highly asymmetrical in virtually every country and region, with greater proportions of individuals or households earning less than the mean income or holding less than the mean wealth. Given this inequality, the median income or wealth are lower than the mean amounts. In these cases (i.e. in all real-life cases!), which redistributive policy will the median voter choose, if by median voter we mean the voter of median income (or wealth)? She will vote for a proportional income tax that finances a lump-sum transfer. Imagine the vote on an income tax to generate funds for a redistribution programme for people on low income, and let's assume that voters are only interested in maximising their income. Individuals on low incomes will vote for a 100 per cent tax on the income of the rich; they will not be taxed and will receive a transfer. In contrast, individuals on earnings above the median income will oppose the proposal, for they will be taxed without receiving anything in exchange. Surely, with inequality in the income distribution, such a proposal will have majority support. Needless to say, despite redistributive policies have been enacted worldwide, income (and wealth) inequality has not been erased—and, in many places, they are growing. Why? Is it not true that according to the median voter theorem the proposal should gain a majority? Yes, but spokes could be put in the wheels of this model: a unidimensional matter could be turned multidimensional. No, you may object: this is the only proposal under discussion; it is unidimensional all the way. Yes, in that sense it is, but I mean becoming multidimensional through the backdoor: the added dimension or dimensions would not be open to voting, but would divide the electorate so that each sub-group would have different income distributions and median incomes. These cross-cutting dimensions may include ethnic background, religion, area of residence, and, among others...age group.

How salient these additional dimensions for the voters are is key. The salience of a dimension depends on the topic in question—for example, religion may be more salient for a vote on abortion than on an extension of an airport runway, but environmental concern may be more salient than religion during a campaign on hydraulic fracturing (i.e. 'fracking'). Roemer (1998b) showed that a salient additional dimension may overthrow the voting result stemming from the median voter theorem before its introduction. The basic idea is that the right additional dimension will strike a chord among a sub-group of the electorate, whose median does not coincide with that of the median voter. Those voters not mobilised by the issue thus reframed—that is, the voters for whom the additional dimension is not salient—may be less likely to turn out and vote. Therefore, the median position would vary. This changed median is the option that political parties or candidates should target in order to win, not the position corresponding to the median voter in the electorate. In his model, Roemer discusses a redistributive policy that consists of levying a uniform tax rate on wealth or income above the median and assumes that the median voter corresponds with the voter earning an income equivalent to the median income or holding wealth equivalent to the median wealth. With the introduction of a bundle of income and a salient characteristic, the new median (income or wealth) lies to the right of (i.e. is greater than) the median for the whole electorate.

Figure 10.5 presents the situation. The median (say, income) for the whole electorate is denoted by μ, and that for the electorate of sub-voters with a particular characteristic is μ^*. With only the redistributive policy under discussion, the median voter theorem predicted that the uniform tax was going to be chosen. However, after the introduction of the bundle of distribution and the additional salient dimension, the predicted median income of the actual voters who are going to turn out is greater, which means that the chosen tax rate will be lower and that the redistribution option might not even be chosen at all.

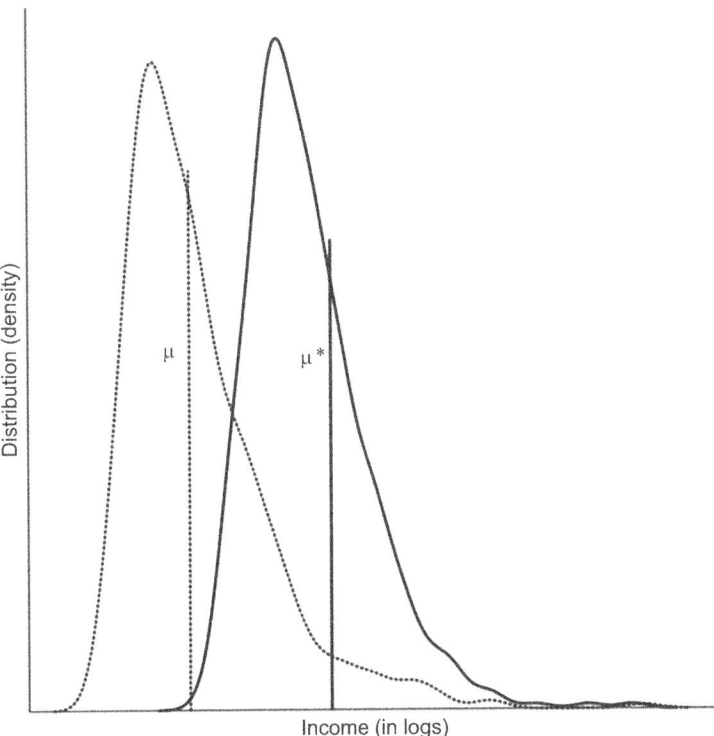

Fig. 10.5 Median before and after the introduction of a salient dimension. *Source: Figure is illustrative, prepared with mock data*

According to Iversen and Goplerud (2018), it is usually ignored that the introduction of such a cross-cutting issue in a bundle does not necessarily lead to less redistribution: if the median income of the mobilised voters lies to the left of the median for the whole electorate, the equilibrium position to which both parties will converge would be pro-redistribution.

Casamatta (2003) modified the model by Roemer to consider the vote on the level of the contribution rate to a pay-as-you-go (PAYG) pension system (see Part II in Volume III) when pensioners are a minority. In PAYG schemes, working-age adults make compulsory contributions on their earnings at a fixed rate against the promise that they will receive an income and other benefits in retirement out of future earmarked taxes. Casamatta presented a two-period overlapping generations model (see Volume I, Chap. 9). In the first period, workers earn an income and make a contribution to the pension scheme. In the second period, they retire and earn a pension income. The number of retirees is smaller than the number of workers.

There is an additional issue that is not related with pensions but on which voters also have to decide. This additional issue is salient only to workers; retirees are indifferent to it, being only interested in the contribution rate to the pension system. However, the workers are mobilised and divided about this additional issue. Two groups of workers are formed, one group favours this other policy point whilst the other one is against it. We denote the contribution rate by τ, which can adopt any value between 0 and 1—that is, $\tau \in [0, 1]$. The additional decision is denoted by a and can be equal to either 0 (if opposed) or 1 (if voted in favour). The salience of the additional policy dimension for the workers is equal to $\theta > 0$. All workers earn the same income, so for simplicity we set it equal to 1. Finally, the population grows at the rate $n > 0$. If we denote by f, o, the voters of working age in favour and opposed to the additional policy issue, respectively, the utility functions (compliant with the usual assumptions; see Chap. 8, Volume I) for each of these groups can be expressed thus:

$$V^f = u(1-\tau) + u[\tau(1+n)] + \theta \cdot a$$
$$V^o = u(1-\tau) + u[\tau(1+n)] + \theta \cdot (1-a) = u(1-\tau) + u[\tau(1+n)]$$
(10.1)

Workers expect that their pension income will be $p = \tau \cdot (1+n)$, as, by the time they retire, the population will have grown by $(1+n)$. The assumption that all workers earn the same income not only simplifies the notation: it also implies that they all have the same preferred level of contribution rate; let's denote it by τ^*. Were the contribution rates the only policy dimension, all workers would vote for this level. However, they do not agree on the additional dimension; some are in favour and some against. We assume that the number of workers in favour of this additional policy point exceeds that of those against it (i.e. $N_w^f > N_w^o$, with $N_w^f + N_w^o = N_w$ representing the number of workers and $N_w + N_r = N$ the number of individuals—N_r is the number of retirees).

We assume that the number of workers exceeds the number of retirees, and that workers in favour of the added dimension are more numerous than workers against it. However we assume that none of the group forms a majority. Furthermore, the number of workers in favour may be equal, greater, or smaller than the number of retirees. The government proposes a policy bundle, that is, a combination of contribution rate and either implementing or not the additional policy point. We assume that the government knows the optimal positions of each group, but two of them *may* form a coalition, reject the government's proposal, and do not re-elect its candidate.

We need to introduce a new variable, the contribution rate that a worker would accept if the undesired option regarding the additional policy point is proposed by the government. Let's denote it by $\tilde{\tau} > \tau^*$. To fix ideas, imagine that the additional point is the legalisation of abortion. The assumption is that some workers are in favour, some are against, and the retirees are indifferent about this policy measure. Remember that both groups of workers prefer a contribution rate equal to τ^* and that retirees want the highest possible contribution rate. Imagine that the government announces that the contribution rate will be equal to τ^* and that abortion is to be legalised if re-elected. The workers who are in favour of legalising abortion will maximise their utility, but those against it will not. However, they could be compensated if the contribution rate were higher than τ^*, to a level equal to $\tilde{\tau}$, with $\tilde{\tau} < 1$. Regarding the additional dimension (i.e. abortion), we assume that it is a binary decision—either it is legalised or not (remember that a is equal to either 0 or 1), but the other policy parameter, the contribution rate, can be set at any level between 0 and 1. We assume that there exists a contribution rate equal to $\tilde{\tau}$ that compensates workers opposed to the legalisation of abortion if the government announces it will legalise it, or compensates workers in favour of legalising abortion if the government announces it will not legalise it if re-elected. It is clear that the government cannot please all workers, because they are divided on the additional policy point. The retirees can exploit this division and force the setting of a contribution rate higher than if the additional policy dimension had not been tabled, that is higher than τ^*. Without this additional issue, both groups of workers would be just one group—that is, the workers—which we assumed were more numerous than the retirees. Hence, as a majority, they would have voted for τ^*. But with the additional issue causing a division within the group of workers, the retirees can act strategically and make an alliance with one of the groups of workers. With which one? With those in favour of the legalisation of abortion if the government announces that it will not go ahead if re-elected, or with those against it if the government announces that it will. But will the retirees be able to have the contribution rate raised all the way to 1? No, even if they are the largest group, the equilibrium rate will be equal to $\tilde{\tau}$, which we assumed is lower than 1. (For the proof of this and the previous propositions, see the appendices in Casamatta 2003.)

This is an intriguing conclusion. The workers are the majority, and they prefer a contribution rate equal to τ^*, but the government is forced to choose a higher rate because of the introduction of an additional policy point that is voted simultaneously, even if it is completely unrelated to pension policy.

The unidimensional nature of the collective choice problem is turned bidimensional. As a result, a policy bundle is formed which creates a division within the hitherto majority group so that when the two dimensions are considered at the same time, there is no one majority group. With no operative majority, one originally minority group seizes the opportunity, enters into a coalition, and forces an increase in the contribution rate which is a better outcome from their perspective—and which would not be voted or implemented had it not been for the introduction of the additional dimension. Let's see multidimensionality more in detail.

Multidimensionality

In the late 1990s, Mulligan and Sala-i-Martin pointed out that the then four countries with highest average age in the world—that is, Monaco, Italy, Greece, and Sweden—had 71, 78, 78, and 78 per cent of their population under age 60, respectively (Mulligan and Sala-i-Martin 1999b). The 2017 population estimates by the United Nations show a similar picture[15]: the four countries with highest average age are Japan, Italy, Germany, and Portugal; the shares of the population over 65 are 26.0, 22.4, 21.1 and 20.7 per cent, respectively, and their median age are all under 50 years (46.3, 45.9, 45.9, and 43.9 years, respectively). The empirical evidence would suggest, then, that the median voter model is not relevant to economics and ageing—yet, and for some time: by 2100, the median age in the forty countries with highest average age is projected to be above 50 years—however, whether 56.8 years of age in 2100 in Singapore (expected to be the country with highest average age, with 40 per cent of its population aged 65 or over) will be considered 'old age' is, though a moot point, very unlikely. Gerontocracy may not be such a safe bet after all, even in the long run. Except, of course, that further assumptions are introduced. It goes without saying that economists (never to let a good model pass by!) did precisely that.

One point to consider is that, in real life, political options are hardly unidimensional, and voters consider (in theory) candidates' proposals on anything from foreign policy to childcare support. We can assume 'intermediate preferences' (Grandmont 1978), which means that the conflict of interests or the ordering of preferences along one dimension (e.g. healthcare) are such that the same ordering is obtained along another dimension (e.g. public spending on pre-school vouchers). If this applies, multiple dimensions can be compressed into one single dimension.

We can also assume the formation of a majority coalition led by, or heavily influenced by, older people. The median voter may not be 'older' but older people would rule the policy-making roost as part of a coalition. For example, Browning (1975) introduced a model that assumes that older people would form a majority coalition with the 'middle aged' (regardless their level of income or wealth). In an alternative specification, Tabellini (2000) assumed that older people would close ranks with the 'poor' (irrespective of their chronological age); this coalition would face the 'young' and the 'rich', either as an opposing coalition or as separate interest groups. Gerontocracy through the back door? Not necessary.

As always with these theoretical models, it all depends on the various assumptions. For example, Conde-Ruiz and Profeta (2007) analysed a bidimensional voting model with individuals divided into three different income groups: low, middle, and high income (see also Razin et al. 2005). These agents live for two periods, working in the first period and retiring in the second one, and have to decide, issue-by-issue, on two policy issues: the pension level and the degree of intragenerational transfer in the benefit formula. Individuals on low incomes are in favour of a redistributive social security system, and high-income individuals oppose any public pension system. This median voter model concludes that if income inequality is high, a coalition of extremes is formed and low-income individuals receive high pension levels in retirement, but the pension system is small in size and a private pension pillar emerges into which higher income individuals invest for their retirement. Instead, if income inequality is low, middle-income individuals are the majority and vote for low transfers to low-income individuals, a larger publicly funded pension system, and a smaller private pillar.

Many models divide the population into three groups, younger workers, middle-aged workers, and retirees, as in Conde-Ruiz and Profeta (2007). Would middle-aged workers favour an increase in contribution rates or not? They are still in employment, so their net income will be affected. On the other hand, they are not far from becoming beneficiaries of pension transfers. The intuition is that they would support an increase if the returns on the contributions are greater that the return on capital accumulation. That is, if by making a contribution to the social security system for the (relatively short) time horizon until they retire, they can obtain more than by saving an equal amount for the same period. According to Galasso (2001), between the 1960s and the 1990s, the median voter's social security returns outperformed alternative forms of investment, including the stock market and government bonds, in the United States of America.

From a different theoretical perspective, Otjes and Krouwel (2018) analysed the composition of the electorate of the 50Plus political party in the Netherlands. Several countries have political parties exclusively promoting the interests of older people (so-called pensioners' parties), including Argentina, Brazil, Russia, Ukraine, Belgium, Luxembourg, Portugal, and Serbia. Willets (2010, p. 251) warned that a '55+ party could sweep Parliament' in the United Kingdom. The reality looks different. In Argentina, the Pensioner's White Party did poorly electorally and de-registered. So did the Party of the Nation's Retirees in Brazil, which eventually joined another party. In Australia, the Seniors United Party is on the brink of being de-registered due to lack of enough members. In Norway, the Pensioners' Party and, in Italy, the Pensioners' Party never managed to obtain more than 1 per cent of the votes in general elections. Similar poor electoral results have been recorded for the United Party of Retirees and Pensioners in Portugal, and also for the Dutch 50Plus party. Perhaps the most successful pensioners' party is the Democratic Party of Pensioners of Slovenia, which gained about 10 per cent of votes in 2014 but less than 5 per cent of votes in 2018. In the United Kingdom, there were short-lived, unsuccessful electoral attempts by the Pensioners' Party—eventually de-registered—and the Senior Citizens Party, which merged with another party.

The bulk of the support for these parties comes from individuals close to retirement rather than already retired, which can be explained by the higher risk the former face concerning their future income. Linked to income vulnerability, another characteristic of the supporters is that they are more likely to be unemployed or economically inactive than the average voter of the same age (the 'social strain' hypothesis). Finally, Otjes and Krouwel (2018) reported that the support moves along two main dimensions:

- Welfare state reform. These parties draw support mostly from individuals who are against rises in the retirement age and shifting more risk and responsibility to individuals.
- Globalisation. These parties draw support mostly from individuals who oppose globalisation and immigration.

Fiscal Leakage

Razin et al. (2002) presented a simple model[16] to explain an empirical 'puzzle': in contrast to the predictions of the canonical median voter theorem, data for the United States of America and twelve Western European countries for the period 1965–1992 showed a negative correlation between the dependency ratio and both the labour-income tax rate and social security transfers generosity within basically pay-as-you-go systems. Higher dependency ratios during the period were the result of population ageing; the median voter theorem predicts higher taxes on labour income (levied on working-age adults) and higher levels of generosity of transfers to the older population. However, the evidence showed exactly the opposite.

The key to explain the puzzle was the 'fiscal leakage' hypothesis. With an ageing population, people of working age may start having second thoughts about the future financial viability of the implicit generational contract underlying a PAYG system: we pay taxes on our labour income to fund the pensions of the current retired population but in the expectation that the succeeding generation (this strand of the literature uses this term to mean birth cohorts) will do likewise to fund our pensions, but what if the succeeding generation cannot afford to keep the promise due to the sheer shrinking size of oncoming cohorts? Population ageing may lead the population of working age to expect lower pension benefits, so they are not prepared to pay as much in taxes as required to maintain the generosity levels of the currently retired cohort. Insofar as retirees do not become a majority, population ageing can reduce pension benefits. There is a 'leak' from the median voter (who belongs to the working population) to the net beneficiaries of the social security system (i.e. the retirees).

However, as Tepe and Vanhuysse (2009) reviewed, the empirical evidence has rejected the 'fiscal leakage' hypothesis—although these authors reported some tentative support for increases on pension expenditure per person, though not on generosity levels. Yes, Razin and others did find an inverse relationship between old-age dependency ratios and both labour income taxes and social security transfers, but Tepe and Vanhuysse explained that the statistical techniques applied in most early studies were not appropriate, suggesting the findings could be biased. In fact, several studies have reported opposite findings: population ageing would accompany an increase in public pension spending as a proportion of the gross domestic product (GDP), or in contribution rates, labour income taxes, and so on (Disney 2007; Pampel 1994; Shelton 2008). However, the literature has failed to find any significant association when the dependent variable was public pension spending per person, rather than as a ratio of GDP (Breyer and Craig 1997; Tepe and Vanhuysse 2010).

10.2.4 Interest Group Models

Population ageing may lead to greater political power of older people but may also lead to *reduced* power. That a group may gain greater political power by an increase in its size is fairly intuitive, but that the opposite may also happen needs explaining. This is the insight by Olson (1965) and Becker (1983), who surmised that interest groups are exposed to free riding problems, which get worse as the group gets bigger. The assumption is that each member would rather shirk her obligations, responsibilities, and contributions to the group and its cause, but benefit from its achievements: each member would rather get a free ride. Consequently, groups have to put in place exclusive benefits for members (not only to attract more people but to create opportunity costs to free riders who would not join but benefit from the political activity of the group)[17] and mechanisms to monitor their members and punish free riders—mechanisms that are costly and reduce the efficiency of the group as they deviate resources away from the group's main objective: the exertion of political pressure towards attaining particular goals.

Interest group models consider both eventualities. Here we are interested in models that assume that interest groups are formed along the chronological age divide, such as young versus older or workers versus pensioners, and so on. Usually these models include a 'political pressure function' that depends on the rate of growth of the population. Sometimes, when the activity of more than one pressure group is modelled, a time intensity independent variable is added to the pressure function. Political pressure takes time, and, say, full-time workers can also dedicate so much whereas retirees would, in theory, have more free time to devote to contributing to the pressure groups they are involved in: the assumption is that it is the amount of time devoted to the group and not money that is crucial for an interest group's political success (Mulligan and Sala-i-Martin 1999a).

The rate of population growth is used as a proxy for population ageing: if the population is increasing, there is no population ageing; in contrast, a shrinking population denotes population ageing. For example, Profeta (2004) analyses whether the pressure function exhibits constant, increasing, or decreasing marginal returns with respect to the number of older people in an economy and its impact on the retirement age and the duration of retirement. With constant marginal returns, the ageing of the population has no influence on the duration of retirement; with increasing returns, population ageing extends the duration of retirement and the opposite happens if the number of older people diminishes the returns on pressure group activity.

In addition, a change in the size of the population and its relative composition by age also has direct macroeconomic consequences, such as on the ratio between the stock of capital and the number of workers or on consumption per capita. Consequently, interest group models combine the direct macroeconomic impacts of demographic change with the implications of such change for the political pressure function of each group, which—in case of an increase in group's size—may either augment its clout or reduce its efficiency.

A final point: interest group models (and also extended median voter models) suggest that if older people are more politically active, they will—*ceteris paribus*—be more able to further their interests and achieve their particular goals. However, Hess et al. (2017) looked into data for twenty-seven European countries from 2009 and reported that higher active participation in society by older people was associated with *higher* support for spending on education compared to spending on pensions than in countries where older people's participation was lower: active ageing policies would reduce intergenerational conflict.

Viriot-Durandal (2003) analysed the political role of pressure groups of retirees in France and concluded that their influence has been marginal as retired individuals as a group lack a collective political identity. In Viriot-Durandal (2012), this same author proposed to depart from the politico-lobbying roles and participation open to older people and to focus instead on *empowerment*. By empowerment, he means the mobilisation of material and immaterial resources by individuals to exercise power in given social situations, through formal and informal decision-making processes, insofar as their material and moral interests are concerned. This empowerment can be manifest in interventions on the institutional environment but also on the self (i.e. self-empowerment).

Notes

1. Although, borrowing from Oscar Wilde's quip about truth, I can say that economics 'is rarely pure and never simple'! (From Wilde 2014, p. 12.)
2. See Groenewegen (2008) for an etymological note on the evolution of the term 'political economy'. This sub-discipline is almost indistinguishable from *public choice*. See Mueller (2015) for such conclusion with which I concur; Drazen (2004) identifies subtle distinctions.
3. Some authors prefer the term 'critical gerontology', which would combine the political economy approach and the humanities (especially the scholarly

production around the concept of 'identity')—see, for example, Bernard and Scharf (2007).
4. In relation to social security, one of the areas of policy most researched by political economists, but these considerations are applicable to other topics as well.
5. In his paper, Breyer referred to political economy models of intergenerational redistribution and included other assumptions relevant to this topic, but the assumptions listed in the text are applicable to any orthodox political economy model.
6. See Persson and Tabellini (2002) for a general introduction.
7. For the record, this is Kenneth Boulding's version of possibly the most famous joke about economists:

There is a story that has been going around about a physicist, a chemist, and an economist who were stranded on a desert island with no implements and a can of food. The physicist and the chemist each devised an ingenious mechanism for getting the can open; the economist merely said, "Assume we have a can opener"!

(Boulding 1970, p. 101)

8. Not quite 'new' in fact, as its intellectual antecedents go back to the Spanish philosopher and monk Ramon Llull in the 1200s and the French philosopher and mathematician the Marquis de Condorcet in the 1700s. In its contemporary 'new' form within economics, it all started with Black (1948) (see endnote 9), extended in Black (1958) and the work by Downs (1957) and Anthony (1957).
9. Although Bowen (1943) set out similar ideas.
10. The number of possible preferences is equivalent to the factorial of the number of alternatives. The factorial of an integer n, expressed $n!$, is equal to $n * (n - 1) * (n - 2) * \ldots$.
11. We denote indifference with the symbol $\sim$. We have:

- $A1 \succ A2 \succ A3$
- $A1 \succ A3 \succ A2$
- $A2 \succ A1 \succ A3$
- $A2 \succ A3 \succ A1$
- $A3 \succ A1 \succ A2$
- $A3 \succ A2 \succ A1$
- $A1 \sim A2 \succ A3$
- $A1 \sim A3 \succ A2$
- $A2 \sim A3 \succ A1$
- $A1 \succ A2 \sim A3$

- A1 ≻ A1 ∼ A3
- A3 ≻ A1 ∼ A2
- A1 ∼ A2 ∼ A1.

12. Short-termism among politicians? What was that about the can opener? On this topic (short-termism in politics, not can openers), see MacKenzie (2016).
13. The fancy term for this assumption is that workers have 'nonzero conjectural variations' (Boadway and Wildasin 1989a,b).
14. See Tullock (1967).
15. Source: Nations (2017).
16. See also Razin and Sadka (2007) for an extension of the basic model.
17. See King and Walker (1992) for a comparison between two US interest groups: the American Association of Retired Persons and Common Cause.

References

Alt, James E and Alberto Alesina (1996). "Political Economy: an overview". In: *A New Handbook of Political Science*. Edited by Robert E Goodin and Hans-Dieter Klingemann. Oxford: United Kingdom: Oxford University Press, pages 645–674.

Anthony, Downs (1957). *An Economic Theory of Democracy*. New York, NY: United States of America: Harper and Row.

Atella, Vincenzo and Lorenzo Carbonari (2017). "Is gerontocracy harmful for growth? A comparative study of seven European countries". In: *Journal of Applied Economics* 1.20, pages 141–168.

Backhouse, Roger E and Steven G Medema (2009). "Retrospectives: On the definition of economics". In: *Journal of Economic Perspectives* 23.1, pages 221–33.

Banks, Jeffrey S and John Duggan (2005). "Probabilistic voting in the spatial model of elections: The theory of office-motivated candidates". In: *Social choice and strategic decisions, Essays in Honor of Jeffrey S. Banks*. Edited by David Austen-Smith and John Duggan. Studies in Choice and Welfare. Berlin: Germany: Springer-Verlag, pages 15–56.

Becker, Gary S (1983). "A theory of competition among pressure groups for political influence". In: *The Quarterly Journal of Economics* 98.3, pages 371–400.

Bernard, Miriam and Thomas Scharf, editors (2007). *Critical perspectives on ageing societies*. Bristol: United Kingdom: Policy Press.

Berry, Craig (2012). "Young people and the ageing electorate: breaking the unwritten rule of representative democracy". In: *Parliamentary Affairs* 67.3, pages 708–725.

Black, Duncan (1948). "On the rationale of group decision-making". In: *Journal of Political Economy* 56.1, pages 23–34.

——— (1958). "The theory of committees and elections". In:.

Boadway, Robin W and David E Wildasin (1989a). "A median voter model of social security". In: *International Economic Review* 30.2, pages 307–328.

Boadway, Robin and David Wildasin (1989b). "Voting Models of Social Security Determination". In: *The Political Economy of Social Security*. Edited by Björn A Gustafsson and N Anders Klevmarken. Contributions to Economic Analysis. Amsterdam: The Netherlands: Elsevier Science Publishers BV, pages 29–50.

Bochsler, Daniel (2010). "The Marquis de Condorcet goes to Bern". In: *Public Choice* 144.1-2, pages 119–131.

Boersch-Supan, Axel, Gabriel Heller, and Anette Reil-Held (2011). "Is intergenerational cohesion falling apart in old Europe?". In: *Public Policy and Aging Report* 21.4, pages 17–21.

Boldrin, Michele and Aldo Rustichini (2000). "Political Equilibria with Social Security". In: *Review of Economic Dynamics* 3.1, pages 41–78.

Börsch-Supan, Axel (2013). "Myths, scientific evidence and economic policy in an aging world". In: *The Journal of the Economics of Ageing* 1, pages 3–15.

Boulding, Kenneth Ewart (1970). *Economics as a Science*. New York, NY: United States of America: McGraw-Hill.

Bowen, Howard R (1943). "The interpretation of voting in the allocation of economic resources". In: *The Quarterly Journal of Economics* 58.1, pages 27-48.

Breyer, Friedrich (1994). "The political economy of intergenerational redistribution". In: *European Journal of Political Economy* 10.1, pages 61–84.

Breyer, Friedrich and Ben Craig (1997). "Voting on social security: Evidence from OECD countries". In: *European Journal of Political Economy* 13.4, pages 705–724.

Browning, Edgar K (1975). "Why the social insurance budget is too large in a democracy". In: *Economic inquiry* 13.3, pages 373–388.

Busemeyer, Marius R, Achim Goerres, and Simon Weschle (2009). "Attitudes towards redistributive spending in an era of demographic ageing: the rival pressures from age and income in 14 OECD countries". In: *Journal of European Social Policy* 19.3, pages 195–212.

Casamatta, Georges (2003). "The Political Power of the Retirees in a Two-Dimensional Voting Model". In: *Journal of Public Economic Theory* 5.4, pages 571–591.

Conde-Ruiz, Ignacio J and Paola Profeta (2007). "The redistributive design of social security systems". In: *The Economic Journal* 117.520, pages 686–712.

Conesa, Juan C and Dirk Krueger (1999). "Social security reform with heterogeneous agents". In: *Review of Economic dynamics* 2.4, pages 757–795.

Congleton, Roger D (2004). "The median voter model". In: *The Encyclopedia of Public Choice*. Edited by Charles K Rowley and Friedrich Schneider. New York, NY: United States of America: Kluwer Academic Publishers, pages 707–712.

Cooley, Thomas F and Jorge Soares (1999). "A positive theory of social security based on reputation". In: *Journal of Political Economy* 107.1, pages 135–160.

Cooley, Thomas F, Jorge Soares, et al. (1999). "Privatizing social security". In: *Review of Economic Dynamics* 2.3, pages 731–755.

Coughlin, Peter J (1992). *Probabilistic voting theory*. Cambridge: United Kingdom: Cambridge University Press.

Cox, Gary W and Mathew D McCubbins (2005). *Setting the Agenda: Responsible Party Government in the U.S. House of Representatives*. New York, NY: United States of America: Cambridge University Press.

—— (2016). *Going Grey: The Mediation of Politics in an Ageing Society*. Abingdon: United Kingdom: Routledge.

De Mello, Luiz et al. (2017). "Greying the budget: Ageing and preferences over public policies". In: *Kyklos* 70.1, pages 70–96.

Dekel, Eddie and Michele Piccione (2000). "Sequential voting procedures in symmetric binary elections". In: *Journal of political Economy* 108.1, pages 34–55.

Disney, Richard (2007). "Population ageing and the size of the welfare state: Is there a puzzle to explain?". In: *European Journal of Political Economy* 23.2, pages 542–553.

Downs, Anthony (1957). "An economic theory of political action in a democracy". In: *Journal of Political Economy* 65.2, pages 135–150.

Drazen, Allan (2004). *Political Economy In Macroeconomics*. Princeton, NJ: United States of America: Princeton University Press.

Eisele, Frederick R (1979). "Origins of "Gerontocracy"". In: *The Gerontologist* 19.4, pages 403–407.

Estes, Carroll L., James H Swan, and Lenore E Gerard (1982). "Dominant and Competing Paradigms in Gerontology: Towards a Political Economy of Ageing". In: *Ageing and Society* 2.2, pages 151–164.

Fullerton, Andrew S and Jeffrey C Dixon (2010). "Generational conflict or methodological artifact? Reconsidering the relationship between age and policy attitudes in the US, 1984-2008". In: *Public Opinion Quarterly* 74.4, pages 643–673.

Galasso, Vincenzo (2001). "Social Security: A financial appraisal for the median voter." *Social Security Bulletin*, 64: 57.

Galasso, Vincenzo (2008a). "Ageing and the Skew Risk in Collective Choices". In: *Revue économique* 59.5, pages 1023–1043.

—— (2008b). *The Political Future of Social Security in Aging Societies*. Cambridge, MA: United States of America: The MIT press.

Galasso, Vincenzo and Paola Profeta (2002). "The political economy of social security: a survey". In: *European Journal of Political Economy* 18.1, pages 1–29.

Goerres, Achim and Markus Tepe (2010). "Age-based self-interest, intergenerational solidarity and the welfare state: A comparative analysis of older people's attitudes towards public childcare in 12 OECD countries". In: *European Journal of Political Research* 49.6, pages 818–851.

Gonzalez-Eiras, Martín and Dirk Niepelt (2008). "The future of social security". In: *Journal of Monetary Economics* 55.2, pages 197–218.

Grandmont, Jean-Michel (1978). "Intermediate preferences and the majority rule". In: *Econometrica* 46.2, pages 317–330.

Groenewegen, Peter (2008). "Political Economy". In: *The New Palgrave Dictionary of Economics - Vol. 6*. Edited by Steven N Durlauf and Lawrence E Blume. 2nd. Palgrave Macmillan, pages 476–480.

Hamil-Luker, Jenifer (2001). "The prospects of age war: Inequality between (and within) age groups". In: *Social Science Research* 30.3, pages 386–400.

Helpman, Elhanan and Torsten Persson (2001). "Lobbying and legislative bargaining". In: *Advances in Economic Analysis & Policy* 1.1. URL: https://doi.org/10.2202/1538-0637.1008.

Hess, Moritz, Elias Nauman, and Leander Steinkopf (2017). "Population ageing, the intergenerational conflict, and active ageing policies–A multilevel study of 27 European countries". In: *Journal of Population Ageing* 10.1, pages 11–23.

Holcombe, Randall G (1989). "The Median Voter Model in Public Choice Theory". In: *Public Choice* 61.2, pages 115–125.

Hu, Sheng Cheng (1982). "Social security majority-voting equilibrium and dynamic efficiency". In: *International Economic Review* 23.2, pages 269–287.

Iversen, Torben and Max Goplerud (2018). "Redistribution Without a Median Voter: Models of Multidimensional Politics". In: *Annual Review of Political Science* 21, pages 295–317.

Jenkins, Jeffery A and Nathan W Monroe (2016). "On Measuring Legislative Agenda-Setting Power". In: *American Journal of Political Science* 60.1, pages 158–174.

Kemmerling, Achim and Michael Neugart (2009). "Financial market lobbies and pension reform". In: *European Journal of Political Economy* 25.2, pages 163–173.

King, D. and Walker, J. (1992). "The provision of benefits by interest groups in the United States", *The Journal of Politics*, 54(2): 394–426.

Kohli, Martin and Camila Arza (2011). "The political economy of pension reform in Europe". In: *Handbook of aging and the social sciences*. Edited by R.H. Binstock and L.K. George. 7th. San Diego, CA: United States of America: Elsevier, pages 251–264.

Kotlikoff, Laurence J and Robert W Rosenthal (1993). "Some Inefficiency Implications of Generational Politics and Exchange". In: *Economics and Politics* 5.1, pages 27–42.

Kurrild-Klitgaard, Peter (2017). "Trump, Condorcet and Borda: Voting paradoxes in the 2016 Republican presidential primaries". In: *European Journal of Political Economy*. URL: https://doi.org/10.1016/j.ejpoleco.2017.10.003.

MacKenzie, Michael K (2016). "Institutional design and sources of short-termism". In: *Institutions For Future Generations*. Edited by Iñigo González-Ricoy and Axel Gosseries. Oxford: United Kingdom: Oxford University Press, pages 24–45.

McCarty, Nolan M and Keith T Poole (1995). "Veto power and legislation: An empirical analysis of executive and legislative bargaining from 1961 to 1986". In: *Journal of Law, Economics, & Organization* 2, pages 282–3128.

Montén, Anna and Marcel Thum (2010). "Ageing municipalities, gerontocracy and fiscal competition". In: *European Journal of Political Economy* 26.2, pages 235–247.

Mueller, Dennis C (2003). *Public Choice III*. Cambridge: United Kingdom: Cambridge University Press.

—— (2015). "Public choice, social choice, and political economy". In: *Public Choice* 163.3-4, pages 379–387.

Mullan, Phil (2002). *The Imaginary Time Bomb. Why an ageing population is not a social problem*. London: United Kingdom: I. B. Tauris.
Mulligan, Casey B and Xavier Sala-i-Martin (1999a). *Gerontocracy, retirement, and social security*. NBER Working Paper 7177. Cambridge, MA: United States of America: National Bureau of Economic Research.
—— (1999b). *Social security in theory and practice (I): Facts and political theories*. NBER Working Paper 7118. Cambridge, MA: United States of America: National Bureau of Economic Research.
Nations, United (2017). *World Population Prospects. The 2017 Revision, DVD Edition*. New York, NY: United States of America.
Olson Jr, Mancur (1965). *The Logic of Collective Action. Public Goods and the Theory of Groups*. Cambridge, MA: United States of America: Harvard University Press.
Otjes, Simon, and André Krouwel (2018). "Old Voters on New Dimensions: Why Do Voters Vote for Pensioners' Parties? The Case of the Netherlands." *Journal of Aging & Social Policy* 30(1): 24–47.
Pampel, Fred C (1994). "Population aging, class context, and age inequality in public spending". In: *American Journal of Sociology* 100.1, pages 153–195.
Persson, Torsten and Guido Enrico Tabellini (2002). *Political Economics: Explaining Economic Policy*. Cambridge, MA: United States of America: The MIT Press.
Prinzen, Katrin (2016). "Attitudes toward intergenerational redistribution in the Welfare State". In: *Social Demography. Forschung an der Schnittstelle von Soziologie und Demografie*. Edited by Karsten Hank and Michaela Kreyenfeld. Kölner Zeitschrift für Soziologie und Sozialpsychologie. Wiesbaden: Germany: Springer, pages 349–370.
Profeta, Paola (2004). "Aging, retirement and social security in a model of interest groups". In: *Mathematical Population Studies* 11.2, pages 93–120.
Razin, Assaf and Efraim Sadka (2007). "Aging population: The complex effect of fiscal leakages on the politico-economic equilibrium". In: *European Journal of Political Economy* 23.2, pages 564–575.
Razin, Assaf, Efraim Sadka, and (in cooperation with Chang Woon Nam) (2005). *The Decline of the Welfare State. Demography and Globalization*. CESifo Book Series. Cambridge, MA: United States of America: The MIT Press.
Razin, Assaf, Efraim Sadka, and Phillip Swagel (2002). "The aging population and the size of the welfare state". In: *Journal of Political Economy* 110.4, pages 900–918.
Roemer, John E (1998b). "Why the poor do not expropriate the rich: an old argument in new garb". In: *Journal of Public Economics* 70.3, pages 399–424.
Romer, Thomas and Howard Rosenthal (1978). "Political resource allocation, controlled agendas, and the status quo". In: *Public choice* 33.4, pages 27–43.
Rothschild, Kurt W (1989). "Political economy or economics?: Some terminological and normative considerations". In: *European Journal of Political Economy* 5.1, pages 1–12.
Rowley, Charles K (1984). "The relevance of the median voter theorem". In: *Zeitschrift für die gesamte Staatswissenschaft/Journal of Institutional and Theoretical Economics* 140, pages 104–126.

Schulz, James and Robert Binstock (2006). *Aging Nation. The Economics and Politics of Growing Older in America*. Westport, CT: United States of America: Praeger Publishers.

Shelton, Cameron A. (2008). "The aging population and the size of the welfare state: Is there a puzzle?". In: *Journal of Public Economics* 92.3-4, pages 647–651.

Sinn, Hans-Werner and Silke Uebelmesser (2003). "Pensions and the path to gerontocracy in Germany". In: *European Journal of Political Economy* 19.1, pages 153–158.

Sørensen, Rune J (2013). "Does aging affect preferences for welfare spending? A study of peoples' spending preferences in 22 countries, 1985–2006". In: *European Journal of Political Economy* 29, pages 259–271.

Street, Debra (1997). "Special Interests or Citizens' Rights? "Senior Power", Social Security, and Medicare". In: *International Journal of Health Services* 27.4, pages 727–751.

Street, Debra and Jeralynn Sittig Cossman (2006). "Greatest generation or greedy geezers? Social spending preferences and the elderly". In: *Social problems* 53.1, pages 75–96.

Tabellini, Guido (2000). "A positive theory of social security". In: *Scandinavian Journal of Economics* 102.3, pages 523–545.

Tepe, Markus and Pieter Vanhuysse (2009). "Are Aging OECD Welfare States on the Path to Gerontocracy?: Evidence from 18 Democracies, 1980–2002". In: *Journal of Public Policy* 29.1, pages 1–28.

—— (2010). "Elderly bias, new social risks and social spending: change and timing in eight programmes across four worlds of welfare, 1980–2003". In: *Journal of European Social Policy* 20.3, pages 217–234.

Thornton, Tim B (2016). *From Economics to Political Economy: The problems, promises and solutions of pluralist economics*. New Political Economy. Abingdon: United Kingdom: Taylor & Francis.

Townley, Peter GC (1981). "Public choice and the social insurance paradox: A note". In: *Canadian Journal of Economics / Revue canadienne d'Economique* 14.4, pages 712–717.

Tullock, Gordon (1967). "The general irrelevance of the general impossibility theorem". In: *The Quarterly Journal of Economics* 81.2, pages 256–270.

Van Deemen, Adrian (2014). "On the empirical relevance of Condorcet's paradox". In: *Public Choice* 158.3-4, pages 311–330.

Viriot-Durandal, Jean-Philippe (2003). *Le pouvoir gris: sociologie des groupes de pression de retraités*. Lien social. Paris: France: Presses Universitaires de France.

—— (2012). "Le ńpouvoir grisż du lobbying au pouvoir sur soi". In: *Gérontologie et Société*. 35.4, pages 23–38.

Walker, Alan (1981). "Towards a political economy of old age". In: *Ageing and Society* 1.01, pages 73–94.

—— (1999). "Political participation and representation of older people in Europe". In: *The politics of old age in Europe*. Edited by Alan Walker and Gerhard Naegele. Maidenhead: United Kingdom: Open University Press, pages 7–24.

Wilde, Oscar (2014). *The Importance of Being Earnest and Other Plays*. Enriched Classics. New York, NY: United States of America: Simon & Schuster.

Willets, David (2010). *The Pinch: how the Baby Boomers Took Their Children's Future And Why They Should Give it Back*. London: United Kingdom: Atlantic Books.

11

Gerontological Views

> **Overview**
> This chapter presents the understanding within political science and social gerontology of political economy. Topics include political participation and behaviour, political franchise, realignment, cognitive mobilisation, and regret. The chapter concludes with a presentation of the social gerontology approach to political economy.

11.1 Political Gerontology

Political gerontology looks into the roles of older people as voters and citizens of welfare states (Komp 2011). Studies of older people as voters focus on whether chronological age drives voting behaviour, preferences, turnout, participation, and so on. It also studies the importance of generational membership and period effects on engagement and salience, among other aspects. One relevant question is, as always, to what extent older people can be considered or treated as one homogeneous group with regards to political behaviour. Studies of older people as citizens of welfare states focus on social and political citizenship, with the opportunities, rights, and responsibilities citizenship creates, and how these may (or should) change with chronological age, especially with cognitive decline.

Some of the models we mentioned above assume that not every voter turns out at the ballot box (or casts his/her vote electronically or by post). Abstention rates are relevant if they are concentrated in particular groups. For example, similarly to Roemer's or Casamatta's models, we can expect that the contents of redistribution policies will be affected if people on low incomes are less (or more) likely to vote than better-off individuals. In economics and ageing, we are particularly interested in voting turnout by chronological age: if younger individuals were less likely to vote, the age of the *actual* median voter would increase, and it would do so faster than the ageing of the population or the electorate as a whole (Davidson 2010).

11.1.1 Political Participation

I mentioned political participation earlier in this part. This is a good moment to provide a formal definition of this concept. According to Cox (2016), political participation includes holding political opinions, being an opinion leader or trying to influence public opinion, joining a political party or movement, voting, and holding office.

The determinants of elements of political participation by age have been vastly researched, for 'there are few variables that are as consistently found to affect different aspects of voting as age' (Dassonneville 2016, p. 137). However, according to Ansolabehere et al. (2012, p. 333), chronological age remains 'among the least well understood' predictors of political participation.

The empirical evidence suggests that older people are more actively engaged in most aspects of political participation than younger people. It is more likely that older people hold political opinions than younger people; older people tend to be over-represented among opinion leaders and influencers as well as among legislative and executive institutions; and their registration and voting turnout rates are usually higher than those of younger voters. Only with regards to membership of parties and politically active movements, there is no clear chronological age gradient.

Eijk and Franklin (2009) identified four effects through which individual and population ageing influence electoral turnout and preferences, and political participation:

- The period effect, which refers to the impact of particular events on behaviour, mindsets, and viewpoints of large number of individuals, especially if they take place at political 'formative' years. It also refers to changes in groups of political enfranchised individuals (e.g. extension of franchise to women or 16-year-olds) and in legislation (e.g. abolition of compulsory voting).

- The age effect, which refers to changes in behaviours and attitudes as individuals grow older. From reduced cognitive or physical ability impairing some older people from accessing voting booths (to say nothing of joining demonstrations, etc.), to changing perspectives and allegiances over time.[1] This age effect is not relevant if the number of people who leave the electorate (e.g. by death) is replaced by voters with the same characteristics. However, it may be influential if there is a change in the age composition of the electorate: the composition effect.
- The composition effect, which refers to changes in the proportion of individuals in the electorate with certain relevant characteristics; in our case, higher proportion of people of a given age group.
- The generational replacement effect, which refers to the entering in the electorate and the political fray of members of generations with different political and social views than leaving generations. Generational replacement may be accompanied by a composition effect or not—that is, the age profile of the electorate and politicians may vary or not; the crucial point for generational replacement to happen is the extent to which the political complexion between individuals leaving and entering politics and the electorate differs.

In addition, Goerres (2009) looked into other aspects of political participation and also identified four age-related effects, which he grouped into age and cohort effects. Age effects include life-cycle and individual ageing effects. Cohort effects include political generation and socio-economic effects.

- The life-cycle effect, which influences the availability of resources and the motivation to participate
- The individual ageing effect, which influences the alternatives and costs due to previous experience but also the adherence to social norms
- The political generation effect, which influences the preferences for activities and the perceived availability of opportunities
- The socio-economic effect, which influences widespread social characteristics and mores (e.g. pro-environmentalist attitudes stemming from post-materialism)

Goerres, p. 161 explained the relative importance of each effect and their interplay with participation and chronological age thus:

> Political generation and socio-economic effects belong to the category of cohort effects. If all differences between older and younger citizens' political behaviour

in a society were due to these effects, that society would experience an unstable difference in participation between these two age groups across time. At one point in time, a particular cohort of older people would demonstrate a certain participatory behaviour vis-à-vis the behaviour of a specific cohort of younger individuals. Ten years later, for example, the groups of older and younger people would consist of different cohorts that again carry different sets of relevant experiences. In contrast to cohort effects, life cycle and individual ageing effects belong to the larger category of age effects. If all differences between older and younger individuals were due to this type of effect, ageing societies would experience long-lasting changes in the participatory process. Older people would represent a growing group of individuals exposed to a specific set of age effects that makes their participatory behaviour different from the shrinking group of younger people. In essence, both cohort and age effects can be found to be at work across all types of participation, making the impact of demographic ageing on the participatory process not only complex, but also unstable across time.

Komp (2011) classified older people into third-agers and fourth-agers. The main difference between both categories is that the third-agers are in relatively good health, and either engaged in or capable of doing paid employment or voluntary work, whilst fourth-agers face health and disability conditions that prevent them from engaging in such activities.[2] Employment, pension, and welfare policies—to name but three relevant areas—aimed at one group would differ from those aimed at the other group. In Komp's conceptualisation, which is applicable to democratic systems, mediating the contributions of older people and the policy outcomes is a 'transformation' process that changes the roles of older people as political actors. For the third-agers, political participation takes the shape of voting and membership of political parties and interest groups; political participation for fourth-agers, in this model, consists of voting. The model proposes that public policies are the 'black box' that transforms older people's inputs (their contributions to social insurance, their engagement in productive activities, etc.) into policy outcomes such as employment regulation and incentives, or the pension, health and long-term care systems. Mediating the inputs and outputs, inside the black box, is the political participation of older people, shaping and influencing policy decisions.

11.1.2 Age-Related Franchise Limits

Another consideration is the limits to political franchise by age. All democratic countries have implemented minimum age limits below which individuals are not allowed to vote or run for office (Grover 2011). However, there are no maximum age limits to vote. Many countries and regions have, instead, maximum age thresholds in place to hold office in their legislative, executive, and judiciary systems.

Two main arguments in favour of setting upper age limits to political franchise are the diminishing mental capability of older people and the short-termism of older people's policy preferences. The first argument is part of a wider discussion about the restrictions of political franchise to mentally handicapped individuals (regardless of chronological age), in place in most democracies. See Sonnicksen (2016) for views against limiting franchise due to dementia and also the papers in the special issue of the *McGeorge Law Review* (Volume 38, Issue 4, from 2016). Mueller (2002) discussed the rationale for setting a maximum voting age that would mirror mental and physical capability tests for driving beyond a certain age that are in force in many countries. This author concludes that setting an upper age limit per se would not be legitimate, but that testing for mental capacity would be legitimate.

The other argument in favour of upper age limits is framed around intergenerational relations. Barron (1953) noted that calls for setting a maximum voting age come from voices who present population ageing as a menace. Stewart (1970, p. 22) argued for an upper voting age limit given that the oldest old would be 'actuarially unlikely to survive, and pay the bills for, the politician or party [they] may help elect' and, more bluntly, '[t]he old, having no future, are dangerously free from the consequences of their own political acts'.

Similarly, Longman (1987) suggested that as older people would be more focused on the short term (given their shorted remaining lifespan), policies would not reflect the interests of children, so he advocated for giving parents an additional vote for each child under voting age. This author also mentioned other proposals, such as giving diminishing weights to votes with a voter's advancing chronological age or remaining life expectancy, or that the electorate would be asked to cast a vote separately for parties (not candidates) so that the public funding to each party, which is usually proportional to the number of votes each one gets in an election, could reflect the age structure of voters supporting each party. Along the same lines, Demeny (1986) proposed allowing custodial parents exercise their children's voting rights until they come of voting age (see also Sanderson and Scherbov 2007).

To prevent any interest groups from gaining influence and power over elected representatives, Hayek (1996) proposed that representatives to legislative assemblies could be elected for long periods without re-election and with a constant fraction renewed each year. Hayek suggested that these representatives should be all from one same age group and that only individuals belonging to that age group would be eligible to vote. As Parijs (1998) explained, this proposal would fix the median age of voters and elected politicians—although Parijs favoured extending franchise to a small number of age groups rather than to only one such group as Hayek proposed. Recently, in the South Korean context, Seo (2017) suggested the redrawing of electoral districts from geographic area to age-cohort to redress the intergenerational imbalance against younger voters in an ageing society.

11.1.3 Realignment, Cognitive Mobilisation, and Regret

Apart from these political influences of individual and population ageing, we have to consider that older people also exhibit political realignment and cognitive mobilisation and that their voting decisions may be driven by minimal regret.

Political realignment (and dealignment) refers to the fact that political allegiances and views are not set in stone after a given chronological age, never to change for the rest of an individual's lifetime (Marsh and McElroy 2016; Wasfy et al. 2017). Generational effects can partially explain these changes at a macro level. For example, Dalton and Flanagan (2017) surmised that generational replacement was at the heart of political changes in the United States of America after the 1960s, and pointed to a macro, aggregate effect: younger cohorts exhibiting lower partisan loyalty than previous generations.

Furthermore, starting with Glenn and Grimes (1968) in the United States of America, the evidence shows that political interest is on the rise among older people in most democracies (Goerres 2007)—a notion sometimes known as 'cognitive mobilisation' (Dalton 2004)—although it would diminish, depending on the country, after ages 60 or 80 (Bhatti and Hansen 2012; Bhatti et al. 2012; Dassonneville 2016). Cognitive mobilisation is affected by the effects listed above—for example, generational influence. Dalton presented evidence from the United States of America about levels of trust in politicians among four different cohorts using cross-sectional data from surveys between 1958 and 2000. He reported that in 1958 there was a negative association between trust and age, which reversed over time. By 2000, the pre-1910 generation had become relatively less cynical of government and younger generations had become more distrustful. Similar increases in trust by age were found in Australia, Canada, Finland, Germany, Sweden, and Norway (with Japan and Switzerland being exceptions in the countries investigated: low levels of trust remained over the life cycle). Dalton concluded [p. 93]:

These differences are admittedly modest, but the incremental effects of generational change can have large cumulative effects, because over time older and more trustful citizens are gradually replaced by younger and more cynical individuals.

'Minimal regret' (Geys 2006) refers to the notion that voters choose not so much for their preferred candidate or outcome but against their perceived worst candidate or outcome. Preferences are related to two variables: position and valence (Davidson 2016). Position refers to the fact that voters make an assessment of their own views on a policy issue in relation to the variations in the positions of the various candidates or political parties. Valence, in turn, refers to a shared agreement among voters of the importance of the issue and a perception of how efficiently each candidate or political party will deliver or perform in this policy domain.

11.2 Social Gerontology and the Political Economy of Ageing

As I mentioned at the beginning of this part, there is a critical approach to later-life issues known as the 'political economy of ageing' that is unrelated to the orthodox economics literature also known as 'political economy' and which is also separate from the political gerontological approach. This final section briefly reflects upon the academic output from this gerontological approach regarding political behaviour and participation. (See Part V in this volume for other considerations from the 'political economy of ageing' gerontology school. A thorough treatment can be found in Díaz-Tendero Bollaín (2012) (in Spanish).)

Of the many theoretical points raised by this strand of the literature, it is worth reflecting—in connection with political participation—on the duality and tension created in the contemporary landscape of ageing in developed economies by, on the one hand, the dual increasing diversity in post-modern, consumer-based lifestyles and, on the other, responsibility (for health, income, long-term care, employability, social participation, etc.), above all in the 'third age' stage of the life cycle that is being placed upon the individual, all accompanied by heightened risk and precariousness, and weakening institutional support. As discussed in Part II in this volume, social inequalities show no signs of abating in later life; moreover, there is evidence that disparities among older people (in work, income, health, cultural, social and political participation, etc.) have been growing since the 1970s (Formosa and Higgs 2015).

Older people are the subject and target of contemporary 'dominating framing' discourses (Kildal and Nilssen 2013), sometimes originating in government and policy circles, that promote the 'independent' and 'active' pursuing of individual goals of choice as a definition of 'successful' ageing. Would it not be contradictory that independent and active individuals in seek of life ambitions and dreams did not at the same time seek to influence policy? The flip side of those older people ageing 'successfully' is the much larger group of experiencing, or being on the verges of, various forms of social exclusion and precariousness (Grenier et al. 2017)—and, in some societies, powerlessness (Kam 2003).[3] Would there be any better instances in democratic societies for them to have their voices heard than through political engagement, either as voters or members of interest groups? As Bernard and Scharf (2007, p. 21) pointed out: 'To be heard in a way that influences actions is a critical, but often neglected, element in a just society.'

Given the mainstream narrative and how the less economically fortunate fit into it, would the political participation of older people not be a sign of a healthy democratic system despite growing varieties of exclusion and numbers of excluded people? After all,

> ...the active participation and inclusion of older people and their representative organizations in the policy process is an important component of the active ageing agenda, which gives due recognition to the rights of older people to contribute to the policy decisions which have a direct impact on their lives
>
> (Doyle and Timonen 2013, p. 79).

These are questions that the critical political economy of ageing set about to provide answers to—and which it answers *affirmatively*. However, not only these authors do not endorse, promote, or acquiesce in a situation of intergenerational strife in which older people have to fight, and are fighting, for their particular interests in the political arena as an organised, monolithic bloc: they do not accept that this is an accurate description of reality. In fact, they abide by the evidence that political engagement is (still?) more greatly fragmented by class, gender, and ethnicity than chronological age and driven by a sense of self-efficacy and empowerment than extrinsic motivations: '[c]ohesiveness of attitudes and voting among older people is relatively weak compared to cohesiveness among classes, races or religions' (Timonen 2008, p. 160). In turn, they emphasise the importance of intergenerational solidarity and cooperation and of restoring meaning and dignity in later life (especially in the so-called old-old age) (Phillipson 2015b). Here, perhaps, lies one explanation as to why 'pensioner parties' perform so badly despite the growing proportion of older people both in the populations and in the electorates of developed democracies.

Review and Reflect

1. Consider the following epistemological defence of the critical gerontological political economy of ageing approach:

   A thoroughgoing approach based on political economy would combine sociological, economic and political analyses. For example, an examination of differential command over resources, would include the exercise of professional power over the lives of elderly people, the application of segregative social policies, benefits and services and the social creation of dependent status.
   (Walker 1981, p. 90)

   How would such examination differ if tackled from the orthodox economics' political economy approach?

2. Comment on the following dictum regarding the median voter theorem. If we accept the conclusion that governments will make decisions so that the median voter's utility is maximised, then:

   ...we can bury politics in [the model's] assumptions ...and use the individual utility maximizing model applied to the median income family to analyze governmental fiscal performance!
   (Inman 1978, p. 46)

3. Discuss the assumptions of the median voter theorem and the 'elderly power' and 'grey peril' hypotheses in the light of the following assertion:

   *...presenting issues in terms of younger **versus** older generations will frustrate positive solutions to the needs of young **and** older people ...the worker versus pensioner perspective is especially unhelpful in that it ignores fundamental changes of the distribution of labour across the life course. ...For many workers the predictability of continuous employment is replaced by insecurity in middle and later life.*
   (Phillipson 2013, p. 27)

4. Discuss the following conclusion from a paper on gerontocracy in OECD countries:

   In sum, it is the logic of retrenchment politics and double fiscal electoral straitjackets, not gerontocracy, which reigns supreme in the political economy of pensions today. More than in previous decades, contempo-

(continued)

> rary public pension politics is driven by tighter macro-fiscal constraints, caused by the increasing budgetary pressures accompanying the growth of elderly cohorts. Larger pensioner populations have indeed led to larger overall spending commitments. At the same time, the concomitant budgetary pressures appear to have stabilized or reduced real pension expenditures per elderly person, as if governments were forced to cut smaller slices out of larger cakes.
>
> (Tepe and Vanhuysse 2009, p. 23)

5. Eisele (1979) remarked that lay people tend to use the term 'gerontocracy' to refer to age-related decision-making and behaviour, whilst specialists tend to use it to describe certain political organizations.

    Do you find a 'lay' use of the term in some of the more academic and scholarly works mentioned above? Do you think these two uses are warranted or introduce needless ambiguity?

6. Gordon Tullock once wrote:

    > I read an article in a political science journal long ago which recommended a maximum voting age of 65. The author, although on the left in politics, apparently wanted lesser pensions and other advantages for the old.
    >
    > (Tullock 2002, p. 248)

    Why would it be surprising that a left-leaning author favoured reduced pensions and benefits to older people? Considering the scant evidence in favour of the median voter hypothesis, do you think that setting a maximum voting age would be detrimental to older people? Could we talk of 'older people's interests' as such anyway? Why or why not?

Notes

1. The famous dictum of either leaning to the left at 20 or not having heart, but changing by 40 or not having a sound mind, which is often wrongly attributed to Winston Churchill, captures this.
2. Note that chronological age is not a classification criterion here.
3. For human rights and older people, see Townsend (2007) and the special issue (Volume 2, Issue 4, December 2017) of the *Journal of Human Rights and Social Work*.

References

Ansolabehere, Stephen, Eitan Hersh, and Kenneth Shepsle (2012). "Movers, Stayers, and Registration: Why Age is Correlated with Registration in the U.S. In: *Quarterly Journal of Political Science* 7.4, pages 333–363.

Barron, Milton L (1953). "Minority group characteristics of the aged in American society". In: *Journal of Gerontology* 8.4, pages 477–482.

Bernard, Miriam and Thomas Scharf, editors (2007). *Critical perspectives on ageing societies*. Bristol: United Kingdom: Policy Press.

Bhatti, Yosef and Kasper M Hansen (2012). "Retiring from voting: Turnout among senior voters". In: *Journal of Elections, Public Opinion & Parties* 22.4, pages 479–500.

Bhatti, Yosef, Kasper M Hansen, and Hanna Wass (2012). "The relationship between age and turnout: A roller-coaster ride". In: *Electoral Studies* 31.3, pages 588–593.

Cox, Harold G (2016). *Later Life: The Realities of Aging*. Abingdon: United Kingdom: Routledge.

Dalton, Russell J (2004). *Democratic Challenges, Democratic Choices: The Erosion of Political Support in Advanced Industrial Democracies: The Erosion of Political Support in Advanced Industrial Democracies*. Comparative Politics. Oxford: United Kingdom: Oxford University Press.

Dalton, Russell J. and Scott E Flanagan (2017). *Electoral Change in Advanced Industrial Democracies: Realignment or Dealignment?* Princeton Legacy Library. Princeton, NJ: United States of America: Princeton University Press.

Dassonneville, Ruth (2016). "Age and voting". In: *The SAGE Handbook of Electoral Behaviour*. Edited by Kai Arzheimer, Jocelyn Evans, and Michael S Lewis-Beck. New York, NY: United States of America: Routledge, pages 137–58.

Davidson, Scott (2010). *Quantifying the Changing Age Structure of the British Electorate 2005-2025*. Report. London: United Kingdom: Age UK.

—— (2016). *Going Grey: The Mediation of Politics in an Ageing Society*. Abingdon: United Kingdom: Routledge.

Demeny, Paul (1986). "Pronatalist policies in low-fertility countries: Patterns, performance, and prospects". In: *Population and Development Review* 12, pages 335–358.

Díaz-Tendero Bollain, Aída (2012). *La teoría de la economía política del envejecimiento. Un nuevo enfoque para la gerontología social en México*. Tijuana: México: El Colegio de la Frontera Norte.

Doyle, Martha and Virpi Timonen (2013). "Powerless observers? Policy-makers' views on the inclusion of older people's interest organizations in the ageing policy process in Ireland". In: *The Making of Ageing Policy. Theory and Practice in Europe*. Edited by Rune Ervik and Tord Skodegal Lindén. Cheltenham: United Kingdom: Edward Elgar, pages 78–97.

Eijk, Cees van der and Mark N Franklin (2009). *Elections and Voters*. Political Analysis. Basingstoke: United Kingdom: Palgrave Macmillan.

Eisele, Frederick R (1979). "Origins of "Gerontocracy"". In: *The Gerontologist* 19.4, pages 403–407.
Formosa, Marvin and Paul Higgs, editors (2015). *Social class in later life: Power, identity and lifestyle*. Bristol: United Kingdom: Policy Press.
Geys, Benny (2006). "'Rational' theories of voter turnout: a review". In: *Political Studies Review* 4.1, pages 16–35.
Glenn, Norval D and Michael Grimes (1968). "Aging, voting, and political interest". In: *American Sociological Review* 4, pages 563–575.
Goerres, Achim (2007). "Why are older people more likely to vote? The impact of ageing on electoral turnout in Europe". In: *The British Journal of Politics and International Relations* 9.1, pages 90–121.
––––– (2009). *The Political Participation of Older People in Europe. The Greying of Our Democracies*. Basingstoke: United Kingdom: Palgrave Macmillan.
Grenier, Amanda et al. (2017). "Precarity in late life: Understanding new forms of risk and insecurity". In: *Journal of Aging Studies* 43, pages 9–14.
Grover, Sonja C (2011). *Young Peoples Human Rights and the Politics of Voting Age*. Ius Gentium: Comparative Perspectives on Law and Justice 6. Dordrecht: The Netherlands: Springer Science+Business Media B.V..
Hayek, Friedrich August von (1996). "Economic Freedom and Representative Government. The Wincott Lectures". In: *Explorations in Economic Liberalism*. Edited by Geoffrey E Wood. Basingstoke: United Kingdom: Macmillan Press, pages 50–63.
Inman, Robert P (1978). "Testing political economy's as if' proposition: is the median income voter really decisive?" In: *Public Choice* 33.4, pages 45–65.
Kam, Ping-Kwong (2003). "Powerlessness of older people in Hong Kong: A political economy analysis". In: *Journal of Aging & Social Policy* 15.4, pages 81–111.
Kildal, Nanna and Even Nilssen (2013). "Ageing policy ideas in the field of health and long-term care. Comparing the EU, the OECD and the WHO". In: *The Making of Ageing Policy. Theory and Practice in Europe*. Edited by Rune Ervik and Tord Skodegal Lindén. Cheltenham: United Kingdom: Edward Elgar, pages 53–77.
Komp, Kathrin (2011). "The Political Economy of the Third Age". In: *Gerontology in the Era of the Third Age. Implications and Next Steps*. Edited by Dawn C Carr and Kathrin Komp. New York, NY: United States of America: Springer Publishing Company, pages 51–66.
Longman, Phillip (1987). *Born to Pay. The New Politics of Aging in America*. Boston: MA: United States of America: Houghton Mifflin.
Marsh, Michael and Gail McElroy (2016). "Voting Behaviour: Continuing De-alignment". In: *How Ireland Voted 2016: The Election that Nobody Won*. Edited by Michael Gallagher and Michael Marsh. Cham: Switzerland: Springer International Publishing, pages 159–184.
Mueller, Dennis (2002). "The Political Economy of Aging Societies". In: *Economic Policy for Aging Societies*. Edited by Horst Siebert. Heidelberg: Germany: Springer, pages 269–284.

Parijs, Philippe van (1998). "The disfranchisement of the elderly and other attempts to secure intergenerational justice". In: *Philosophy & Public Affairs* 27.4, pages 292–333.

Phillipson, Chris (2013). *Ageing*. Polity Press.

— (2015b). "The political economy of longevity: developing new forms of solidarity for later life". In: *The Sociological Quarterly* 56.1, pages 80–100.

Sanderson, Warren C and Sergei Scherbov (2007). "A near electoral majority of pensioners: Prospects and policies". In: *Population and Development Review* 33.3, pages 543–554.

Seo, Yongseok (2017). "Democracy in the ageing society: Quest for political equilibrium between generations". In: *Futures* 85, pages 42–57.

Sonnicksen, Jared (2016). "Dementia and representative democracy: Exploring challenges and implications for democratic citizenship". In: *Dementia* 15.3, pages 330–342.

Stewart, Douglas J (1970). "Disfranchise the Old". In: *New Republic (29/8/1970)*, pages 20–22.

Tepe, Markus and Pieter Vanhuysse (2009). "Are Aging OECD Welfare States on the Path to Gerontocracy?: Evidence from 18 Democracies, 1980–2002". In: *Journal of Public Policy* 29.1, pages 1–28.

Timonen, Virpi (2008). *Ageing Societies: A Comparative Introduction*. Maidenhead: United Kingdom: McGraw-Hill Open University Press.

Townsend, Peter (2007). "Using human rights to defeat ageism: dealing with policy-induced 'structured dependency' ". In: *Critical perspectives on ageing societies*. Edited by Miriam Bernard and Thomas Scharf. Bristol: United Kingdom: Policy Press, pages 27–44.

— (2002). "Undemocratic Governments". In: *Kyklos* 55.1, pages 247–264.

Walker, Alan (1981). "Towards a political economy of old age". In: *Ageing and Society* 1.01, pages 73–94.

Wasfy, Jason H, Charles Stewart III, and Vijeta Bhambhani (2017). "County community health associations of net voting shift in the 2016 US presidential election". In: *PloS one* 12.10. URL: https://doi.org/10.1371/journal.pone.0185051.

Part V

The Silver Economy

12

The Silver Economy

Overview

This chapter—the only one in this part—covers the emergence of the so-called grey or silver economy or markets in the context of consumer societies. It reviews the influence of population ageing on the changes in consumer demand across economic sectors. Topics include the notion of the 'ageing' consumer and the retirement-consumption 'puzzle'. It also reviews the 'successful ageing' conceptual framework and the notion of affluenza and later life.

12.1 Introduction

The 'silver' economy and markets is an expression that originated in Japan, where *shirubā* initially meant to refer to goods and services for older people. There is no agreed or accepted definition of silver economy (other expressions apart from 'silver' also used in the literature are 'grey', 'older', 'ageing', or 'mature'). Moody and Sasser (2018) included the 'new ageing marketplace' as one of the twelve controversies surrounding ageing. However, older consumers have been the focus of extensive scholarly research in marketing and consumer studies since the early 1970s, mainly in the United States of America, although Dodge (1958) analysed how to reach to older consumers. In 1960, the US business magazine *Business Week* featured an article about the 'old age market' (Week 1960), and later in that decade some authors already referred to the senior market as 'elusive' (Samli 1967) and 'overlooked' (Morse 1964) and wondered whether it was a fact or a fiction (Reinecke 1964).[1]

The European Commission defined the silver economy as

> ...a wide range of age-related products and services in many existing sectors, including information and communication technologies, financial services, housing, transport, energy, tourism, culture, infrastructures, and local services as well as long-term care.
>
> (European Commission 2007, p. 96)

The Commission (2015) identified three silver generations in the silver economy market development for member states: (a) the older working age (age 50 to the senior retirement age in Europe, i.e. age 67 plus 7 further years) whose experience and abilities make an increasing large actual and percentage contribution to corporate turnover and profit and to their national economies and by offering training, mentoring, and skills transfer to the young, (b) the active pensioners (65 to 75/80) who look forward to and require the supply for recreational and active retirement and the maintenance of good health and interactivity with society, and (c) the older and frail elderly (80+), requiring a comfortable and extended life being consumers in the market development for home-based long-term care services and institutional care. However, other authors point out that the silver economy and market go beyond particular chronological age groups and include 'ideas for the "universal design" and the "transgenerational design" (or the "intergenerational design") and "barrier-free products and services" for those who are frail, handicapped, or disabled' of all ages (Klimczuk 2015, p. 77).

Even though the focus of this part is on older people as consumers, it is worth noting that the notion of the silver economy is sometimes extended to include older workers, volunteers, investors, or entrepreneurs as well—in other words, older people as actors in the manifold roles in modern economies rather than solely as consumers. In turn, the silver economy is also viewed as workers, volunteers, investors, and entrepreneurs of all ages whose activities are tailored for or targeting at older consumers. Another approach includes 'age brands'—that is, goods and services targeted to older consumers and the 'new ageing enterprises' and charitable organisations focused on older people, which according to Klimczuk (2015, p. 79) would increase the 'sense of agency and activity' of older people.

The increasing importance of the silver economy is seen by some authors as an opportunity for furthering the quality of life of older people. In this regard, Enste et al. (2008, pp. 337–338) put forth the following recommendations 'for a sociopolitically sensible, and at the same time economically positive, further development of the silver market':

- *Customer-oriented enhancement of the range of products as well as of services and a differentiated market development*
- *Sensitisation and co-ordination of the actors*
- *Further development and increased deployment of senior marketing*
- *Provision for the consumer needs of poor older people*
- *Empowerment and a better representation of the interests of older consumers*
- *Dialogical product and services development*
- *Enhancement and expansion of the existing products and services*
- *Further development of user-friendly and seniors-oriented design*
- *Promotion of consumer protection for older people*

Despite the growing presence of older consumers and the potential for further growth of the silver economy, these markets face localised barriers that may hamper their development in specific regions or cities. Klimczuk (2015) identified the following five:

- lack of firms with a well-developed research base and marketing know-how
- viewing the silver economy as only relevant to affluent and influential older people
- different consumption patterns stemming from other sets of values than those prevalent in developed countries
- migration patterns among older people that tend to reinforce particular negative images of these regions
- the negative influence and impact that the promotion of certain products and markets developed as a result of population ageing (e.g. anti-ageing medicine) would have on consumers and ageism

12.2 The 'Ageing' Consumer

Many studies in marketing research on older consumers delimit the 'senior' markets by means of thresholds defined by chronological age, so that 'ageing' or 'silver' consumers are adults aged 60 or over, or 55 or over, and so on. However, as I described in Chap. 1 in Volume I, gerontologists distinguish between three conceptualisations of 'age' apart from chronological age: biological, psychological, social. These notions of age (and ageing) are relevant to how consumption patterns change in later life, though they play a minor role in consumer research (Zniva and Weitzl 2016). Changes in biological age—not in the sense of a theoretical construct based on biomarkers but of the onset and progression of physiological and cognitive decline—as well as changes in psychological and social age are behind some of the main motivations to focus on, and drivers of, the silver economy and markets. For example, in Japan, due to the demographic change its population is experiencing, the market for adult nappies (i.e. diapers) surpassed in size (both volume and revenue) that of babies. Other new and growing age-related markets in Japan and elsewhere include those of assistive or healthcare robots and of pet-related products. Regarding the latter, according to Usui (2008), the market for pets and related products would be booming in Japan because pets are increasingly considered as family members rather than animals, to the extent that there are firms offering amusement parks and hot springs for dogs. This change in attitudes towards pets is driven by a decline in household size, an increase in 'empty nests', loneliness, and reduced social interaction. Furthermore, concerning loneliness in later life, markets for talking dolls, robotic pets, and companion robots have also emerged. Hence, population ageing—though not exclusively in terms of chronology—is behind processes of creative destruction and of market, product, and service development, changing several economic or industrial sectors. In addition, it is also changing the type of consumer. It may come as no surprise, then, that two main approaches are applied to the study of the silver economy. One is based on sectors; its objective is to discern trends in demand driven by population ageing. The second approach is based on the identification of sub-groups of older consumers, usually in specific markets.

12.2.1 Demand-Driven Market Segmentations

According to Enste et al. (2008, p. 330), the silver economy 'should not be regarded as an own economic sector but rather as a cross-section market, in which numerous industrial sectors are involved'. Different authors have come up with different enumerations of sectors that would be in a state of flux due to the ageing of the population. Serrière (2018) listed the following main sectors under stark transformation:

- Tourism
- Food and nutrition
- Telecare and home automation
- Long-term care
- Robotics

- Fitness
- Video games
- Smartphones
- Technology
- Self-driving cars
- E-health
- Home improvement

Enste et al. (2008, p. 332) included these same sectors in their own list, but added education and culture, clothing and fashion, insurance coverage against age-specific or age-related "risks", and financial services 'sensitive to demography' (i.e. capital protection and decumulation).

Section 9.2 in Volume III presents a discussion on the relationship between population ageing and relative prices, where it is mentioned that one of the mechanisms linking ageing with changes in relative prices is the variation in consumption patterns as people get older. Regardless of its effects on aggregate prices, different age-consumption patterns between older and younger people have been identified for many goods and services in various countries. More generally, consumption patterns vary along the life cycle, so that the evidence suggests the existence of different age-consumption profiles not only between 'younger' and 'older' people but at all ages. Some economists have resorted to a broad classification of goods and services—that is, tradeable and non-tradeable—and asserted that as people grow older, they tend to reduce the demand for the latter (Börsch-Supan 2003; Braude 2000; Ewijk and Volkerink 2012; Hobijn et al. 2003). For example, as it would be expected, older people spend a larger share of their income in health-related goods and services (Hobijn et al. 2003) and leisure (Higgs et al. 2009). However, in their study of consumption patterns by chronological age in the United Kingdom, Julia Twigg and Majima (2014) were surprised to find that women aged over 75 were the most frequent attenders of hairdressers—although these authors ascribed this finding to a cohort effect.

One problem with a number of studies is that they exclusively focused on consumption patterns of older people—however their authors may have set the minimum chronological age threshold to define them rather than, at least, contrast these patterns with those of younger people. Tongren (1988) objected to this approach on the grounds that comparisons would be necessary to ascertain how unique or peculiar older consumers are—see also Zniva and Weitzl (2016). This is a justifiable objection: Higgs et al. (2009, Figure 2), for example, showed for the United Kingdom that the proportion of total household expenditure devoted to food by households whose heads defined themselves as retired declined from over 30 per cent in 1968 to just over 20 per cent in 2004/2005. One would be inclined to conclude that there was a shift in consumption patterns away from essential goods such as food among older people during the period. However, Figure 3 in the same paper shows that a similar though more pronounced negative trend could be found among other groups—for example, the employees and the self-employed. Therefore, a different conclusion could be reached: that there was nothing peculiar about consumption of food (in these aggregate terms) among retired household heads and that the reduction reflected, instead, a more general trend in British society over a period of rising living standards. Or, as the authors put it, that older age had been 'increasingly less an obstacle to engagement with consumption than it was in the past' (Higgs et al. 2009, p. 116).

More generally, Higgs et al. theorised that this confluence in consumption patterns of aggregate categories across birth cohorts could be explained with the help of two notions: generational field and generational habitus.[2] A generational field is a 'cultural field shaped by later life consumption patterns' (Gilleard and Higgs 2009, p. 23) that emerges as a result of changes in the relationship between past and present social spaces and that 'is determined by...a post-scarcity consumption that supports the search for distinction and that implicitly or explicitly rejects, denies or marginalizes "old-age"' (Gilleard and Higgs 2009, p. 25). In turn, Gilleard and Higgs (2005, p. 70) defined generational habitus as 'a set of dispositions that generate and structure individual practices, which emerge and are defined by the forces operating within a particular generational field'. The post-war generation, according to Higgs et al., experienced a generational habitus characterised by rising living standards and the growth of the consumer society and youth culture,[3] which marked the cultural practices of the younger cohorts as manifest in 'the fusing of fashion and identity, in the commodification of lifestyle and in the erosion of status and tradition' (Higgs et al. 2009, p. 106).

Furthermore, consumption and a person's sense of self or identity are linked in multifarious ways in contemporary societies: to some extent, individual identity is forged by consumption (Belk 1988). This is also the case in later life[4] (Grant 2004; Oh and Choi 2015)—what Schau et al. (2009) termed 'consumer identity renaissance'. Consumption goods and services have a symbolic content, which is used both as a social and cultural marker and for social distinction. Furthermore, Csikszentmihalyi and Halton (1981) reported that for people in their late 40s and 50s, identity is less defined by what they do and more by what they *have* and that in later life, possessions are important for their ability to relate to others and to help construct life projects and stories, which are especially prominent at retirement.

How people approach the retirement transition psychologically (i.e. their lifestyle postures) is also important for their consumption patterns at retirement. Lifestyle, incidentally, has been defined as

> distinctive attributes or recognizable patterns of behaviors reflecting shared interests and life situations incorporating related values, attitudes, and orientations that create characteristic social identities
>
> (Hendricks and Hatch 2006, p. 303)

Hornstein and Wapner (1986) distinguished four lifestyle postures towards the retirement transition:

- as a new beginning
- as a continuation of pre-retirement life structure or lifestyle
- as an imposed disruption
- as a transition to old age

Hopkins et al. (2006) remarked that retirees who approached retirement as a new beginning are more likely to increase consumption of outward-oriented, symbolic products that replace their former work-related roles, whilst those who perceive retirement as a transition to old age consume relatively more non-experiential inward-oriented products.

Shopping Habits

Another related point is shopping habits, for not only does the type of product demanded change with population ageing, but average shopping habits also evidence modifications with chronological age (Lumpkin et al. 1985; Martin 1976; Zniva 2016). Retail and marketing studies found that older consumers tend to go on shopping more often and at different times than younger shoppers. They also tend to shop in places they know and where they are known. 'Two-for-one' and other discounts based on quantity are less enticing to older people on average, as they tend to live in smaller households than younger people.

Furthermore, attitudes towards prices change with chronological age. Impulse buying declines in later life (Amos et al. 2014; Wood 1998)—which some evidence from neuroeconomics suggests may be related to changes in grey matter volume in the brain (Yokoyama et al. 2014). Exploration and search for information also tend to decline in later life: older consumers are, on average, less engaged in seeking information about prices and brands than younger consumers (Cole and Balasubramanian 1993; Cole et al. 2008; Faras 2018; Lambert-Pandraud et al. 2005).

12.2.2 Other Market Segmentations

Finally, segmentation exercises have identified sub-groups of older consumers with regard to several characteristics and variables—a reflection of the increasing heterogeneity among individuals as they get older. Market segmentation involves

> the subdivision of the entire market for a product or service into smaller market groups or segments, consisting of customers who are relatively similar within each specific segment and maximally different from customers comprising other segments.
>
> (Moschis et al. 1997, p. 284).

Initially, older consumers were crudely classified according to their chronological age (e.g. below or above pensionable age; 55–64, 65–74, 75 and over). Other criteria were used, but consumers' characteristics merely provide a description of data with no predictive power. Consequently, multiple segmentation techniques have been regularly applied since the mid-1970s focusing on specific products, which has brought new light to the understanding of consumers in later life.

One such approach is known as 'gerontographics'. Introduced by Moschis (1992, 1996), it

> …acknowledges individual differences in aging processes as well as differences in types of aging dimensions that occur in later life…[It is] based on the premise that the observed similarities and differences in the consumer behavior of older adults is the outcome of several social, psychological, biophysical, life-time events, and other environmental factors, all affecting the aged person differently.
>
> (Moschis 1992, pp. 19–20).

Using cluster analysis on data for over 20,000 consumers aged 50 or over in the late 1990s in the United States of America, Moschis and his team identified four sub-groups or segments of older consumers according to differences in preferences for products and services, information needs, patronage motives, industry images or perceptions, and advertising perceptions.

Other segmentation studies are based on lifestyle factors—or 'psychographics'. Its first application to older consumers by Oates et al. (1996) rendered five distinct groupings based on data on activities, interests, and opinions of older consumers. Both approaches, gerontographics and psychographics, have been applied to anything from cosmetics to housing, from health insurance to magazines, and from online shopping for food to travel.

Benefit segmentation is another popular technique; it is based on the premise that 'the benefits which people are seeking in consuming a given product are the basic reasons for the existence of true market segments' (Haley 1968, p. 31)—see also Haley (1984). Koubaa et al. (2017) presents an application to senior markets in France.

12.3 The Retirement-Consumption Puzzle

Economic activity is a source of variation in the consumption patterns in later life as well as social location, income, wealth and cultural background are also relevant. Moreover, chronological age is a significant factor: Abdel-Ghany and Sharpe (1997) reported substantial differences in the United States of America between households headed by a person aged between 65 and 74 years and those headed by someone aged 75 years or older in the consumption of products such as food away from home, alcohol, housing, transportation, healthcare, personal insurance, and so on. However—and, yes, once again—there is disagreement in the literature regarding these findings, though mostly because the various studies refer to different chronological ages, periods, and sometimes goods.[5] Of the changes in economic activity, the transition into retirement is likely to have some bearing on consumption patterns in later life.

A vast literature is devoted to exploring a finding that has puzzled economists for years: that consumption levels fall among retirees, what is known as the retirement-consumption puzzle. The decline in consumption in retirement is puzzling because the life-cycle hypothesis predicts that economic agents would seek to smooth their consumption over the life cycle, precisely to avoid cliff-edge experiences in consumption levels. Aggregate data tend to confirm the existence of a reduction in consumption among retirees (Bullard and Feigenbaum 2007; Hamermesh 1984a; Thurow 1969),[6] whereas disaggregated data paints a more nuanced picture.

A drop in consumption in retirement is part of wider research efforts into the determinants of hump-shaped age-consumption profiles. Again, against the predictions of the life-cycle hypothesis, several authors confirmed that age-consumption profiles are far from flat and would exhibit, instead, a gradual increase until the mid- to late-40s and 50s to fall gradually thereafter (Attanasio and Browning 1995; Carroll and Summers 1991; Fernández-Villaverde and Krueger 2007; Gourinchas and Parket 2002). Different explanations have been proposed for the finding that retirees would experience a pronounced reduction in consumption levels (Aguiar and Hurst 2013; Haider and Stephens 2007):

- Time-inconsistent preferences and hyperbolic discounting (Angeletos et al. 2001)—see Chap. 8 in Volume I and Chap. 8 in this volume.
- Unanticipated income shocks: before retirement, individuals may expect a larger income when they retire than what they actually receive in retirement (Banks et al. 1998; Hurd and Rohwedder 2003).
- Unanticipated health shocks (Hurd and Rohwedder 2003). This view posits that it is not an inadequacy of income and wealth that can explain the drop in consumption at retirement but health shocks that agents have failed to anticipate.
- Precautionary savings and impatience (Gourinchas and Parket 2002). The hypothesis is that economic agents change behaviour regarding savings and consumption at ages 40–45:

> The neoclassical representative-agent model of aggregate consumption is incorrect precisely because of this changing behavior over the working life.
> (Gourinchas and Parket 2002, p. 82)

During younger years, agents would adopt a 'buffer-stock' behaviour: despite their preference for consuming out of future income, younger people save for precautionary purposes until they reach a given target. In contrast, agents older than 40–45 years of age but in paid employment would save actively towards retirement.
- The existence of an unfunded social security system (Hansena and Imrohoroglu 2008).
- A change in household bargaining power between spouses (Lundberg et al. 2003). The hypothesis is that as women tend to live longer than men and married women tend to be younger than their husbands, wives would exhibit a higher preference for saving than their spouses. Combined with this difference in life expectancy, there is another hypothesis related to in-household decision-making: the relative bargaining power of each household member would depend on the private and public resources each one of them holds. The main household income in many countries is

earned by the male, so that his retirement would reduce his bargaining power within the household, leading to a fall in household consumption as a reflection of the wife's increasing power.
- Home production (Hurd and Rohwedder 2003; Lührmann 2010; Stancanelli and Van Soest 2012): with additional leisure, retirees may substitute consumption of purchased market goods and services with home production.
- Expectations about consumption in retirement Ameriks et al. (2002).
- Stepping 'outside the framework of rational, farsighted optimization' (Bernheim et al. 2001, p. 855), poor or imperfect planning, and the use of heuristic rules of thumb to make saving decisions during working years.

In actuality, the consensus is that certain consumption goods and services do show a sharp decline as people retire, but that in aggregate terms there is no such thing as a substantial reduction in consumption levels between and after the retirement transition. See, among other studies,

- Aguiar and Hurst (2005, 2013), Aguila et al. (2011), Blau (2008), Fisher et al. (2008), Haider and Stephens (2007), Hamermesh (1984b), Hurd and Rohwedder (2013), Hurst (2008), and Laitner and Silverman (2005) (United States of America)
- Barrett and Brzozowski (2010, 2012) (Australia)
- Li et al. (2015, 2016) (China)
- Moreau and Stancanelli (2015) (France)
- Lührmann (2010) and Schwerdt (2005) (Germany)
- Battistin et al. (2009) and Miniaci et al. (2003) (Italy)
- Stephens and Unayama (2012) and Wakabayashi (2008) (Japan)
- Cho (2012) (Korea)
- Kolasa et al. (2017) (Poland)
- Luengo-Prado and Sevilla (2013) (Spain)
- Banks et al. (1998) and Smith (2006) (United Kingdom)

Detailed studies of disaggregated consumption patterns of people in employment and in retirement cast some surprising findings along with more obvious ones. Some goods and services would be less demanded in retirement whilst others would be more so. It would be striking not to find that work-related consumption (e.g. non-durable transportation and clothing) falls substantially in retirement. In turn, it would be more difficult to predict that retirees would significantly reduce their food consumption compared to people in employment of similar chronological age, but this has been also widely reported. However, more detailed studies shed additional light on this issue: for example, Plessz et al. (2015) reported an increase in vegetable consumption among retirees in France—a behaviour that would be mediated by place of lunch which, at retirement, is almost exclusively at home; Appleton et al. (2017) also found an increase in vegetable consumption at retirement in France, Italy, and the United Kingdom.

Hurst (2008) identified five stylised effects in the consumption patterns of people in retirement. Apart from differences in work-related consumption from people in employment and a decline in food expenditure, Hurst mentioned that despite the reduction in food consumption, food intake stays at similar levels after retirement, that there is a greater heterogeneity among retirees that among people in employment—especially among retirees with low pension wealth—and that there is a sharp decline in consumption among retirees who experienced involuntary retirement, which is even more pronounced when accompanied by a health shock.

12.4 Ageing and the Consumer Society

Offe (1985) contended that in contemporary advanced societies consumption has replaced labour as a key vital interest. In this same vein, some social thinkers talked of consumption as the central tool for making lifestyle a life project in modern societies (Featherstone 1987, 2007). It is in this context that the 'consumer society' appears in sociological and philosophical works. Sometimes the same expression in one language loses a subtle difference in meaning when it is translated into another language. 'Consumer society' is the topic of an influential essay by French philosopher Jean Baudrillard,[7] except that the French expression for 'consumer society' is 'société de consommation'—literally, "society of consumption". In English, the focus is on the economic actor (i.e. the consumer), whereas in French the spotlight is on the practice, the act of consumption. This semantic distinction is relevant because sociological approaches to the consumer society see the consumer through a different lens from mainstream economics, according to which a consumer is the 'central figure in microeconomic theory' (Kreps 1990, p. 17), an actor that exerts her subjectivity by means of choosing between goods and services in order to maximise her utility subject to a budget constraint. One line of sociological and philosophical thought on the consumer society understands that the actors become commodities in the act of consuming: in the words of Offe (1985, p. 12), no subject 'can keep his or her subjectness secure without perpetually resuscitating, resurrecting and replenishing the capacities expected and required of a sellable commodity.' Consumers in consumer societies would become indistinguishable from the commodities they consume—a consumer society would blend into a society of consumption.

Lewis (2008, p. 35) asserted that 'contemporary developed societies are characterized by':

- *increasing differentiation and diversity (in cultural interaction, ways of living and acting, etc.)*
- *increasing consumerism*
- *increasing density of televisual product and modes of knowing* (i.e. ways of acquiring beliefs and knowledge, including reason, language, emotion, perception, intuition, etc.)

The thesis is that consumption is intertwined in these three characteristics, as it satisfies—albeit partially and fleetingly—the desire for social differentiation and distinction (Bourdieu 1979). Such desire as well as the values and social imaginary it stems from are transmitted and reproduced via social and other media, which are increasingly reliant on the image as their medium *par excellence*.[8]

These images convey the message of the superiority of certain lifestyles attainable via consuming particular bundles of goods and services thus creating a propensity to purchase any of those lifestyles. As Hendricks and Hatch (2006, p. 301) explained, it is through lifestyles that 'social differentiation is made apparent, group recognition made possible, and differential outcomes fashioned'. In this regard, the United Nations refer to the 'global consumer' (United Nations Population Fund 2004)—a new 'consumer class' of around 1.7 billion people (half of whom would live in developing countries) with one common denominator: a 'lifestyle and culture that became common in Europe, North America, Japan, and a few other pockets of the world in the twentieth century [that] is going global in the twenty-first' (Halweil and Mastny 2004, p. 4). Here lies the key element of the consumer/consumption society: a new social imaginary is forged from lifestyles to be achieved by means of consuming, and in the process a new subjectivity is formed, which is in fact the dissolution of the subject into the commodities she consumes.

12.4.1 Successful Ageing and the Consumer Society

Rowe and Kahn (1997) introduced the concept of successful ageing as 'the many factors which permit individuals to continue to function effectively, both physically and mentally, in old age' (Rowe and Kahn 1998, p. xii). It consists of three main elements (Rowe and Kahn 1997):

- maintaining a low probability of disease and disability
- maintaining high cognitive and physical ability and functioning
- maintaining high levels of engagement in social activities and productive activities

The emphasis, as Kahn (2004, p. 4) indicated, is in 'what individuals can do for themselves to maintain vitality in old age'. In order to complement this individualist approach with a more macro focus, a number of 'success criteria' at a societal level have been proposed, including (Rowe and Kahn 2015, p. 594):

- productivity and engagement both in paid employment and voluntary work
- cohesion between generations and socio-economic strata
- balance between the risks and benefits of demographic change
- resilience—that is, the capacity to respond effectively to and bounce back and even flourish from stress
- sustainability—that is, the capacity to maintain high functioning levels over time

The most promising approach to the consumer society is the development of a critical framework to interrogate the concept of 'successful ageing' and related terms such as 'productive ageing' or 'new ageing'. Successful, productive, new ageing and others are a set of contemporary narratives that inform consumption patterns in later life, particularly in their prescriptive and normative elements related to the 'right' choices regarding future fitness, performance, health and well-being, and, ultimately, agency and self (Higgs and Gilleard 2015). These narratives link ageing with decline and are one of the most powerful drivers of silver markets. Inserted in a consumer culture that constantly promotes and idolises youth and youthfulness, they turn ageing—as Twigg (2004, p. 61) commented—into 'a project to be worked upon, fashioned and controlled'. According to Estes et al. (2003, p. 70), successful ageing is 'a remedial intervention approach, masking a decline model of ageing' that 'presents a literal response to age as a natural, manageable problem'. These authors, comparing the successful ageing and productive ageing frameworks, commented: 'Productive ageing is the active justification of "I can still work" and successful ageing the passive justification of "Look, I'm trying hard not to be a burden on others"' (Estes et al. 2003, p. 72).

Besides, these discourses are related to what Gilleard and Higgs (2009) termed 'active agentic consumerism'—that is, 'consumption that most expresses choice, autonomy, pleasure, and self-expression' [p. 26], and, especially, an 'ageless' self (Kaufman 1994). For example, 'new ageing'— 'a buoyant and optimistic cultural imagery around which marketing and consumerism have rallied' (Katz and Marshall 2003, p. 4)—is such an agentic concept: 'agentic', in this context, means to be able to 'influence intentionally one's functioning and life circumstances' (Bandura 2008, p. 16).[9] Beyond the macro or societal dimension, concepts such as successful ageing, with their strong focus on an older person's agency, would—to be begin with—only be applicable to, admittedly, the minority of older people (even in developed countries) with enough autonomy and agency to be able to choose between an array of lifestyle choices. As Barnes et al. (2018, p. 50) remarked: "'Choice' is not a means to empowerment when you are feeling frail, vulnerable and facing decisions you do not want to make.'. And, as also Vincent (1995, p. 118) commented: 'Real unalienated choice is created by proper pensions and also fair prices'. Furthermore, no structural elements are brought into play, despite the inextricable relationship between agency and social structure and despite the various structural factors that would impede the realisation of many a 'chosen' lifestyle.

There is nothing intrinsically 'wrong' with the promotion of 'successful' ageing and other lexical concepts of similar meaning such as 'healthy', 'positive', 'optimal', 'independent', 'active', 'productive', or 'effective' ageing; after all, 'successful ageing' is 'one of gerontology's most successful ideas', as Katz and Calasanti (2014, p. 26) playfully remarked! But we must be wary of the less desirable attitudes that may be subtly lurking behind policies and narratives that put the emphasis *and onus* on older individuals, as such policies and discourses may reinforce and perpetuate pernicious outcomes. The successful ageing framework presents a static 'ideal' that neglects the 'historical and cultural context, social relationships, and structural forces' that have a bearing on how older people live their (later) lives (Stowe and Cooney 2014). The two main authors behind the notion of 'successful ageing' asserted that their 'main message is that we can have a dramatic impact on our own success or failure in aging. Far more than is usually assumed, successful aging is in our hands.' (Rowe and Kahn 1998, p. 18). Consumer society and successful ageing coalesce in the notion of the 'silver economy' and its conceptual cousins, in which firms are encouraged to find 'gold in gray' (Minkler 1989), and sweep under the carpet the poverty and inequality and deprivation that many people suffer in later life, including in the wealthiest countries and regions in the world—see Chap. II in this volume.

12.4.2 Affluenza

'Affluenza' is a portmanteau word derived from the blend of affluence and influenza. It was popularised by James (2007), in a book that proposed the theory that contemporary capitalism increases the affluence in an economy and society at the expense of inequality and neglect of interpersonal relationships. These trends would be paramount in materialistic consumer societies and would create and deepen emotional and psychological stress (see also Graaf et al. 2001; Hamilton and Denniss 2005). It is characterised by a plethora of persuasive arguments and propositions whose goal is to mount evidence in favour of theses that are ultimately untestable. It is part of a wider movement of criticism of consumerism and capitalism, which largely ignores inequality and deprivation (see Part II in this volume) in general, not only among older people. In fact, it has little to say specifically about later life.

Some empirical research is at odds with these arguments, but there is a risk of cherry-picking findings to falsify or to validate conjectures that are actually unfalsifiable. With this in mind, I simply want to touch upon some trends in social networks in later life that seem to be discordant with the corollaries from those supportive of the affluenza thesis.

The composition of personal social networks varies over the life course (Youm et al. 2018), but the affluenza thesis suggests that the quantity and quality of interpersonal relationships would be eroding over time, not as an age or cohort effect, but as a period effect, and that the reduction would be more marked, the more developed or affluent were the country or the individual. However:

- Tomini et al. (2016) investigated the size of social networks among older people in sixteen European countries. They reported *larger* networks in the *more affluent* Western and Northern countries compared to the Southern and Eastern countries.
- Schwartz and Litwin (2018) found an *expansion* in the size of the social networks of people aged 65 or over in Europe, with continuous additions of new members, especially among women. Conway et al. (2013) reached a similar conclusion with data from older people in the United States of America.

The main merit of a line of reasoning such as this is not so much the production of testable hypotheses but the posing of challenging questions at a more general level about the economic systems in place in almost all the countries in the world.

Notes

1. For a review of the literature before 1980, see Meadow et al. (1981).
2. 'Field' and 'habitus' are concepts introduced by French social thinker Pierre Bourdieu—see Bourdieu and Wacquant (1992), Hilgers and Mangez (2015), and Medvetz and Sallaz (2018).
3. It has to be noted that this hypothesis would apply mostly to the middle classes in developed countries and some developing countries.
4. '…aging is not about the inevitable end but rather about the evolving self' (Schau et al. 2009, p. 256).
5. For example, Fareed and Riggs (1982) failed to find any differences between people aged under and over 65 years in the United States of America.
6. In fact, Hamermesh (1984a) found that consumption a few years into retirement does not fall but that given these levels were not sustainable, households reduced their consumption gradually over time giving way to a J-shaped age-consumption profile in retirement.
7. Baudrillard (1974).
8. Sartori (2000) opined that contemporary societies are characterised by 'post-thinking': the individual would have been transformed into a *homo videns* who consumes so many images that her capacity to think is ultimately annulled.
9. See also Bandura (2001).

References

Abdel-Ghany, Mohamed and Deanna L Sharpe (1997). "Consumption patterns among the young-old and old-old". In: *Journal of Consumer Affairs* 31.1, pages 90–112.

Aguiar, Mark and Erik Hurst (2005). "Consumption versus Expenditure". In: *Journal of Political Economy* 113.5, pages 919–948.

—— (2013). "Deconstructing life cycle expenditure". In: *Journal of Political Economy* 121.3, pages 437–492.

Aguila, Emma, Orazio Attanasio, and Costas Meghir (2011). "Changes in consumption at retirement: evidence from panel data". In: *Review of Economics and Statistics* 93.3, pages 1094–1099.

Amos, Clinton, Gary R Holmes, and William C Keneson (2014). "A meta-analysis of consumer impulse buying". In: *Journal of Retailing and Consumer Services* 21.2, pages 86–97.

Ameriks, John, Andrew Caplin, and John Leahy (2002). *Retirement consumption: Insights from a survey*. NBER Working Paper No. w8735. Cambridge, MA: United States of America. National Bureau of Economic Research.

Angeletos, Laibson, Repetto, Tobacman, and Angeletos, George-Marios et al. (2001). "The hyperbolic consumption model: Calibration, simulation, and empirical evaluation". In: *Journal of Economic Perspectives* 15.3, pages 47–68.

Appleton, Dinnella, Spinelli, Morizet, Saulais, Appleton, Katherine M et al. (2017). "Consumption of a High Quantity and a Wide Variety of Vegetables Are Predicted by Different Food Choice Motives in Older Adults from France, Italy and the UK". In: *Nutrients* 9.9. DOI: https://doi.org/10.3390/nu9090923.

Attanasio, Orazio P and Martin Browning (1995). "Consumption over the life cycle and over the business cycle". In: *The American Economic Review* 85.5, page 1118.

Bandura, Albert (2001). "Social cognitive theory: An agentic perspective". In: *Annual review of psychology* 52.1, pages 1–26.

—— (2008). "Toward an agentic theory of the self". In: *Advances in self research* 3, pages 15–49.

Banks, Blundell, and Banks, James, Richard Blundell, and Sarah Tanner (1998). "Is There a Retirement-Savings Puzzle?". In: *The American Economic Review* 88.4, pages 769–788.

Barnes, Marian, Beatrice Gahagan, and Lizzie Ward (2018). *Re-imagining old age: Wellbeing, care and participation*. Series in Sociology Wilmington, DE: United States of America: Vernon Press.

Barrett, Garry F and Matthew Brzozowski (2010). *Involuntary Retirement and the Resolution of the Retirement-Consumption Puzzle: Evidence from Australia*. Social and Economic Dimensions of an Aging Population Research Papers 275. McMaster University.

—— (2012). "Food expenditure and involuntary retirement: Resolving the retirement-consumption puzzle". In: *American Journal of Agricultural Economics* 94.4, pages 945–955.

Battistin, Erich et al. (2009). "The retirement consumption puzzle: evidence from a regression discontinuity approach". In: *American Economic Review* 99.5, pages 2209–26.

Baudrillard, Jean (1974). *La société de consommation: ses mythes, ses structures*. Idées. Sciences humaines. Paris: France: Gallimard.

Belk, Russell W (1988). "Possessions and the extended self". In: *Journal of Consumer Research* 15.2, pages 139–168.

Bernheim, Skinner, and Bernheim, B Douglas, Jonathan Skinner, and Steven Weinberg (2001). "What accounts for the variation in retirement wealth among US households?" In: *American Economic Review* 91.4, pages 832–857.

Blau, David M (2008). "Retirement and consumption in a life cycle model". In: *Journal of Labor Economics* 26.1, pages 35–71.

Börsch-Supan, Axel (2003). "Labor market effects of population aging". In: *Labour* 17.s1, pages 5–44.

Bourdieu, Pierre (1979). *La distinction: critique sociale du jugement*. Collection Le sens commun. Edition de Minuit.

Bourdieu, Pierre and Loic Wacquant (1992). *An Invitation to Reflexive Sociology*. Sociology (University of Chicago). Chicago, IL: United States of America: University of Chicago Press.

Braude, Jacob (2000). *Age structure and the real exchange rate*. Discussion Paper 2000.10. Jerusalem: Israel: Bank of Israel.

Bullard, James and James Feigenbaum (2007). "A leisurely reading of the life-cycle consumption data". In: *Journal of Monetary Economics* 54.8, pages 2305–2320.

Carroll, Christopher D and Lawrence H Summers (1991). "Consumption growth parallels income growth: Some new evidence". In: *National saving and economic performance*. Edited by B Douglas Bernheim and John B Shoven. Chicago, IL: United States of America: University of Chicago Press, pages 305–348.

Cho, Insook (2012). "The retirement consumption in Korea: evidence from the Korean Labor and Income Panel Study". In: *Global Economic Review* 41.2, pages 163–187.

Cole, Catherine A and Siva K Balasubramanian (1993). "Age differences in consumers' search for information: Public policy implications". In: *Journal of Consumer Research* 20.1, pages 157–169.

Cole, Laurent, Drolet, Ebert, Gutchess, Cole, Catherine et al. (2008). "Decision making and brand choice by older consumers". In: *Marketing Letters* 19.3-4, pages 355–365.

Commission, European (2015). *Growing the European silver economy*. Background Paper 23. Brussels: Belgium.

Conway, Francine et al. (2013). "A six-year follow-up study of social network changes among African-American, Caribbean, and US-born Caucasian urban older adults". In: *The International Journal of Aging and Human Development* 76.1, pages 1–27.

Csikszentmihalyi, Mihaly and Eugene Halton (1981). *The Meaning of Things: Domestic Symbols and the Self*. Cambridge: United Kingdom: Cambridge University Press.

Dodge, Robert E (1958). "Selling the older consumer". In: *Journal of Retailing* 34.2, pages 73–81.

Enste, Peter, Gerhard Naegele, and Verena Leve (2008). "The discovery and development of the silver market in Germany". In: *The Silver Market Phenomenon*. Edited by Florian Kohlbacher and Cornelius Herstatt. Heidelberg: Germany: Springer, pages 325–339.

Estes, Carroll L., Simon Biggs, and Chris Phillipson (2003). *Social Theory, Social Policy and Ageing. A Critical Introduction*. Open University Press.

European Commission (2007). *Europe's demographic future: Facts and figures on challenges and opportunities*. Technical report. Luxembourg: Luxembourg.

Ewijk, Casper van and Maikel Volkerink (2012). "Will ageing lead to a higher real exchange rate for the Netherlands?" In: *De Economist* 160.1, pages 59–80.

Fareed, AE and GD Riggs (1982). "Old-Young Differences in Consumer Expenditure Patterns". In: *Journal of Consumer Affairs* 16.1, pages 152–160.

Faras, Pablo (2018). "Determinants of knowledge of personal loans' total costs: How price consciousness, financial literacy purchase recency and frequency work together". In: *Journal of Business Research*. DOI: https://doi.org/10.1016/j.jbusres. 2018.01.047.

Featherstone, Mike (1987). "Lifestyle and consumer culture". In: *Theory, Culture & Society* 4.1, pages 55–70.

—— (2007). *Consumer Culture and Postmodernism*. Published in association with Theory Culture & Society London: United Kingdom: SAGE Publications.

Fernández-Villaverde, Jesús and Dirk Krueger (2007). "Consumption over the life cycle: Facts from consumer expenditure survey data". In: *The Review of Economics and Statistics* 89.3, pages 552–565.

Fisher, Jonathan D et al. (2008). "The retirement consumption conundrum: Evidence from a consumption survey". In: *Economics Letters* 99.3, pages 482–485.

Gilleard, Chris and Paul Higgs (2005). *Contexts of Ageing: Class, Cohort and Community*. Cambridge: United Kingdom: Polity Press.

—— (2009). "The third age: field, habitus or identity". In: *Consumption and generational change: The rise of consumer lifestyles*. Edited by Ian Jones, Paul Higgs, and David J Ekerdt. New Brunswick, NJ: United States of America: Transaction Publishers, pages 23–36.

Gourinchas, Pierre-Olivier and Jonathan A. Parket (2002). "Consumption over the Life Cycle". In: *Econometrica* 70.1, pages 47–89.

Graaf, John de, David Wann, and Thomas H Naylor (2001). *Affluenza: The All-Consuming Epidemic*. San Francisco, CA: United States of America: Berrett-Koehler Publishers.

Grant, Bevan C (2004). "A new sense of self and a new lease of life: Leisure in a retirement village". In: *Annals of Leisure Research* 7.3-4, pages 222–236.

Haider, Steven J and Melvin Stephens Jr (2007). "Is there a retirement-consumption puzzle? Evidence using subjective retirement expectations". In: *The review of economics and statistics* 89.2, pages 247–264.

Haley, Russell I (1968). "Benefit segmentation: A decision-oriented research tool". In: *The Journal of Marketing* 32, pages 30–35.

—— (1984). "Benefit segmentation - 20 years later". In: *Journal of Consumer Marketing* 1.2, pages 5–13.

Halweil, Brian and Lisa Mastny (2004). *State of the World, 2004. Special Focus: The Consumer Society*. State of the World. Washington, DC: United States of America: Worldwatch Institute.

Hamermesh, Daniel S (1984a). "Consumption during retirement: The missing link in the life cycle". In: *The Review of Economics and Statistics* 66.1, pages 1–7.

—— (1984b). "Life-cycle effects on consumption and retirement". In: *Journal of Labor Economics* 2.3, pages 353–370.

Hamilton, Clive and Richard Denniss (2005). *Affluenza: when too much is never enough*. Crows Nest, NSW: Australia: Allen & Unwin.

Hansena, Gary D and Selahattin Imrohoroglu (2008). "Consumption over the life cycle: The role of annuities". In: *Review of Economic Dynamics* 11, pages 566–583.

Hendricks, Jon and Laurie Russell Hatch (2006). "Lifestyle and aging". In: *Handbook of Aging and the Social Sciences (Sixth Edition)*. Edited by Robert Binstock et al. Burlington, MA: United States of America: Academic Press, pages 301–319.

Higgs, Paul and Chris Gilleard (2015). "Fitness and consumerism in later life". In: *Physical Activity and Sport in Later Life*. Edited by Emmanuelle Tulle and Cassandra Phoenix. Basingstoke: United Kingdom: Palgrave Macmillan, pages 32–42.

Higgs, Paul et al. (2009). "From passive to active consumers?: Later life consumption in the UK from 1968 to 2005". In: *The Sociological Review* 57.1, pages 102–124.

Hilgers, Mathieu and Eric Mangez (2015). *Bourdieu's Theory of Social Fields: Concepts and Applications*. Routledge Advances in Sociology Abingdon: United Kingdom: Routledge.

Hobijn, Bart, David Lagakos, et al. (2003). "Social security and the consumer price index for the elderly". In: *Current Issues in Economics and Finance* 9.5.

Hopkins, Christopher D, Catherine A Roster, and Charles M Wood (2006). "Making the transition to retirement: appraisals, post-transition lifestyle, and changes in consumption patterns". In: *Journal of Consumer Marketing* 23.2, pages 87–99.

Hornstein, Gail A and Seymour Wapner (1986). "Modes of experiencing and adapting to retirement". In: *The International Journal of Aging and Human Development* 21.4, pages 291–315.

Hurd, Michael D and Susann Rohwedder (2013). "Heterogeneity in spending change at retirement". In: *The journal of the economics of ageing* 1, pages 60–71.

Hurd, Michael and Susann Rohwedder (2003). *The retirement-consumption puzzle: Anticipated and actual declines in spending at retirement*. NBER Working Paper 111639586. Cambridge, MA: United States of America: National Bureau of Economic Research.

Hurst, Erik (2008). *The retirement of a consumption puzzle*. NBER Working Paper 13789. Cambridge, MA: United States of America: National Bureau of Economic Research.

James, Oliver (2007). *Affluenza: How to be Successful and Stay Sane*. London: United Kingdom: Vermilion.

Julia Twigg, Julia and Shinobu Majima (2014). "Consumption and the constitution of age: Expenditure patterns on clothing, hair and cosmetics among post-war baby boomers'". In: *Journal of Aging Studies* 30, pages 23–32.

Kahn, Robert L (2004). *Successful aging: Myth or reality. The 2004 Leon and Josephine Winkelman Lecture*. Technical report. Ann Arbor, MI: United States of America.

Katz, Stephen and Toni Calasanti (2014). "Critical perspectives on successful aging: Does it "appeal more than it illuminates"?" In: *The Gerontologist* 55.1, pages 26–33.

Katz, Stephen and Barbara Marshall (2003). "New sex for old: Lifestyle, consumerism, and the ethics of aging well". In: *Journal of Aging Studies* 17.1, pages 3–16.

Kaufman, Sharon R (1994). *The Ageless Self: Sources of Meaning in Late Life*. Madison, WI: United States of America: The University of Wisconsin Press.

Klimczuk, Andrzej (2015). *Economic Foundations for Creative Ageing Policy: Volume I Context and Considerations*. New York, NY: United States of America: Palgrave Macmillan US.

Kolasa, Aleksandra et al. (2017). "Life Cycle Income and Consumption Patterns in Poland". In: *Central European Journal of Economic Modelling and Econometrics* 9.2, pages 137–172.

Koubaa, Yamen, Rym Srarfi Tabbane, and Manel Hamouda (2017). "Segmentation of the senior market: how do different variable sets discriminate between senior segments?" In: *Journal of Marketing Analytics* 5.3-4, pages 99–110.

Kreps, David Marc (1990). *A course in microeconomic theory*. Harlow: United Kingdom: Pearson Education Limited.

Laitner, John and Dan Silverman (2005). *Estimating Life-Cycle Parameters from Consumption Behavior at Retirement*. NBER Working Paper 11163. Cambridge, MA: United States of America: National Bureau of Economic Research.

Lambert-Pandraud, Raphaëlle, Gilles Laurent, and Eric Lapersonne (2005). "Repeat purchasing of new automobiles by older consumers: empirical evidence and interpretations". In: *Journal of Marketing* 69.2, pages 97–113.

Lewis, Jeff (2008). *Cultural Studies: The Basics*. London: United Kingdom: SAGE Publications.

Li, Hongbin, Xinzheng Shi, and Binzhen Wu (2015). "The retirement consumption puzzle in China". In: *American Economic Review* 105.5, pages 437–41.

—— (2016). "The retirement consumption puzzle revisited: Evidence from the mandatory retirement policy in China". In: *Journal of Comparative Economics* 44.3, pages 623–637.

Luengo-Prado, Mara José and Almudena Sevilla (2013). "Time to cook: expenditure at retirement in Spain". In: *The Economic Journal* 123.569, pages 764–789.

Lührmann, Melanie (2010). "Consumer expenditures and home production at retirement—new evidence from Germany". In: *German Economic Review* 11.2, pages 225–45.

Lumpkin, James R, Barnett A Greenberg, and Jac L Goldstucker (1985). "Marketplace needs of the elderly: Determinant attributes and store choice". In: *Journal of Retailing* 61.2, pages 75–105.

Lundberg, Shelly Richard Startz, and Steven Stillman (2003). "The retirement-consumption puzzle: a marital bargaining approach". In: *Journal of public Economics* 87.5-6, pages 1199–1218.

Martin Jr, Claude R (1976). "A transgenerational comparison the elderly fashion consumer." In: *Advances in Consumer Research* 3.1, pages 453–456.

Meadow, H Lee, Stephen C Cosmas, and Andy Plotkin (1981). "The elderly consumer: past, present, and future". In: *Advances in Consumer Research*. Edited by Kent B Monroe. Volume 08. Ann Arbor, MI: United States of America: Association for Consumer Research, pages 742–747.

Medvetz, Thomas and Jeffrey Sallaz (2018). *The Oxford Handbook of Pierre Bourdieu*. New York, NY: United States of America: Oxford University Press.

Miniaci, Raffaele, Chiara Monfardini, and Guglielmo Weber (2003). *Is there a retirement consumption puzzle in Italy?* IFS Working Papers. London: United Kingdom: The Institute for Fiscal Studies. URL: http://hdl.handle.net/10419/71551.

Minkler, Meredith (1989). "Gold in gray: reflections on business' discovery of the elderly market". In: *The Gerontologist* 29.1, pages 17–23.

Moody, Harry R and Jennifer R Sasser (2018). *Ageing. Concepts and Controversies* 9th. Thousand Oaks, CA: United States of America: Sage Publications, Inc.

Moreau, Nicolas and Elena Stancanelli (2015). "Household consumption at retirement: A regression discontinuity study on French data". In: *Annals of Economics and Statistics/Annales d'Économie et de Statistique* 117/118, pages 253–276.

Morse, Leon (1964). "Old Folks: An Overlooked Market?" In: *Duns Review and Modern Industry* 83, pages 45–46.

Moschis, George P (1992). "Gerontographics: a scientific approach to analyzing and targeting the mature market". In: *Journal of Services Marketing* 6.3, pages 17–26.

— (1996). *Gerontographics: Life-stage Segmentation for Marketing Strategy Development*. Westport, CT: United States of America: Quorum Books.

Moschis, George P Euehun Lee, and Anil Mathur (1997). "Targeting the mature market: opportunities and challenges". In: *Journal of Consumer Marketing* 14.4, pages 282–293.

Oates, Barbara, Lois Shufeldt, and Bobby Vaught (1996). "A psychographic study of the elderly and retail store attributes". In: *Journal of Consumer Marketing* 13.6, pages 14–27.

Offe, Claus (1985). *Disorganized Capitalism: Contemporary Transformations of Work and Politics*. Studies in contemporary German social thought. Cambridge, MA: United States of America: The MIT Press.

Oh, EunHye and HyeKyung Choi (2015). "Study on the meaning of elderly consumers' consumption after retirement". In: *Advanced Science and Technology Letters (Bioscience and Medical Research)* 91, pages 71–74.

Plessz, Marie et al. (2015). "Ageing, retirement and changes in vegetable consumption in France: findings from the prospective GAZEL cohort". In: *British Journal of Nutrition* 114.6, pages 979–987.

Reinecke, John A (1964). "The "Older" Market. Fact or Fiction?" In: *The Journal of Marketing* 28.1, pages 60–64.

Rowe, John W and Robert L Kahn (1997). "Successful aging". In: *The Gerontologist* 37.4, pages 433–440.

— (1998). *Successful Aging*. New York, NY: United States of America: Dell Publishing.

— (2015). "Successful aging 2.0: conceptual expansions for the 21st century". In: *The Journals of Gerontology: Series B* 70.4, pages 593–596.

Samli, A Coskun (1967). "The elusive senior citizen market". In: *Dimensions* 1, pages 7–16.

Sartori, Giovanni (2000). *Homo videns: Televisione e post-pensiero*. Bari: Italy: Editori Laterza.

Schau, Hope Jensen, Mary C Gilly and Mary Wolfinbarger (2009). "Consumer identity renaissance: the resurgence of identity-inspired consumption in retirement". In: *Journal of Consumer Research* 36.2, pages 255–276.

Schwartz, Ella and Howard Litwin (2018). "Social network changes among older Europeans: the role of gender". In: *European Journal of Ageing*, pages 1–9.

Schwerdt, Guido (2005). "Why does consumption fall at retirement? Evidence from Germany". In: *Economics Letters* 89.3, pages 300–305.

Serrière, Frédéric (2018). *Le Guide Silver Eco 2018*. Grenoble: France: Frédéric Serrière Consulting.

Smith, Sarah (2006). "The retirement-consumption puzzle and involuntary early retirement: Evidence from the British Household Panel Survey". In: *The Economic Journal* 116.510, pages C130–C148.

Stancanelli, Elena and Arthur Van Soest (2012). "Retirement and home production: A regression discontinuity approach". In: *American Economic Review* 102.3, pages 600–605.

Stephens Jr, Melvin and Takashi Unayama (2012). "The impact of retirement on household consumption in Japan". In: *Journal of the Japanese and International Economies* 26.1, pages 62–83.

Stowe, James D and Teresa M Cooney (2014). "Examining Rowe and Kahn's concept of successful aging: Importance of taking a life course perspective". In: *The Gerontologist* 55.1, pages 43–50.

Thurow, Lester C (1969). "The optimum lifetime distribution of consumption expenditures". In: *The American Economic Review* 59.3, pages 324–330.

Tomini, Florian, Sonila M Tomini, and Wim Groot (2016). "Understanding the value of social networks in life satisfaction of elderly people: a comparative study of 16 European countries using SHARE data". In: *BMC Geriatrics* 16.1. URL: https://doi.org/10.1186/s12877-016-0362-7.

Tongren, Hale N (1988). "Determinant behavior characteristics of older consumers". In: *Journal of Consumer Affairs* 22.1, pages 136–157.

Twigg, Julia (2004). "The body gender, and age: Feminist insights in social gerontology". In: *Journal of Aging Studies* 18.1, pages 59–73.

United Nations Population Fund (2004). *UNFPA State of World Population, 2004: The Cairo Consensus at Ten: Population, Reproductive Health and the Global Effort to End Poverty*. New York, NY: United States of America: United Nations Population Fund.

Usui, Chikako (2008). "Japan's Demographic Changes, Social Implications, and Business Opportunities". In: *The Silver Market Phenomenon*. Edited by Florian Kohlbacher and Cornelius Herstatt. Heidelberg: Germany: Springer, pages 71–82.

Vincent, John (1995). *Inequality and Old Age*. London: United Kingdom: UCL Press.

Wakabayashi, Midori (2008). "The retirement consumption puzzle in Japan". In: *Journal of Population Economics* 21.4, pages 983–1005.

Week, Business (1960). "How the Old Age Market Looks". In: *Business Week*, pages 37–38.

Wood, Michael (1998). "Socio-economic status, delay of gratification, and impulse buying". In: *Journal of economic psychology* 19.3, pages 295–320.

Yokoyama, Ryoichi et al. (2014). "Association between gray matter volume in the caudate nucleus and financial extravagance: Findings from voxel-based morphometry". In: *Neuroscience letters* 563, pages 28–32.

Youm, Yoosik, Edward O Laumann, and Keunbok Lee (2018). "Changes of Personal Network Configuration Over the Life Course in the USA: A Latent Class Approach". In: *Social Networks and the Life Course. Integrating the Development of Human Lives and Social Relational Networks*. Edited by Duane F Alwin, Diane H Felmlee, and Derek A Kreager. Frontiers in Sociology and Social Research. Cham: Switzerland: Springer International Publishing, pages 367–387.

Zniva, Robert (2016). *Ältere Konsumenten in Handel und Marketing: Empirische Überprüfung der Bedeutung von Convenience*. Wiesbaden: Germany: Springer Fachmedien.

Zniva, Robert and Wolfgang Weitzl (2016). "Its not how old you are but how you are old: A review on aging and consumer behavior". In: *Management Review Quarterly* 66.4, pages 267–297.

Part VI

Postscript

Deleuze and Guattari opined that a book is like an assemblage with 'lines of articulation or segmentarity, strata and territories; but also lines of flight, movements of deterritorialization and destratification'. Exploring this idea further they wrote that a book has two sides:

> One side of a machinic assemblage faces the strata, which doubtless make it a kind of organism, or signing totality, or determination attributable to a subject; it also has a side facing a body without organs, which is continually dismantling the organism, causing asignifying particles or pure intensities to pass or circulate, and attributing to itself subjects that it leaves with nothing more than a name as the trace of an intensity.
>
> (Deleuze and Guattari 2004, p. 4)

Along its four volumes, this textbook has described many conjectures, hypotheses, models, and theories about economics and ageing, not all from economics or by economists. It has tried to articulate various strata and territories but has also offered avenues towards de-stratification and uprooting of the received wisdom in economics *of* ageing, lines of flight into a deeper and richer assemblage of ideas, the ever higher 'plateau'—that is, the ever higher 'multiplicity connected to other multiplicities by superficial underground stems' (Deleuze and Guattari 2004, p. 22)—of economics *and* ageing.

Another endeavour was to present empirical findings without shying away from conflicting views and interpretations. As Max Weber said,

> The primary task of a useful teacher is to teach his students to recognize "inconvenient" facts–I mean facts that are inconvenient for their party opinions.

And for every party opinion there are facts that are extremely inconvenient, for my own opinion no less than for others. I believe the teacher accomplishes more than a mere intellectual task if he compels his audience to accustom itself to the existence of such facts.

(Weber 1991, p. 147)

In closing, I leave you with the following reflection by the US economist Prof. Peter Arthur Diamond, who was awarded the Nobel Prize in Economic Sciences in 2010, which would be advisable for all of us to take heed of:

I am concerned that too many economists take the findings of individual studies literally as a basis for policy thinking, rather than seeking inferences from an individual study to be combined with inferences from other studies that consider other aspects of a policy question, as well as with intuitions about aspects of policy that are not in the models. To me, taking a model literally is not taking the model seriously. It is worth remembering that models are incomplete—indeed that is what it means to be a model. We construct multiple models to highlight different aspects of an issue, so, thinking thoroughly about policy calls for thinking through multiple models, and requires recognizing issues that have not made it into any of the available models.

(Diamond 2009, pp. 2–3)

And, now, along with Thomas Stearns (T.S.) Eliot, I say to you

Not fare well,
But fare forward, voyagers.

('The dry salvages', in *Four Quartets*, T.S. Eliot, 1943)

References

Deleuze G, Guattari F (2004) A Thousand Plateaus. Capitalism and Schizophrenia. Continuum
Weber M (1991) Science as a Vocation, Routledge, pp 129–156
Diamond P (2009) Taxes and pensions. Southern Economic Journal 76(1):2–15

Glossary: Volume IV
(Numbers Refer to Chapter)

Absolute poverty '[A] condition of severe deprivation of basic human needs, including food, safe drinking water, sanitation facilities, health, shelter, education and information' (United Nations 1996, p. 38). 5

Ageing in place '[T]he ability of older people to live in their own home and community safely, independently, and comfortably, regardless of age, income or level of intrinsic capacity' (WHO 2015, p. 36). 7

Anchoring effect A behavioural characteristic by which, when given options that have values attached, economic agents tend to choose the first of those values. 8

Asset-based welfare A poverty eradication policy based on the premise that it would be better for the poor to accumulate assets than to receive income support. 7

Behavioural life cycle A life-cycle model in which the variable 'wealth' is broken down into as many asset types as an agent's mental accounting processes are assumed to be able to distinguish. 8

Diminishing sensitivity A behavioural characteristic by which, above a reference point, the marginal value of increases in monetary values for an economic agent diminishes with their magnitude, and similarly for the marginal value of decreases below the same reference point. 8

Easterlin paradox The finding that in the short term happiness and national income are positively correlated but in the long term there is no statistical association and happiness does not grow despite increases in national income. 2

Emotional well-being The frequency and intensity of positive and negative emotions in everyday life. 1

Endowment effect A behavioural characteristic by which an economic agent demands more to give up an object than she would be willing to pay to acquire it. 8

Eudaimonic happiness The notion that happiness consists of functionings, fulfilment, interactions, meaning and purpose. 1

Five factor personality model A psychology model that posits that the structure of personality can be described in terms of five factors or traits: openness, conscientiousness, extraversion, agreeableness, and neuroticism. 2

Framing effect A behavioural characteristic by which economic agents' decisions depend on how choices between alternative options are presented. 8

Happiness A subjective evaluation of how satisfied a person is with her life in all its dimensions and facets as a result of the degree that her needs are met. 1

Happiness-adjusted life years See happy life expectancy. 3

Happy life expectancy The product of life expectancy at birth and a happiness score. HLE is also known as happy life years (HLY) and happiness-adjusted life years. 3

Happy life years See happy life expectancy. 3

Headcount rate The ratio between the number of people or households below the poverty line and the total population. 5

Hedonic happiness The notion that happiness consists of avoiding pain and maximising pleasure. 1

Hedonic treadmill The theory that individuals adapt to pleasurable goods, experiences, and activities so that these things cease to be a source of happiness. 2

Human development The process in which individuals live a long and healthy life, are educated, and have access to resources needed for a decent standard of living (UNDP 1990). 1

Implicit price The hypothesis that economic agents derive utility or happiness from the ranked position in a reference group rather than from the absolute level of income. 1

Inequality-adjusted happiness A measure that combines the average and the standard deviation of happiness scores. 3

Inertia The same as the status quo bias. 8

Intergenerational income elasticity The percentage difference in income in one generation divided by the percentage difference in income in the previous generation. 5

Intergenerational mobility The degree to which an individual is in the same social position as their parents'. 6

Lack of willpower A behavioural characteristic by which an agent who prefers x to y does y. 8

Life-cycle model of residential mobility A model that posits that changes in life stages are main drivers to move house. 7

Life satisfaction What an individual thinks about or how she assesses her life. 1

Loan-to-value ratio The ratio between the amount of the mortgage loan on the borrowing date and the purchase value of the property. 7

Loss aversion A situation in which an economic agent seeks risk when facing a loss prospect but exhibits risk aversion for gains. 8

Mental accounting A behavioural characteristic by which an agent values some amount of money less or more than the same amount from a different source. 8

Moral desert Ethical theory that holds that 'the moral value of achieving a one unit gain of well-being is greater, the greater the individual's level of deservingness' (Arneson 1997, p. 334). 4

Mortality paradox The statistical artefact that renders lower poverty rates the worse the survival conditions of the poor. 5

Myopia A behavioural characteristic by which an agent neglects the future and gives undue importance to short-term gains and pleasures at the expense of receiving larger benefits in the long run. 8

Objective well-being Well-being in relation to living conditions and life chances and events. 1

Poverty gap How far each unit classified as in relative poverty is below the poverty line. 5

Poverty line Between 40 per cent and 70 per cent of the median household income in the country. 5

Priming A behavioural characteristic similar to anchoring in which the anchor may be a word, image, colour, and so on. 8

Quality of life '[T]he set of non-monetary attributes of individuals [that] shapes their opportunities and life chances, and has intrinsic value under different cultures and contexts' (?, p. 5). 1

Range-frequency theory The theory that category judgements depend on how the individual mentally divides the range between the maximum and minimum categories and how often each category is presented. 2

Rank-income hypothesis The hypothesis that economic agents derive utility or happiness from the ranked position in a reference group rather than from the absolute level of income. 2

Reference dependence A behavioural characteristic by which an economic agent's valuation of monetary changes depends not only on the amount of the variation but also of a reference point. 8

Relative poverty Not being able to afford a living standard deemed socially acceptable in a country or region where a person lives. 5

Retirement-consumption puzzle The decline in consumption in retirement. 12

Reverse mortgage A loan against a property owned by the debtor and in which she resides that allows the person to release equity without having to leave the house. 7

Status quo bias A behavioural characteristic by which economic agents would value more any expected losses from a current situation than any expected gains. 8

Stress threshold model of residential mobility A model that posits that mismatches between preferences and the utility derived from actual and perceived amenities of a property are the main drivers to move house. 7

Subjective well-being Emotional well-being and life satisfaction. 1

Sunk cost effect The *'tendency to continue an endeavor once an investment in money, effort, or time has been made'* (Arkes and Blumer 1985, p. 124). 8

Sustainable development The process in which economic, human, social, and environmental capital grow at a time-consistent rate. 1

Utility The satisfaction that an economic agent derives from the goods and services she consumes. 1

Welfare Either the total amount of utility experienced or achieved from the consumption of goods or services or the set of health, education, housing, employment, and social security public subsidies and transfers. 1

Index

Page numbers followed by 'n' refer to notes.

A

Absolute poverty, 179, 182–187, 192, 198, 203, 207, 230
Ageing in place, 312, 313, 322
Anchoring effect, 349–350
Asset-based welfare, 318, 322

B

Behavioural life cycle, 352, 353, 380

D

Diminishing sensitivity, 342

E

Easterlin paradox, 45, 66–67, 72, 73
Emotional well-being, 11, 13, 79, 108, 111
Endowment effect, 51, 342, 345, 372
Eudaimonic happiness, 18

F

Five factor personality model, 74
Framing effect, 346–349

H

Happiness, 3–6, 8, 9, 11, 13, 17–20, 23, 28, 32n1, 32n3, 45–48, 50–53, 59, 60, 63–71, 73–76, 78–84, 85n15, 101–115, 125, 175, 204, 393, 400
Happiness-adjusted life years, 109, 110
Happy life years, 109–111
Headcount rate, 188
Hedonic happiness, 18
Hedonic treadmill, 67–74, 102
Human development, 3, 4, 6, 18–19, 77, 105, 322

I

Implicit price, 15, 21
Inequality-adjusted happiness (IAH), 111

Inertia, 345, 374–376, 404
Intergenerational income elasticity, 223–227, 229, 266
Intergenerational mobility, 77, 229, 255, 261–272, 289

L

Lack of willpower, 217, 220, 354–356, 360n10, 360n11
Life-cycle model of residential mobility, 306
Life satisfaction, 4, 5, 10, 11, 13, 14, 16, 17, 20, 21, 23, 28, 46, 49, 50, 58–60, 62–67, 69, 72–76, 78, 79, 83, 84, 103, 104, 106–110, 112, 114, 115, 125, 234
Loan-to-value ratio, 309, 315
Loss aversion, 208, 211, 341, 343–345, 375, 376

M

Mental accounting, 352–353
Moral desert, 126–127, 272
Mortality paradox, 202
Myopia, 220, 354, 379

O

Objective well-being, 10–15

P

Poverty gap, 188–192, 202, 203, 208, 213, 214
Poverty line, 183, 184, 186–192, 195, 198, 200, 204, 205, 209–213, 239n6, 314, 315
Priming, 349, 350

Q

Quality of life, 3, 4, 6, 9, 12, 14–17, 19–31, 45, 46, 52, 65, 77, 78, 101, 105, 109, 113, 314, 393, 446

R

Range-frequency theory, 62
Rank-income hypothesis, 60–64
Reference dependence, 342
Relative poverty, 179, 181, 187–192, 194, 198, 200, 203–204, 210, 217, 218
Retirement-consumption puzzle, 445, 453–456

S

Status quo bias, 345, 346, 370, 374, 375
Stress threshold model of residential mobility, 306, 307
Subjective well-being, 4–6, 10–16, 50–53, 59, 62, 63, 69, 70, 72–76, 79–81, 83, 102, 104, 108, 160, 178
Sunk cost effect, 345, 350–351
Sustainable development, 19

U

Utility, 3, 4, 6–10, 17, 29, 48, 51, 54–56, 60, 61, 63, 71, 79, 108, 115, 124–126, 128, 129, 220, 225, 265, 266, 272, 301, 307, 309, 310, 340–343, 355, 357, 378, 381, 392, 393, 395–399, 404, 412, 413, 437, 456

W

Welfare, 3, 4, 6, 9–10, 17, 51, 55, 56, 65, 70, 71, 101, 102, 105, 111, 124, 126–129, 135, 176, 258, 261, 262, 265, 266, 272, 283, 289, 307, 312, 316, 318, 319, 322, 373, 381, 392, 393, 398, 400, 401, 416, 429, 432

CPI Antony Rowe
Eastbourne, UK
January 31, 2020